S0-CBU-269

SOIL BEHAVIOR AND SOFT GROUND CONSTRUCTION

PROCEEDINGS OF THE SYMPOSIUM

October 5–6, 2001
Cambridge, Massachusetts

SPONSORED BY
The Geo-Institute of the American Society of Civil Engineers

EDITED BY
John T. Germaine
Thomas C. Sheahan
Robert V. Whitman

 **American Society of Civil Engineers**
1801 ALEXANDER BELL DRIVE
RESTON, VIRGINIA 20191–4400

Abstract: This proceedings, *Soil Behavior and Soft Ground Construction*, consists of papers presented at a symposium that was organized to pay tribute to and celebrate the career of Charles C. "Chuck" Ladd. The symposium was held at the Massachusetts Institute of Technology, Cambridge, Massachusetts from October 5 to 6, 2001. The volume contains state-of-the-art papers covering our knowledge base in fundamental areas of soft ground mechanical behavior, geochemistry, and links between the two. In addition, there are several papers on the state of the practice, historical overviews, and case studies in soft ground construction. The papers are divided into four focus areas that encompass Professor Ladd's primary technical interests. Papers in the area of Soil Physics and Chemistry address the fundamental micro-level factors controlling macroscopic mechanical and chemo-mechanical soil behavior. Papers on Mechanical Properties and In Situ Measurements deal with the measurement and characterization of soil physico-mechanical properties and behavior for engineering. The last set of papers, on Engineering Applications, presents historical reviews and case studies on soft ground engineering. In all of the papers, Professor Ladd's influence on the many aspects of this field is clear, as is the ways in which he has shaped our understanding of these difficult soils and our ability to engineer them.

Library of Congress Cataloging-in-Publication Data

Soil behavior and soft ground construction : proceedings of the symposium October 5-6, 2001, Cambridge, Massachusetts, sponsored by The Geo-Institute of the American Society of Civil Engineers / edited by John T. Germaine, Thomas C. Sheahan, Robert V. Whitman.
    p. cm. -- (Geotechnical special publication ; no. 119)
    Includes bibliographical references and index.
    ISBN 0-7844-0659-6
    1. Soil mechanics--Congresses. I. Germaine, John T. II. Sheahan, Thomas C. III. Whitman, Robert V., 1928- IV. American Society of Civil Engineers. Geo-Institute. V. Series.

TA710.A1 S5224 2002
624.1'5136--dc21   2002038440

Preface

This special publication contains papers presented at a symposium to pay tribute to and celebrate the career of Charles C. "Chuck" Ladd. The symposium was held at the Massachusetts Institute of Technology from October 5 to 6, 2001 and included one full day of technical presentations by invited speakers followed by social events and an open house in the MIT geotechnical engineering laboratories.

The organizing committee members for the symposium were: Lewis Edgers, John T. Germaine, Deidre A. O'Neill, Thomas C. Sheahan, Robert V. Whitman, and Andrew J. Whittle. Alise Kalemkiarian provided administrative support for the committee.

The symposium was divided into four parts, each encompassing a technical area of soft ground engineering in which Professor Ladd has made a significant impact. Dr. Anwar Wissa moderated the session *Soil Physics and Chemistry*; Dr. Jacques Levadoux moderated the *Mechanical Properties* session; Dr. John Christian moderated the session *In Situ Measurements*; and Dr. Edward Kinner moderated the last session, simply titled *Applications*.

The papers in this volume have been peer reviewed and revised as necessary by the authors. Each paper was sent to three established experts in the respective areas. The editors appreciate the efforts of the following reviewers:

Akram Alshawabkeh	Sandra Houston	Juan Pestana-Nascimento
Charles Aubeny	Roman Hryciw	John Powell
Jean Benoit	Thomas Ingra	Anand Puppala
John Burland	Edward Kinner	Rodrigo Salgado
John Christian	Michael Long	Charles Shackelford
Patricia Culligan	Thomas Lunne	Satoru Shibya
Lewis Edgers	Allan Marr	Gilliane Sills
Richard Finno	Paul Mayne	Francisco Silva-Tulla
Patrick Fox	Gerald Miller	Timothy Stark
Nadim Fuleihan	Thomas O'Rourke	Gioacchino Viggiani
John Garlanger	Samuel Paikowsky	

The volume contains state-of-the-art papers covering our knowledge base in fundamental areas of soft ground mechanical behavior, geochemistry, and links between the two. In addition, there are several papers on the state of the practice and case studies in soft ground construction. In all of the papers, Professor Ladd's influence on the many aspects of this field is clear. The symposium as well as the papers illustrate how this influence shaped our understanding of these complex soils and our ability to engineer them.

John T. Germaine
Thomas C. Sheahan
Robert V. Whitman

Charles Cushing Ladd III

Professor Ladd has been internationally recognized for his work as a researcher, practitioner, and teacher of geotechnical engineering. He has been especially interested in compacted and natural clay behavior, laboratory and field testing innovations, soft ground construction, frozen soil mechanics, and offshore engineering. His mastery of these diverse areas is a testament to his intellect and his well-known thoroughness and determination to understand completely any topic he is studying (this extends even to his leisure activities such as golf).

Chuck Ladd was born in Brooklyn, New York, and lived in Connecticut, Arizona, and Rhode Island during his youth. He graduated from Providence Country Day High School in 1950. He then moved on to study under a 3-2 plan at Bowdoin College and MIT. In 1955, he received both the AB (cum laude) in Mathematics and Physics from Bowdoin and the BS degree in Building Engineering and Construction from MIT. As has been pointed out, this makes him one of the few engineers having membership in both Tau Beta Pi and Phi Beta Kappa. He then earned an SM in Civil Engineering and the ScD in Soil Engineering from MIT in 1957 and 1961, respectively. His doctoral thesis was supervised by Professor T.W. Lambe and entitled "Physico-Chemical Analysis of the Shear Strength of Saturated Clays."

Chuck's entire career has been spent at MIT, in the department now known as Civil and Environmental Engineering. He was an Instructor from the time he received his SM in 1957 until 1961, and then became an Assistant Professor. In 1964, he became an Associate Professor and attained the rank of Professor in 1970. In 1983, he was a Visiting Senior Scientist at the Norwegian Geotechnical Institute, and from 1983 through 1994, he was Director of the Center for Scientific Excellence in Offshore Engineering at MIT. In 1 994, Chuck became the Edmund K. Turner Professor – a title that he held until his official retirement in 1996. He has remained active in teaching and other departmental activities since then.

A major focus of his research has been the improved understanding of the behavior of soft, cohesive soils. He developed the SHANSEP method for overcoming the effects of sample disturbance on laboratory stress-strain-strength properties. His state-of-the-art papers at the 1977 and 1986 international conferences are c onsidered landmark synopses of research and practical developments in soft ground construction issues. In particular, he emphasized what he felt were the three major factors that influenced our ability to engineer these soils: stress-strain-strength anisotropy, sample disturbance and s train r ate e ffects. H e c ontinually t aught t hese t hemes t o a cademic c olleagues, students, and perhaps most importantly, practitioners doing the day-to-day engineering. His influence is evident in the papers contained in this volume, ranging from geochemical aspects of clay behavior to case studies involving large-scale construction works on soft clays.

Like Karl Terzaghi and Ralph Peck, Chuck's research contributions have been integrally linked to professional practice, and his impacts on the practice have been enormous. His research has been driven by practical concerns, he has always striven to present results in terms that are useful in practice, and he continually tries to improve the state of practice. Two outstanding examples are his Terzaghi Lecture (presented in 1986 and published in 1991), in which he combined virtually all of his soft ground engineering knowledge into the complex topic of staged embankment construction; and the paper describing the engineering design of the St. James Bay embankments. His research and consulting work for B oston's Central Artery/Third Harbor Tunnel project was instrumental in understanding how Boston Blue Clay would respond to the engineered conditions that "pushed the envelope" of our design capabilities.

Chuck has been devoted in his service to the American Society of Civil Engineers. At the Society level, he served on the Professional Activities Committee and chaired the Committee on Curricula and Accreditation. Within the Geotechnical Engineering Division, Chuck was a member of the Publications Committee, chaired the Soil Properties a nd Awards Committees, a nd t hen se rved o n a nd chaired t he E xecutive Committee. It was during the time of Chuck's chairmanship that the Executive Committee took steps that led to the establishment of the Geo-Institute, and Chuck was a member of the founding Board of Governors. At the local level, Chuck served a term as president of the Boston Society of Civil Engineers Section of ASCE, and contributed his experience as a member of the Board of Commissioners for the Department of Public Works in his hometown of Concord, Massachusetts.

In addition to all these major accomplishments, Chuck is remembered by several generations of students for his teaching, both in the classroom and in one-to-one work with research students. His subjects *1.361 Advanced Soil Mechanics* and *1.322 Soil Behavior* were for decades the centerpieces of MIT's graduate program in geotechnical engineering. His lectures were meticulously prepared, generally accompanied by hand written notes (both figures and text) bearing his trademark date and initials in the upper right hand corner of each and every sheet. His thesis supervision is often remembered by students as both the most difficult and the most fulfilling experience of their careers. Like a legendary coach, Chuck has challenged his students to rise to new levels of intellect and determination to advance the state of knowledge.

For his contributions, Chuck has received numerous awards and recognitions. From ASCE, he received the Walter L. Huber Civil Engineering Research Prize, the Middlebrooks Award, the Karl Terzaghi Award, and two of the highest recognitions given by the Society: the Croes Medal in 1973 and the Norman Medal in 1976. In 1995, he was made an Honorary Member of the Society. Chuck also received the Hogentogler Award of the American Society of Testing materials. In 1983, he was elected to the National Academy of Engineering.

Through all of this professional accomplishment, Chuck's family has remained an important part of his life. His wife, Carol, c ontinues to put up with his l ong work hours (even when he is technically "retired"), and he is a proud father and grandfather.

The 200 practitioners, former students and faculty colleagues who came together for this symposium are symbolic of the extended "family" that Chuck has established during his distinguished, highly productive career. His selfless dedication to teaching what he has learned, through research and engineering practice, has influenced a generation of geotechnical engineers and has truly changed the way we think about soil behavior and soft ground engineering.

Contents

Up and Down Soils – A Reexamination of Swelling Phenomena in Earth Materials

James K. Mitchell[1], Hon.M., ASCE

Abstract

In his first published paper in 1960 Professor Charles C. Ladd stated that several soil composition, state, and environmental factors govern the swelling behavior of a soil. He summarized the theoretical basis for analysis of interparticle forces and swelling in terms of diffuse double layer theory and osmotic pressure. He contributed data on the influence of compaction conditions and pore water chemistry on the swelling behavior of compacted fine-grained soils. When examined from the perspective of today's understanding of soil behavior, the principles he stated 40 years ago still apply. Theoretical and experimental studies carried out in the intervening years, and summarized in this paper, have provided further clarification of the most important factors controlling swelling and the qualitative and quantitative applicability of double layer and water adsorption theories of swelling. The most important single factor influencing the swell potential of a soil is the effective specific surface. The smectite clays (montmorillonite) have the greatest potential for expansion owing to weak interlayer bonding and, therefore, the large specific surface available for water adsorption. In the presence of high concentration salt solutions and/or potassium or divalent adsorbed cations interlayer swelling is suppressed, and the amount of swell is much less than when sodium is the dominant adsorbed cation.

Current research on the fundamentals of water adsorption and clay swelling employs Monte Carlo and Molecular Dynamics modeling to obtain knowledge of the distribution and energy states of water and cations adjacent to clay particle surfaces. There may be applications of the results of these current studies to practical problems in geotechnical engineering in the future.

Introduction

On this very special occasion, organized to honor and pay tribute to our distinguished friend and colleague, Professor Charles C. Ladd, the organizers have suggested that we reflect on significant developments in our topic area during the past 50 years. The topic that I have chosen – soil swelling – provides an excellent opportunity to do just

---

[1] University Distinguished Professor, Emeritus, Virginia Tech, 109B Patton Hall, Blacksburg, VA 24061-0105; jkm@vt.edu

1

that. The first paper on Chuck Ladd's publications list is *Mechanisms of Swelling by Compacted Clay*, published in 1960. Although this is not quite 50 years ago, the paper provides a significant bench marking point, especially since much of the fundamental understanding of swelling processes at that time had been elucidated only in the preceding 10 years, and Ladd identified and explained virtually all the key concepts in a manner useful for engineering purposes, while providing new test data to demonstrate their validity.

The annual economic losses associated with failure to understand and properly evaluate and treat expansive soils rival those of virtually all other types of natural disasters (e.g., floods, earthquakes, hurricanes). Proper identification of expansive soils and dealing with them to avoid unanticipated volume changes and disruptive ground movements remain one of the most important challenges in geotechnical engineering. Knowledge of the fundamentals of earth material expansion provides a sound basis for meeting these challenges.

My objectives in this paper are to summarize and demonstrate the continuing applicability of the ideas and understanding in Ladd (1960) and to review some of the developments in our understanding of swelling and shrinkage phenomena in clays since 1960. The focus is on the physics and chemistry of soil swelling, recognizing that swelling and compression are coupled chemo-mechanical processes in which the details of composition, mineralogy, soil fabric, pressure, and temperature each play a role in the process.

Background and Context

Three general classes of mechanical behavior of earth materials are usually of interest in any problem or project in geotechnical engineering:
1. *Volume change* in response to changed conditions; e.g., exposure to water, change in applied stress, change in chemical and electrical environment, change in temperature, and mechanical disturbance.
2. *Deformation and strength* under different loading conditions.
3. *Conductivity* of the soil or rock to fluids, gases, heat, chemicals, and electricity.

Any specific property within each of these classes is likely to have some dependency on time and the rates of any imposed loading or other stimulus. Furthermore, changes in a property in one class will mostly likely produce changes in the properties and behavior within another class. For example, if an expansive soil is allowed to swell, its strength will probably go down, and the saturated hydraulic conductivity will go up; if a soil is deformed extensively in shear, its conductivity properties and subsequent shrink-swell behavior will change; and if the hydraulic conductivity is reduced the rate of swelling will be slowed. Volume change response; i.e., swelling, compression, and shrinkage, is of primary interest herein, so the discussion will focus on the mechanisms of these processes without consideration of the associated strength and conductivity property changes.

It is well known that certain soil minerals swell and shrink more than others; e.g., Smectite (montmorillonite) >>> Illite > Kaolinite, owing to their different compositions and mineral structures, which leads to significant differences in specific surface, cation exchange, and colloidal characteristics. Details of clay mineral composition and structure are not considered here, but can be found in Mitchell (1993) and references cited therein.

What Ladd told us in 1960[2]

Governing factors. *The swelling behavior of compacted clay will be governed by the following factors:*

1. *Composition of the clay – composition and amount of clay minerals, nature and amount of exchangeable cations, proportions of sand and silt in the clay, and presence of organic matter and cementing agents.*
2. *Compaction conditions – molded water content, dry density, degree of saturation, and type of compaction.*
3. *Chemical properties of the pore fluid – both that during compaction and that which is imbibed during swelling.*
4. *Confining pressure during swelling.*
5. *Time allowed for swelling.*

This listing is important, as it illustrates that many factors determine what will happen; in essence, a given soil may represent a nearly infinite variety of materials, each with its specific response to stress and environmental changes. How many times in the years since these factors have been recognized has failure to take one or more of them into account led to problems?

The Diffuse Double Layer and Swelling. At the time Ladd (1960) was written, newly developed concepts of colloidal chemistry, most notably double layer theory, had recently been introduced into soil mechanics (the term *geotechnical engineering* had not yet been coined) in an effort to better understand and explain fine-grained soil behavior, especially swelling and compression and the role of soil fabric; e.g., Kruyt (1952), Lambe (1953), Bolt (1956), Mitchell (1956), Lambe (1958). Application of these concepts for explanation of the swelling of compacted clays was a logical extension.

Owing to the net negative charge of clay particles, a "double layer" composed of the particle surface charge and the — *portion of the water surrounding the particle in which there is a negative electric field requiring an excess of positive charges relative to negative charges* is formed. A schematic of the double layer and charge distributions within it is shown in Fig. 1.

When particles approach each other, double layers overlap and generate repulsions, which acting in conjunction with interparticle attractions and any applied compressive stresses, determine the equilibrium water content and void ratio[3]. For ideal systems of very small platy particles with a fabric that remains essentially unchanged during compression and swelling; e.g., parallel plates, calculations based on double layer theory have given compression and swelling response that is in quite

---

[2] Text in this section that is in italics is taken directly from Ladd (1960).

[3] A more complete description of attractive and repulsive interparticle forces and their relationship to effective stress is given in Mitchell (1993).

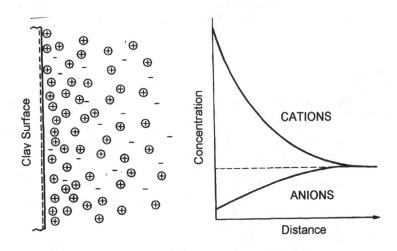

Fig. 1 - Schematic diagram of the clay-water-electrolyte system double layer.

good agreement with experimental observations, as demonstrated, for example by Warkentin, et al., (1957) and shown in Fig. 2. That the equilibrium interparticle spacing at any pressure during the first compression is greater than that during subsequent compression and swelling is a reflection of mechanical particle interactions that prevent development of parallel platy particle orientations throughout the micro-fabric of the clay.

Key predictions of the theory are that the higher the salt concentration in the pore water and the higher the valence of the adsorbed cations, the lower will be the swell and swell pressure. However, *most natural soils are far from these ideal systems so that one can only use such theories in a qualitative manner when dealing with soils.* A complicating factor in many systems is the presence of positive edge charges, especially in kaolinite, that *may cause an additional attractive force between particles that can lead to a nonparallel particle orientation.* Ladd (1960) also points out that the theory will apply only for particle spacing greater than 1 to 2 nm. He provides a lucid description of the concept of osmotic pressure and how it has been used to compute swelling pressure in an ideal system.

Other Factors and Mechanisms. Ladd (1960) discusses several other factors that may influence swelling. These include van der Waals attractive forces acting on the water surrounding clay particles, the effect of the negative electric field on the double layer water, water of hydration of the adsorbed cations, and the absence of osmotic repulsive pressures for interparticle spacings less than 1 to 2 nm. Of these, cation water of hydration may be the most important and is discussed later. In addition, *Mention should be made of swelling due to elastic rebound and "unbending" of soil particles.* The importance of the latter in most natural soils is considered subsequently.

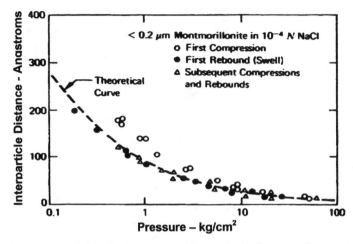

Fig. 2 - Relationship between particle spacing and pressure for montmorillo-nite (modified from Warkentin, et al., 1957) (1.0 kg/cm^2 = 98.06 kPa)

Swelling in Compacted Clays. Direct application of the foregoing concepts and theory in their simple form to swelling of compacted clay is complicated by the facts that:

1. *The dry density of the soil and thickness of the double layer water around clay particles vary.*
2. *The particle orientation varies.*
3. *The pore water tensions vary.*
4. *The degree of saturation, and hence the amount of the air in the sample varies. The pressure in the air may also vary.*

The idea of double layer water deficiency (Lambe, 1958) as a contributing factor to compacted clay swelling is also discussed by Ladd. *The thickness of double-layer water on compacted clay particles is nearly always less than that which the particles would like to have if given free access to water.* Contained in the discussion is the very important concept that most of the water in a typical soil is associated with the clay phase, owing to the much greater specific surface of the clay relative to the silt and sand fractions.

After consideration of the above factors, along with discussion of capillarity and pore water tensions in a partially saturated soil, Ladd concludes with: *the clay micelles expand their double-layers and swelling occurs between clay particles, just as for the saturated clays, until the repulsive pressure minus the attractive pressure between particles is in equilibrium with any applied effective stress.*

The Role of Air. In addition to its effects on pore water tensions, air in a partly saturated soil influences behavior on exposure of the soil to water in another way.

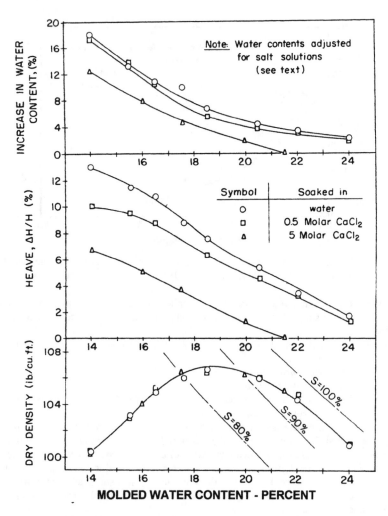

Fig. 3 - Effect of salt concentration on swelling behavior (from Ladd, 1960)
($1 \text{ lb/ft}^3 = 0.157 \text{ kN/m}^3$).

During water absorption from the outer boundary of a soil mass, air trapped internally
is compressed. The increased pressure creates a hoop stress type of tension on the
solids surrounding the pore. This can cause disruption of the grain fabric. Such dis-
ruptions, if near the surface of the soil mass and unconstrained, can manifest themselves as rapid slaking.

Experimental Data. Ladd (1960) reported the results of swelling tests on compacted Vicksburg Buckshot Clay, a highly plastic (CH) clay, with a clay size (<0.002 mm) content of 36 percent. About 50 percent of the material by weight was an interstratified illite-montmorillonite, and the cation exchange capacity was 30 meq./100 gm. Samples were compacted into fixed ring type consolidometer rings using a dynamic method and an effort of 28,000 ft-lb/cu ft. Free swell in water and in 0.5 and 5 M calcium chloride was measured under a surcharge of 200 lb/sq ft (0.1kg/sq cm). The results of these tests are shown in Fig. 3. From this figure and related observations Ladd concluded that:

1. *The amount of heaving and water pickup decreases with increasing molded water content.*
2. *The degree of saturation after soaking is less than 100 percent. The volume of water pickup also exceeds the volume of expansion.*
3. *The soaking of compacted samples in salt solutions produces a marked decrease in the amount of fluid pickup and heaving. -------- the 5 molar salt solution prevents swelling for a sample compacted 2 percent wet of optimum water content.*
4. *The initial rate of swelling is practically unaffected by the salt concentration in the soaking solution.*

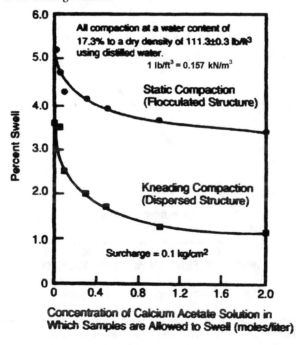

Fig. 4-Effect of structure and electrolyte concentration of absorbed solution on swell of compacted clay (adapted from Seed, et al., 1962)

Seed, et al. (1962) reported similar results, as shown in Fig.4, for tests on an expansive sandy clay, compacted wet of optimum water content using both static and kneading compaction, and allowed to swell under a surcharge pressure of 0.1 kg/sq cm (205 psf) in calcium acetate solutions of different concentrations. There are, however, two important differences. First, high solution concentration was insufficient to prevent all swelling. Second, the sample prepared by static compaction, producing a flocculent soil fabric, swelled considerably more than the sample prepared by kneading compaction, which produced a dispersed fabric. Note, also, that these differences in swell amount are essentially the same irrespective of the swelling solution concentration. These differences are believed due to mechanical effects caused by bending and distortion of platy particles and particle aggregates in the soil structure. A large mechanical contribution to swelling was demonstrated by Terzaghi (1931) and is shown in Fig. 5, where swelling index, C_s, for sand-mica mixtures is given as a function of mica content.

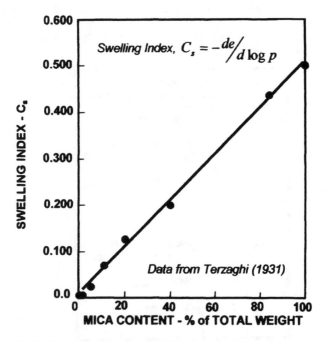

Fig. 5 - Swelling index as a function of mica content for coarse-grained mixtures.

Another finding by Seed, et al. (1960) was that the amount of swell of compacted clay, and presumably also by saturated clay, is unloading stress path dependent, as shown in Fig. 6. Three samples were compacted at the same water content to the same density by the same method. Each was then exposed to water and allowed to swell under different initial confining stress, but unloaded ultimately to 1 psi, as

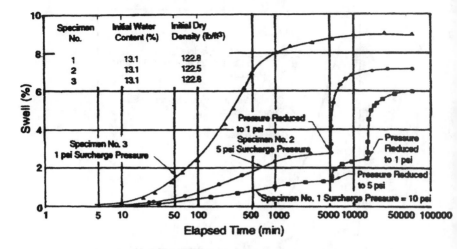

Fig. 6 - Effect of unloading stress path on swelling of compacted sandy clay
(Seed, et al., 1962) (1psi = 6.895 kPa; 1 lb/ft^3 = 0.157 kN/m^3)

shown in the figure. The amount of swell was significantly less when it occurred under a series of reducing stresses than when it was allowed to develop under the lowest stress from the outset. These results suggest that the internal structural adjustments that accompany swelling are dependent on the increment of stress change and that two samples of the same material will have different fabrics and interparticle force balances at the same effective stress at the end of the expansion process.

Air Permeability. As a part of his test program on Vicksburg Buckshot clay, Ladd (1960) determined the rate of air flow through samples compacted at different water contents and obtained the results shown in Fig. 7. Air voids are discontinuous through the material at water contents equal to optimum or greater, as indicated by an air flow rate of zero. As compaction results from soil densification by removal of air, once the air voids become discontinuous further increases in density are not possible, and increasing the water content only serves to dilute the solids concentration; hence, the point where air voids become discontinuous corresponds to optimum water content.[4]

Assessment. This first of C.C. Ladd's many important published contributions is remarkable for its insights into the fundamentals of soil swelling and for elucidating principles that are as useful today as they were then. Among the most important of these are:

---

[4] Ladd (personal communication, July 2001) views this finding as probably the most important finding in the paper.

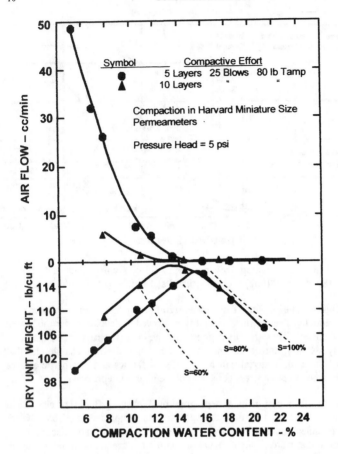

Fig. 7 – Air flow rates through compacted Vicksburg Buckshot Clay (redrawn from Ladd, 1960) (1psi = 6.895 kPa; 1 lb/ft³ = 0.157 kN/m³)

1. Double layer theory and the osmotic pressure concept, while not always quantitatively applicable to natural soils, can explain why clays swell when unloaded in the presence of water.
2. Pore water chemistry greatly influences the amount of swelling.
3. Factors in addition to osmotic pressure influence the amount of swell of compacted clay. *These other factors may be: the effect of negative electric and London van der Waals force fields on water, cation hydration and the attraction of the particle surface for water molecules, elastic rebound, particle orientation, and the presence of air.*
4. Although all details of the influence of air are not yet known, capillary forces, the effects of air compression during water absorption, and continuity of air voids all impact the swelling behavior of a soil.

An understanding of swelling in the 1990's

I now do a fast-forward to the 1990's and provide a summary review of our under-standing of the fundamentals of soil swelling that takes into account developments over the preceding 30+ years. Much of this summary draws heavily on the relevant sections of Chapter 13 in Mitchell (1993).

Applicability of Osmotic Pressure Concepts

A reasonably clear understanding of the applications and limitations of double layer and osmotic pressure theories for soil swelling had developed by 1990. Good agree-ment between theoretical and measured compression curves for sodium montmorillo-nite particles finer than 0.2μm in dilute salt solution was shown in Fig. 2. Compres-sion curves illustrating the influence of cation valence are shown in Fig. 8, also for montmorillonite composed of particles finer than 0.2μm. These data, as well as the results of other tests show generally good qualitative agreement between the predic-tions of theory and experiment, except that the measured curves consistently fall above the theoretical curves. Probable reasons for this are "dead" volumes of liquid associated with terraced particle surfaces (Bolt, 1956) and non-parallel plate particle associations.

Agreement between osmotic pressure theory for swelling and experiment has not generally been good for clays containing particles larger than about 0.2μm, which, of course, represents almost all natural soils. Whereas the very fine fraction (< 0.2μm) of two bentonites studied by Kidder and Reed (1972) gave swelling pres-sures close to theoretical, a coarser fraction (0.2 to 2μm) gave pressures that were less than predicted, even though the particle surface charge densities of the two fractions were the same. Compression curves for three size fractions of sodium illite are shown in Fig. 9. Rebound curves are shown for the two coarsest fractions. The ex-perimental compression curves for the < 0.2μm fraction, Fig. 9(a), are in the correct relative positions with respect to the influence of salt concentration, even though they are displaced significantly from the theoretical curves. On the other hand, the curves are in the reverse order to theoretical predictions in Figs. 9(b) and 9(c). Furthermore the slopes of the compression curves in Figs. 9(b) and 9(c) are essentially independ-ent of salt concentration. This is because the compression of the larger particle size clay was controlled more by initial particle orientations and physical interactions be-tween the particles than by osmotic repulsive pressures[5].

---

[5] Comprehensive summary tables that list differences between kaolinite and montmorillonite with re-spect to the effects of different pore fluid types, solution concentrations, and cation valences on several engineering properties are given by Santamarina and Fam (1995) and Santamarina, et al., (2001).

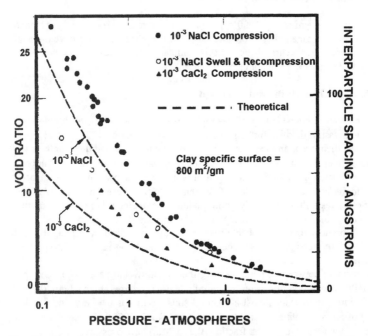

Fig. 8 - Compression curves for Na-montmorillonite and Ca-montmorillonite composed of particles finer than 0.2μm in calcium and sodium chloride solutions. (adapted from Bolt, 1956).

Kaolinite and illite clay particles of the size typically found in clays of engineering interest assume a flocculated arrangement in high electrolyte, high valence, or low dielectric constant environments. Thus on first compression the non-parallel particle assemblages have a greater resistance to stress than would be the case in a low electrolyte, low valence, or high dielectric constant solution, where the particles remain deflocculated and dispersed and are more easily moved into denser packing under stress. This accounts for the reverse order of the curves in Figs. 9(b) and 9(c).

Overall these, and other data presented and discussed in Mitchell (1993) show that physical and fabric factors related to particle size cause sufficient deviations from the conditions and assumptions of the osmotic pressure (double layer) theory of swelling and that the theory cannot be applied directly for analysis of the first compression of most natural clays of the type encountered in geotechnical practice.

However, for compression of clays with a very high specific surface resulting from very small particles, such as bentonite, and for swelling of most other clays from a precompressed state, the theory gives a reasonable prediction of behavior. For example, Madsen and Müller-Vonmoos (1985, 1989) show the successful application of the osmotic pressure theory of swelling for Opalinum shale, illustrated in Fig. 10.

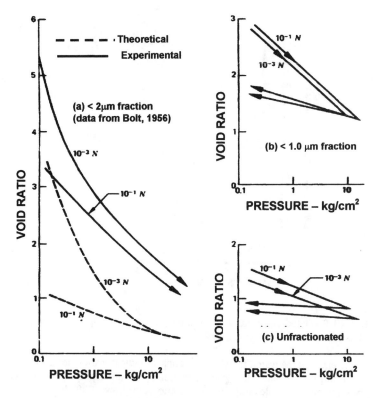

Fig. 9 - Influence of NaCl concentration and particle size on compression and swelling behavior of illite (from Mitchell, 1993). (1.0 kg/cm^2 = 98.06 kPa)

Particle spacing was calculated from specific surface area and water content, and the swelling pressure, P, was calculated using the van't Hoff equation for osmotic pressure:

$$P = RT \sum (c_{iA} - c_{iB}) \tag{1}$$

in which R is the gas constant per mole, T is the absolute temperature, c_{iA} is the ionic concentration at the midplane between particles, and c_{iB} is the concentration of the free pore fluid solution.

Water Adsorption Theory of Swelling

Low (1987, 1992), from the results of many carefully done experiments and analyses, developed an alternative to the osmotic pressure theory for clay swelling based on the hydration of clay surfaces. Interaction of water with clay surfaces reduces the chemi-

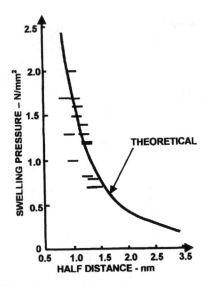

Fig. 10 Predicted and measured swelling pressures for Opalinum shale (data from Madsen and Müller-Vonmoos, 1989)

cal potential of the water, thereby generating a gradient that causes flow of additional water into the system. The swelling pressure, p, in atmospheres, for pure clays was shown to obey the following empirical relationship (Low, 1980):

$$(p+1) = B\exp(\alpha/w) = B\exp[k_i/(t\rho_w)] \qquad (2)$$

in which B and α are clay parameters, w is water content, ρ_w is the density of water, t is the average thickness of water layers, and $k_i = \alpha/(\rho_w A_s)$, where A_s is the specific surface of the clay.

Equation (2) shows that the lower the water content and the higher specific surface, and therefore the thinner the water layer thickness, the higher the swelling pressure. Low (1987, 1992) demonstrates that this theory can explain the swelling of pure clays quite accurately in situations where the osmotic pressure theory cannot. However, the influences of surface charge density, cation valence, electrolyte concentration, and dielectric constant, all known to have major effects on swelling, are not accounted for directly by the water hydration theory, unless adjustments in B, α, and k_i are made.

Reconciliation of Osmotic Pressure and Surface Hydration Swelling Theories

The osmotic pressure theory of swelling is built on the concepts of interparticle repulsions generated by interactions between the diffuse parts of electrical double layers developed by negatively charged clay particles. The surface hydration theory of swelling assumes that the direct attraction of water to particle surfaces is responsible for swelling; the larger the specific surface, the more water is needed to satisfy hydration forces, and the greater the swell and swell pressure.

The proportions of fully expandable and partially expandable clay layers in an expansive soil are determined by surface charge density and cation type. For example, calcium montmorillonite does not swell to interplate distances greater than about 0.9 nm (Norrish, 1954; Blackmore and Miller, 1962). At this spacing the particle structure is stabilized by attractive interactions between the basal planes of the unit layers as influenced by exchangeable cations and adsorbed water (Sposito, 1984). If the electrolyte concentration is high or the pore fluid dielectric constant is low, then interlayer swelling is suppressed, and the effective specific surface is greatly reduced relative to the case where interlayer expansion occurs. Thus the water required for surface hydration is reduced.

A hydration water layer thickness on smectite (montmorillonite) surfaces of about 10 nm is needed to reach a distance beyond which the water properties are no longer influenced by surface forces (Low, 1987). The swelling pressure is about 100 kPa for a water layer thickness of 5 nm. For a fully expanding smectite having a specific surface of 800 sq m/gm the corresponding water content would be 400 percent. Thus, this material would be expected to be expansive over a wide range of water contents, and experience shows clearly that it is. On the other hand, a smectite in an environment that does not allow full interlayer expansion might have an effective specific surface that is much less, perhaps of the same order as that of illite. As the exposed surface structures of both illite and smectite are the same (the bases of silica sheet tetrahedral), it would be expected that the hydration forces should be similar, and an adsorbed water layer thickness of 5nm at a swelling pressure of 100 kPa would be reasonable for both. As the specific surface areas of the illite and nonexpanded smectite are only about 100 sq m/gm, the corresponding water content is only 50 percent. For a kaolinite having a specific surface of 15 sq m/gm, the water content would only need to be 7.5 percent for a 5 nm thick water layer.

Therefore, the specific surface most directly determines the amount of water required to satisfy the forces of hydration. Except for heavily over-consolidated clays and soils that contain significant amounts of expandable smectite, there is enough water present, even at low water contents, to satisfy surface hydration forces, and swelling is small. For those soils in which the clay content is high and particle dissociation into unit layers is extensive, the effective specific surface area is large, and swelling can be significant. The extent of smectite dissociation into unit layers can be evaluated in terms of double layer interactions, with conditions favoring the development of high repulsive forces and, therefore, high osmotic pressures, leading to greater dissociation. Accordingly, both the osmotic pressure theory and the surface hydration theories of swelling have applicability for understanding swelling and shrinkage phenomena in expansive soils. However, neither by itself is suitable quantitatively for explaining all the observations. Overall, the effective specific surface area is the most important single property determining plasticity, swelling behavior, compressibility, and importance of time-dependent deformations in a fine-grained soil. As a simple illustration of this point, Farrar and Coleman (1967) found that the liquid limit (LL) of nineteen British clays was related to their specific surface area, A_s according to

$$LL = 19 + 0.56A_s \qquad\qquad (3)$$

to an accuracy of ±20 percent.

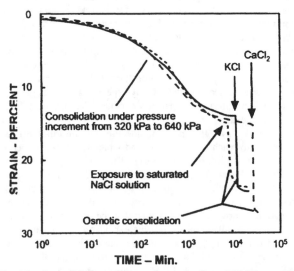

Fig. 11 - Consolidation of Ponza bentonite from 320 kPa to 640 kPa in water and saturated salt solutions (from Di Maio, 1996)

Osmotic and Mechanical Effects in Bentonite. The experimental results reported by Di Maio (1996) are illustrative of the significantly different compression and swelling behavior of bentonite after alternate exposures to high concentration salt solutions and different cation types. Ponza bentonite, having a plascticity index of 320 percent, containing 80 percent clay size material, and composed mainly of Na-montmorillonite, was used for the tests. The liquid limit of the bentonite was about 400 percent when mixed with distilled water, but it dropped to about 70 percent when mixed with concentrated (> 2 molar) solutions of NaCl, KCl, and $CaCl_2$.

Di Maio (1996) mixed the bentonite with distilled water at about the liquid limit and consolidated samples in one dimension in increments, doubling the pressure for each successive increment, under pressures from 40 kPa to 2500 kPa. Curves of vertical strain vs. time for the pressure increment from 320 kPa to 640 kPa are shown in Fig. 11. After sustained periods of consolidation in distilled water, different samples were exposed to saturated solutions of NaCl, KCl, and $CaCl_2$ without change in the applied vertical stress. The data in Fig. 11 show that substantial further consolidation developed as the high concentration salt diffused into the clay, thereby reducing the osmotic swelling pressure. As would be expected, the magnitude of strain that accompanied the change in pore solution concentrations decreased with increasing vertical stress at which the concentrated solution was introduced owing to the lower void ratios at higher stress.

The osmotic consolidation is in many ways similar to mechanical consolidation, with an "equivalent" mechanical consolidation pressure being defined for any increment of osmotic consolidation in the manner shown in Fig. 12. The "equivalent" consolidation pressure is the increment of applied stress that would be required to produce the same void ratio reduction as caused by the osmotic pressure reduction generated by the increase in NaCl concentration. DiMaio's data showed that while the rates of consolidation by application of pressure and reduction in osmotic swelling pressure were about the same, there was a time lag after the increase in salt concentration before osmotic consolidation started attributed to the time necessary for salts to diffuse into the clay.

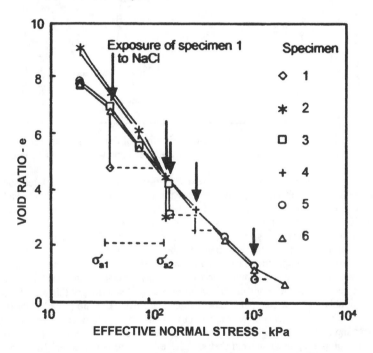

Fig. 12 - "Equivalent" mechanical consolidation pressure developed by exposure of Ponza bentonite to concentrated NaCl solution (from DiMaio, 1996)

Samples that had been allowed to equilibrate in concentrated salt solutions were then re-exposed to distilled water. Those that had been saturated by KCl and CaCl₂ did not change thickness significantly, whereas those previously saturated with NaCl solution swelled a large amount. Fig. 13 shows vertical strain for successive cycles of water and NaCl replacement after mechanical consolidation in water under a vertical stress of 40 kPa. The magnitudes of osmotic swelling and compression resulting from repeated cycles of exposure to water and concentrated NaCl solution decreased with increasing initial consolidation pressure. These magnitudes of swelling became progressively smaller proportions of the original osmotic consolidation as the applied stress at which osmotic consolidation was initiated increased.

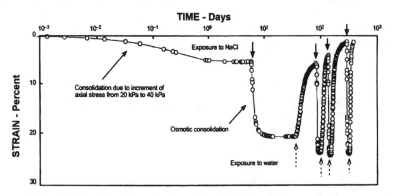

Fig. 13 - One-dimensional volumetric strain of Ponza bentonite accompanying repeated exposure to concentrated NaCl and water (from Di Maio, 1996)

The failure of samples to re-expand after exposure to concentrated solutions of KCl and CaCl₂ solutions suggests that the original Na-montmorillonite had been converted, by cation exchange, to K-montmorillonite and Ca-montmorillonite. That the Ca-montmorillonite did not swell in the presence of water is a consequence of both the lower osmotic swelling pressure associated with divalent cations and limitations on the interplate spacing, as noted above. The failure to swell of the samples saturated with concentrated KCl solution and then re-exposed to water is less readily explained. The Gouy-Chapman model for the double layer cannot explain the difference between the behavior in NaCl solution and KCl solution, as this model assumes the double layer cations to be point charges. However, DiMaio (1996) notes that the smaller hydrated radius of K compared to Na may be a significant factor, and this is accounted for in the Stern (1924) model for the double layer. Scanning electron microscope photographs showed that samples exposed to concentrated NaCl solutions and then re-exposed to water consisted of thin, deformable particles, similar to the bentonite prior to testing. On the other hand, the samples that had been exposed to concentrated KCl or CaCl₂ solutions exhibited fabrics, dependent on experimental conditions, ranging from that in samples exposed only to water to fabrics with large and stiff particles.

Accordingly, it is clear that cation type has an influence on the swelling behavior of clay that is well beyond what is predicted solely by the simple double layer theory. When the physical influences of cation type on fabric and effective specific surface are taken into account, however, the behavior can be better understood.

Some Mineralogical and Geo-chemical Considerations in Soil Expansion

The details of expansive mineral structure, the nature of interlayer materials and the presence of certain chemical compounds in the soil may both limit expansive behavior or enhance it. Examples include hydroxy interlaying and the presence of sulfates, and they are described briefly here.

Hydroxy Interlayering. Hydroxy interlayers composed of Fe-OH, Al-OH, and Mg-OH can form between the unit layers of otherwise expansive clay particles (Rich, 1968). Conditions favoring formation of these layers include a supply of Al^{3+} ions, moderately acid pH, low oxygen content, and frequent wetting and drying. Hydroxy-aluminum and Fe-OH tend to form in acid soils, and $Mg(OH)_2$ is the predominant material in alkaline soils. Randomly distributed islands of interlayer material bind adjacent layers and thereby reduce the swelling potential. Even though the amount of interlayering is likely to be small (10 to 20 percent), it is sufficient to fix the basal spacing of montmorillonite at 1.4 nm, with consequent reduction in both the cation exchange capacity and the swelling.

Sulfate Induced Swelling of Cement- and Lime-Stabilized Soils. Some fine-grained soils, especially in arid and semi-arid areas contain significant amounts of sulfate and carbonate. Sodium sulfate, Na_2SO_4, and gypsum, $Ca\ SO_4 \cdot 2H_2O$, are the common sulfate forms, and calcium carbonate, $CaCO_3$, and dolomite, $MgCO_3$, are the usual carbonate forms. The dominant clay minerals in these soils are expansive smectites. Delayed expansion following admixture stabilization of these soils using portland cement and lime has developed at several sites (Mitchell, 1986). Although test programs have shown suppression of swelling and substantial strength increase at short times (days) as a result of the incorporation of the stabilizer, subsequent heave of magnitude sufficient to destroy pavements has developed as a result of exposure to water at some later time. The mechanism associated with this process appears to be as follows.

When cement or lime is mixed with soil and water, there is a pH increase to about 12.4, some calcium goes into solution and exchanges with sodium on the expansive clay. This ion exchange, along with light cementation by carbonate and gypsum, if present, suppresses the swelling tendency of the clay. The mixed and compacted soil is non-expansive and has higher strength than the untreated material. If sodium sulfate is present, then available lime is depleted according to

$$Ca(OH)_2 + Na_2SO_4 \rightarrow CaSO_4 + 2\ NaOH$$

Silica (SiO_2) and alumina (Al_2O_3) dissolve from the clay in the high pH environment and/or they may be present in amorphous form initially. These compounds can then combine with calcium, carbonate, and sulfate to form ettringite, $Ca_6[Si(OH)_6]_2(SO_4)_3 \cdot 26H_2O$, and/or thaumasite, $Ca_6[Si(OH)_6]_2(SO_4)_2(CO_3)_2 \cdot 24H_2O$ which are very expansive materials (Mehta and Hu, 1978). In addition, in the case of lime-treated soil, if the available lime is depleted, the pH will drop and the further dissolution of SiO_2 from the clay will stop. As silica is needed for formation of the cement (CSH) that is the desired end product of the pozzolanic lime stabilization reaction, long-term strength gain is prevented. Consequently, when the treated material is given access to water, a large amount of swell may occur. Further details concerning lime-sulfate heave reactions in soils are given in Dermatis and Mitchell (1992).

Some Current Studies of Swelling Behavior of Soils

An exhaustive review of the literature on soil swelling published during the past eight years has not been made; however, the contributions to a recent workshop, *Clay Behaviour:Chemo-Mechanical Coupling, from nano-structure to engineering applications*, held in Maratea, Italy, June 2001 provides an excellent insight to recent contributions and current research[6]. Some of them are described below.

Skipper (2001) is using Molecular Dynamics and Monte Carlo simulations to study ions and solvent at hydrated clay surfaces. His aim is to develop a molecular scale insight into fluid-dependent processes in clay such as hydration-dehydration, solute transport, compression, and swelling. Molecular Dynamics analysis requires solving the equations of motion for all particles in the system in the force fields of interactions with the other particles. Both time-independent and time-dependent quantities can be calculated. Modeling a clay-fluid system requires development of six interaction potential functions: fluid-fluid, fluid-counterion, clay-fluid, clay-counterion, counterion-counterion, and clay-clay.

Delville (2001) considers two regions adjacent to a clay surface. At particle spacing less than 2 nm the swelling is dominated by crystalline association of water molecules and ions. Osmotic interactions are dominant for greater interparticle spacing. The swelling of montmorillonite in the crystalline regime is related to the hydration enthalpy of the sodium and the surface. Monte Carlo and Molecular Dynamics simulations are used to determine the equilibrium properties at the clay-water interface. The Monte Carlo simulations are useful to describe phenomena at different scale lengths; for example, solvent molecules confined between two smooth surfaces, condensation of neutralizing counterions in an electrostatic well near a charged surface, and the self organization of charged clay plates. Delville (2001) states that "numerical simulations help understand charged colloid behavior and offer new keys to modify and improve properties."

Moyne and Murad (2001) have derived a macroscopic model for expansive montmorillonite clays by scaling up the pore scale description using electrohydrodynamics coupled with the convection-diffusion equations governing fluid movement and ion transport. The result is a model in which the macroscopic quantities are correlated to micro-structural behavior. According to Moyne and Murad, the procedure "provides a natural derivation of electro-osmotic phenomena along with micromechanical representation for the electrokinetic coefficients (e.g., electro-

---

[6] It is anticipated that published proceedings of this workshop will be available.

osmotic permeability) and swelling pressure which appear in the macroscopic model."

Hueckel, Loret and Gajo (2001) define Representative Elementary Volumes (REV) for saturated clay as closed (exchanging of mass internally between components or open (external mass exchanges with the surroundings) systems. Three scale phenomena within the REV are considered: nano-scale, micro-scale, and macro-scale. Water adsorption and interlayer and external cation adsorption occur at the nano-scale, interphase mass exchange occurs at the micro-scale, and chemo-mechanical coupling occurs at the macro-scale. Reversible and irreversible effects of chemical changes on mechanical behavior are assumed to result from chemo-elastic and chemo-plastic softening resulting from inter-phase mass transfers and reactions. The derived relationships are considered by Hueckel, et al., (2001) to "form a basis for developing constitutive relationships for reversible and irreversible processes of deformation and mass transfer of species of water and salt during chemical swelling and collapse." These relationships are derived in Gajo, et al., (2001)

Commentary. There seems to be consensus that for small interparticle spacing, of the order of 1-2 nm, clay swelling is controlled by surface hydration forces, with osmotic diffuse double layer forces of interaction important only at greater spacing. Monte Carlo and Molecular Dynamics methods are being used to develop a better fundamental understanding of the many interactions in a clay-water-electrolyte system. These techniques allow determination of the nano-scale distribution and energetics of ions and molecules in relation to particle surfaces. Direct links from the results of these studies to practical applications for better geotechnical engineering of expansive soils may be forthcoming, but are not yet available.

Conclusion

The publication of Professor Charles C. Ladd's first paper in geotechnical engineering in 1960 came at a time when new developments in colloid chemistry and clay mineralogy made better understanding of the how's and why's of expansive soil behavior possible. Drawing on these new developments, he was able to identify the key factors controlling the swelling of compacted expansive clays in earthwork construction. Subsequent experimental and theoretical research on diffuse double layer theory, osmotic pressure, and adsorption of water on clay particle surfaces has established the qualitative influences of soil and pore fluid composition, soil structure, stress history and time on behavior, and the quantitative limits of the theories have been reasonably determined. Many of these developments have been reviewed in this paper. Having done so, it is clear that, although some details and emphases in Ladd (1960) might be somewhat changed were the paper to be written today, the essential concepts are as valid now as they were more than 40 years ago.

References

Blackmore, A. V. and Miller, R. D. (1962), "Tactoid Size and Osmotic Swelling in Calcium Montmorillonite," *Soil Science Society of America Proceedings,* Vol. 25, pp. 169-173

Bolt, G. H. (1956), "Physico-Chemical Analysis of the Compressibility of Pure Clay," *Geotechnique,* Vol. 6, No. 2, pp. 86-93.

Delville, A. (2001), "The Influence of Electrostatic Forces on the Stability and the Mechanical Properties of Clay Suspensions," *Proceedings of Clay Behavior: Chemo Mechanical Coupling,* Maratea, Italy, June 2001, Vol. I.

Dermatis, D. and Mitchell, J. K. (1992), "Clay Soil Heave Caused by Lime-Sulfate Reactions," *Innovations and Uses for Lime, ASTM STP 1135,* D. D. Walker, Jr., T. B. Hardy, D. C. Hoffman, and D. D. Stanley, Eds., American Society for Testing and Materials, Philadelphia, pp. 41-64.

Di Maio, C. (1996), "Exposure of Bentonite to Salt Solution: Osmotic and Chemical Effects," *Geotechnique,* Vol. 46, No. 4, pp. 695-707.

Farrar, D. M. and Coleman, J. D. (1967), "The Correlations of Surface Area with Other Properties of Nineteen British Clay Soils," *Journal of Soil Science,* Vol. 18, No. 1, pp. 118-124.

Gajo, A., Loret, B., and Hueckel, T. (2001), "Electro-Chemo Mechanical Couplings in Honoionic and Heteroionic, Elastic-Plastic Expansive Clays," *Proceedings of Clay Behavior: Chemo Mechanical Coupling,* Maratea, Italy, June 2001, Vol. III.

Hueckel, T., Loret, B., and Gajo, A. (2001), "Swelling Materials as Reactive, Deformable, Two-Phase Continua: Basic Modeling Concepts and Options," *Proceedings of Clay Behavior: Chemo Mechanical Coupling,* Maratea, Italy, June 2001, Vol. II.

Kidder, G. and Reed, L. W. (1972), "Swelling of Characteristics of Hydroxy-Aluminum Interlayered Clays," *Clays and Clay Minerals,* Vol. 20, pp. 13-20.

Kruyt, H.R. (1952), *Colloid Science I, Irreversible Systems,* Elsevier Pub. Co., new York.

Ladd, C.C. (1960), "Mechanisms of Swelling by Compacted Clay," *Water Tensions; Swelling Mechanisms; Strength of Compacted Soil,* Highway Research Board Bulletin 245, pp. 10-26.

Lambe, T.W. (1953), "Structure of Inorganic Soil," ASCE Separate No. 315.

Lambe, T.W. (1958), "The Structure of Compacted Clay" and " The Engineering Behavior of Compacted Clay," *Journal of the Soil Mechanics and Foundations Division*, ASCE, Vol. 84, No. SM2.

Low, P. F. (1987), "Structural Component of the Swelling Pressure of Clay," *Langmuir*, Vol. 3, pp. 18-25.

Low, P. F. (1992), "Interparticle Forces in Clay Suspensions: Flocculation , Viscous Flow, and Swelling," *Proceedings of the 1989 Clay Minerals Workshop on the Rheology of Clay/Water Systems*.

Madsen, F. T. and Müller-Vonmoos, M. (1985), "Swelling Pressure Calculated from Mineralogical Properties of a Jurassic Opalinum Shale, Switzerland," *Clays and Clay Minerals*, Vol. 33, No. 6, pp. 501-509.

Madsen, F. T. and Müller-Vonmoos, M. (1989), "The Swelling Behaviour of Clays," *Applied Clay Science*, Vol. 4, No. 2, pp. 143-156.

Mehta, P. K. and Hu, F. (1978), "Further Evidence for Expansion of Ettringite by Water Adsorption," *Journal of the American Ceramic Society – Discussions and Notes*, Mar.-Apr. 1978, pp. 179-180.

Mitchell, J. K. (1956), "The Fabric of Natural Clays and its Relation to Engineering Properties," *Proceedings of the Highway Research Board*, Vol. 35, pp. 693-713.

Mitchell, J. K. (1986), "Practical Problems from Surprising Soil Behavior," *Journal, Geotechnical Engineering Division*, ASCE, Vol. 112, No. 3, pp. 255-289.

Mitchell, J. K. (1993), *Fundamentals of Soil Behavior*, 2nd Edition, Wiley Interscience, New York, 437 pp.

Moyne, C. and Murad, M. (2001), "Micromechanical Computational Modeling of the Hydration Swelling of Montmorillonite," *Proceedings of Clay Behavior: Chemo Mechanical Coupling*, Maratea, Italy, June 2001, Vol. II.

Norrish, K. (1954), "The Swelling of Montmorillonite," *Discussions of the Faraday Society, London*, No. 18, pp. 353-359.

Rich, C. I. (1968), "Hydroxy Interlayers in Expansible Layer Silicates," *Clays and Clay Minerals*, Vol. 16, pp. 15-30.

Santamarina, J. C. and Fam, M. A. (1995), "Changes in Dielectric Permittivity and Shear Wave Velocity During Concentration Diffusion," *Canadian Geotechnical Journal*, Vol. 32, pp. 647-659.

Santamarina, J. C., Klein, K. A., and Fam, M. A. *Soils and Waves*, John Wiley and Sons, Inc.,Chichester, 488 pp.

Seed, H. B., Mitchell, J. K. and Chan, C. K. (1962), "Swell and Swell Pressure Characteristics of Compacted Clay," *Highway Research Board Bulletin 313*, pp. 12-39.

Skipper, N. (2001), "Influence of Pore-Liquid Composition on Clay Behaviour: Molecular Dynamics Simulations of Nano-Structures," *Proceedings of Clay Behavior: Chemo Mechanical Coupling*, Maratea, Italy, June 2001, Vol. II.

Sposito, G. (1984), *The Surface Chemistry of Soils*, Oxford University Press, New York, 234 pp.

Stern, O. (1924), "Zur Theorie der Elektrolytischen Doppelschriht, *Zeitschrift Electrochem*, Vol. 30, pp. 508-516.

Terzaghi, K. (1931), "The Influence of Elasticity and Permeability on the Swelling of Two Phase Systems," *Colloid Chemistry, Vol. III*, J. Alexander, Ed., Chemical Catalog Co., New York, pp. 65-88.

Warkentin, B. P., Bolt, G. H., and Miller, R. D., (1957), "Swelling Pressure of Montmorillonite," *Soil Science Society of America Proceedings*, Vol. 25, No. 5, pp. 495-497.

Soil Behavior at the Microscale:
Particle Forces

J. Carlos Santamarina[1]

Abstract

Soils are particulate materials. Therefore, the behavior of soils is determined by the forces particles experience. These include forces due to boundary loads (transmitted through the skeleton), particle-level forces (gravitational, buoyant, and hydrodynamic), and contact level forces (capillary, electrical and cementation-reactive). The relative balance between these forces permits identifying various domains of soil behavior. Furthermore, the evolution of particle forces helps explain phenomena related to unsaturation, differences between drained and undrained strength under various loading modes (including the effect of plasticity), sampling disturbance, and fines migration during seepage. Generally accepted concepts gain new clarity when re-interpreted at the level of particle forces.

Introduction

The limitations with continuum theories for the analysis of soil behavior were recognized early in the twentieth century. Terzaghi wrote "... Coulomb... purposely ignored the fact that sand consists of individual grains, and ... dealt with the sand as if it were a homogeneous mass with certain mechanical properties. Coulomb's idea proved very useful as a working hypothesis for the solution of one special problem of the earth-pressure theory, but it developed into an obstacle against further progress as soon as its hypothetical character came to be forgotten by Coulomb's successors. The way out of the difficulty lies in dropping the old fundamental principles and starting again from the elementary fact that sand consists of individual grains" (Terzaghi 1920 - includes references to previous researchers).

The fundamental understanding of soil behavior begins by recognizing the particulate nature of soils and its immediate implications: the interplay between particle characteristics (e.g., size, shape, mineralogy), inter-particle arrangement and interconnected porosity, inherently non-linear non-elastic contact phenomena, and

---

[1] Professor, School of Civil and Environmental Engineering, Georgia Institute of Technology, Atlanta, GA 30332. E-mail: carlos@ce.gatech.edu.

particle forces. While all these parameters are interrelated, the focus of this manuscript is on particle forces, their relative importance, and the re-interpretation of soil phenomena relevant to engineering applications with emphasis on processes studied by Prof. C. Ladd. Finally, the microscale analysis of forces leads to a re-interpretation of the effective stress principle and previously suggested modifications.

Particle Forces

Particle forces in soils were considered in the seminal paper by Ingles (1962), and later reviewed in Mitchell (1993) and Santamarina et al. (2001-a). Particle forces are classified herein in relation to the location of the generation mechanism:

- *Forces due to applied boundary stresses*: they are transmitted along granular chains that form within the soil skeleton. Capillary effects at high degree of saturation prior to air-entry fall under this category.
- *Particle-level forces*: includes particle weight, buoyancy and hydrodynamic forces. A particle can experience these forces even in the absence of a soil skeleton.
- *Contact-level forces*: includes capillary forces at low degree of saturation, electrical forces, and the cementation-reactive force. The first two can cause strains in the soil mass even at constant boundary loads. Conversely, the cementation-reactive force opposes skeletal deformation.

Mass-related magnetic forces (not relevant in most soils) and contact-level hydrodynamic squirt-flow type forces (that develop during dynamic excitation) are not considered in this review. The emphasis in this section is on recent developments in the understanding of particle forces, and includes simple, order-of-magnitude expressions to estimate these forces for the case of spherical particles.

Skeletal Forces (related to applied boundary stresses)

Early analyses based on spherical particles and regular packings (Deresiewicz 1973), photoelastic models (Durelli and Wu 1984), and the more recent developments in numerical micro-mechanics pioneered by Cundall (Cundall and Strack 1979) have provided unique insight into the distribution and evolution of inter-particle skeletal forces in soils.

Both normal N and tangential T forces develop at contacts when an effective stress σ' is applied at the boundary[2]. The normal force N at a contact is related to the applied state of effective stress σ' and the particle diameter d. The first order approximation $N = d^2\sigma'$ is appropriate for a simple cubic packing of equal size spheres. For a random packing of spheres, the mean normal contact force $\underline{N}$ is related to the void ratio e through some empirical or semi-empirical correction functions, rendering expressions such as (for $0.4 < e < 1$)

$$\underline{N} = \sigma' d^2 \left[\frac{\pi(1+e)^2}{12} \right] \tag{1}$$

---

[2] For convenience, the more common equivalent continuum concept of "stress" is herein used to refer to the distributed force applied at the boundary.

The boundary stress is not supported uniformly by the skeletal forces N and Weibull or exponential distributions apply (Dantu 1968; Gherbi et al. 1993; Jaeger et al. 1996). Further insight is gained from photoelastic studies such as those shown in Figure 1:

- Chains of particles form columnar structures that resist the applied boundary stress (Drescher and de Josselin de Jong 1972; Oda et al. 1985). These chains resemble a fractal-type structure. The smallest scale of chains is a few particle diameters in size. Particles that form part of these chains are loaded in the direction of the applied principal stress.

- Particles that are not part of the main chains play the secondary yet very important role of preventing the buckling of the main chains. Hence, the main forces acting on these particles are transverse to the main chains (Radjai et al. 1998 - see Figure 1-a).

- There are many particles that sit within the granular medium and do not carry skeletal load. These are "movable" particles and, if smaller than the pore throats, may migrate when proper fluid flow conditions develop in the medium.

- When large pores are present, force chains arch around the pores. These arches tend to collapse during shear (see Figure 1-b).

- The stability of the columns is related to the direction of particle movement during loading, so that reversing the direction of loading promotes instability. Load history dependency is manifested even at small deformations as shown in Figure 1c (Duffy and Mindlin 1957): the stiffness contributed by the contact shear resistance would be lost if the loading direction is reversed.

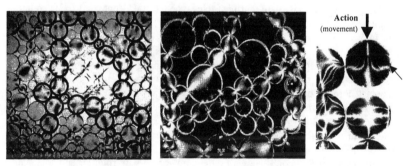

Figure 1: Skeletal force distribution – Photoelastic disks. (a) Random packing and force chains - different force directions along principal chains and in secondary particles (b) arches around large pores - precarious stability around pores. (c) Resistance mobilized during loading - contribution of shear stiffness (Courtesy of J. Valdes, M. Guimaraes, and M. Aloufi).

While the increase in mean stress promotes volume reduction and a higher coordination number (contacts) between particles, the increase in deviatoric load causes internal anisotropy in contact distribution. Inter-particle coordination and its anisotropy restrict the possible axis of rotation eventually leading to rotational arrest or frustration (Figure 2). In general, the probability for rotational frustration increases with increasing coordination, i.e., with decreasing void ratio (the dense system shown on the right is frustrated for any rotation). Therefore, rotational frustration is overcome by either frictional slippage at contacts or by fabric changes that lead to fewer contacts among particles, i.e., decreasing coordination number which is often associated with local dilation. In fact, the higher interparticle friction, the lower the extent of frictional slippage at contacts, and the lower the coordination number that develops during shear (see Thornton 2000). While these concepts are illustrated with monosize spherical particles, rotational frustration is affected by the relative size among neighboring particles, by the ability to attain high densities in soils with high coefficient of uniformity, and by particle shape (sphericity, angularity and roughness).

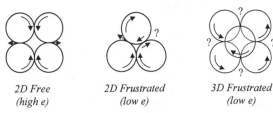

| 2D Free | 2D Frustrated | 3D Frustrated |
| (high e) | (low e) | (low e) |

Figure 2.   Rotational frustration. The lower the void ratio, the higher the number of contacts per particle and the higher the probability of rotational frustration.

These observations gain further relevance in the context of 3D micro-mechanical simulations of axial compression AC (b=0) and axial extension AE (b=1), where $b=(\sigma_2-\sigma_3)/(\sigma_1-\sigma_3)$. The distribution of contacts and average normal and shear contact forces in a given direction θ, herein denoted as $\underline{N}(\theta)$ and $\underline{T}(\theta)$, are depicted in Figure 3 (Chantawarangul 1993; see also Rothenburg and Bathurst 1989, and Thornton 2000). The following observations can be made:

- Contact normals during anisotropic loading become preferentially oriented in the direction of the main principal stress σ_1, in agreement with observations made above (see also Oda 1972).
- The main reduction in inter-particle contacts takes place in the direction of the minor principal stress: σ_2 and σ_3 directions in AC, and σ_3 direction in AE. This situation allows for more degrees of freedom for particle rotation and for chain buckling in AC (even when the total coordination number at failure is about the same in both cases).
- Such volume-average microscale response provides insight into the observed effective peak friction angle (macroscale - numerical results presented in the lower frame of Figure 3): higher friction angle is mobilized in AE than in AC. Furthermore, the lack of particle displacement in the direction of plane strain hinders rearrangement and causes an even higher peak friction angle in plane strain loading. The critical state friction angle obtained in numerical simulations follows a similar trend, but with less pronounced differences.
- Results by Chantawarangul (1993 - not presented here) also show that early volume contraction before the peak strength, is more pronounced in AE than in AC tests – relevant to undrained strength.

3D Distribution of:	(a) Isotropic confinement	At peak deviatoric load	
		(b) Axial Compression b=0	(c) Axial Extension b=1
Contact normals	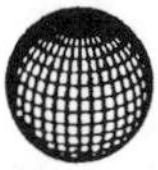	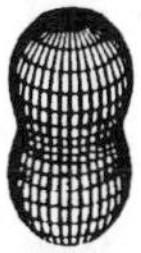	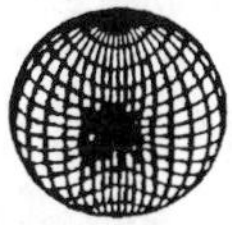
Average Normal Contact Force $\underline{N}(\theta)$	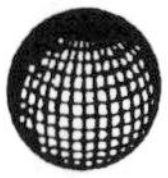	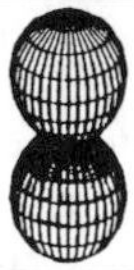	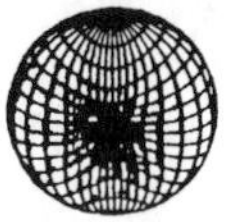
Average Shear Contact Force $\underline{T}(\theta)$ (magnified x5)	·	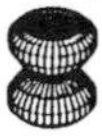	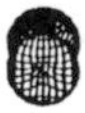
Friction angle	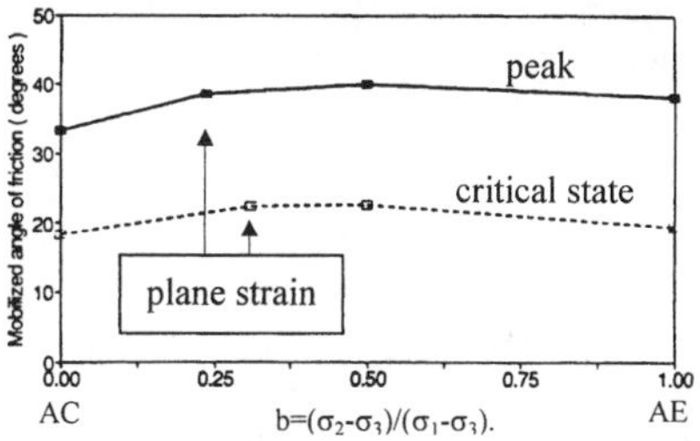		

Figure 3.   Numerical simulation: Evolution of inter-particle contacts and average normal and shear contact forces during axial compression and axial extension loading. Variation in friction angle with the intermediate stress σ_2. Figure compiled from Chantawarangul (1993).

The evolution of anisotropy in contact normals and in contact forces $\underline{N}(\theta)$ and $\underline{T}(\theta)$ reflects the soil response to the anisotropic state of stress that is applied at the boundaries. Ultimately, *the shear strength of a soil is the balance between two competing trends: the reduction in coordination to minimize frictional resistance by freeing particle rotation, and the increase in coordination following the buckling of particle chains.* Therefore, the shear strength of a soil reflects the restrictions to particle motion due to either mutual frustration or boundary conditions.

Gravitational Force: Weight and Buoyancy (Particle-Level)

Newton's fourth law specifies the attraction force between two masses m_1 and m_2 at a distance r: $F_G = Gm_1m_2/r^2$, where $G=6.673\times10^{-11}$ N·m^2/kg^2. This force causes tides as the moon interacts with oceans, and gives rise to the weight of soil grains on the earth (e.g., m_1 is the mass of the earth and m_2 is the mass of a soil grain); for a spherical particle of diameter d is,

$$W = \frac{1}{6}\pi G_s \gamma_w d^3 \qquad\qquad \textit{Weight of a sphere} \qquad\qquad (2)$$

where G_s is the specific gravity of the mineral that makes the particle and γ_w is the unit weight of water. The gravitational attraction F_G between two grains is much smaller than the weight of grains (about 10^{12} times smaller for millimetric particles).

Hydrostatic fluid pressure results from the weight of the fluid above the point under consideration. When a particle is submerged in water (or any other fluid), the water pressure is normal to the particle surface. The integral of the fluid pressure acting on the particle renders the buoyancy force. This force does not change, regardless of the submerged depth because the difference between the water pressure acting at the bottom and at the top of the particle remains the same, $\Delta u = \gamma_w d$ (rigid particles). In Archimedes' words, the buoyant force is equal to the weight of fluid the particle displaces, regardless of depth. For completeness,

$$U = Vol \cdot \gamma_w = \frac{1}{6}\pi \gamma_w d^3 \qquad\qquad \textit{Buoyant force} \qquad\qquad (3)$$

The effective weight of a submerged particle becomes W-U.

A related experiment considers the case of two soil grains press together using a clamp, and submerged into a pond. The particles experience not only the same buoyant force but also the same inter-particle contact force due to the clamp at all depths (Figure 4). Hence, *the local pore fluid pressure around a particle does not alter the effective inter-particle skeletal force.*

Figure 4   Hydrostatic fluid pressure, buoyant force U, weight W, and inter-particle skeletal forces N.

Hydrodynamic Force (Particle-Level)

The pore fluid moving along the interconnected pore network in the soil exerts viscous drag forces and forces resulting from velocity gradients (null average in straight macro-flow). Consider the pore fluid moving with velocity v relative to a particle of diameter d. The magnitude of the viscous drag force F_{drag} is proportional to the viscosity of the fluid μ, as predicted by Stoke's equation (applies to low Reynolds numbers - Graf 1984; Bear 1972):

$$F_{drag} = 3\pi\mu v d \qquad (4)$$

where the viscosity of water at 20°C is $\mu \cong 1$ centiPoise=0.001 N·s/m^2. The velocity of fluid moving through the pores in a soil is related to the hydraulic gradient i, the hydraulic conductivity k, and the porosity n of the soil, v= ki/n. Combining this relation with Equation 4 permits estimating the drag force experienced by a potentially movable particle sitting on a pore wall. Viscous drag also acts on the particles that form the skeleton, and together with the velocity gradient forces alters the effective stress acting on the soil (often referred to as the seepage force).

Capillary Force - Mixed Fluid Phase (Boundary to Contact-Level)

A molecule in a fluid experiences van der Waals attraction to neighboring molecules. As these forces act in all directions, they tend to cancel. However, this is not the case for water molecules at the surface of the fluid: these molecules feel an effective pull-in resultant force normal to the fluid surface. At the macroscale, this effect resembles a membrane that tries to shrink, creating a surface tension T_s which characterizes the interface (T_s=0.0727 N/m for water-air at room temperature). Because this membrane tries to shrink, the fluid inside drops has positive pressure. This pore fluid pressure is computed using Laplace's equation,

$$u = T_s\left(\frac{1}{r_1} + \frac{1}{r_2}\right) \qquad (5)$$

where r_1 and r_2 are the curvature radii of the air-water interface. For a spherical drop, r_1=r_2.

On the other hand, water tends to wet and hydrate hydrophilic mineral surfaces. When a saturated soil mass begins to dry, the gradually shrinking volume of water pulls the membrane in, while the membrane attempts to cling to the mineral surfaces around pore throats. Therefore, suction develops inside the pore fluid. The membrane tension is transmitted onto the skeleton in terms of effective stress, as in a triaxial specimen surrounded by a thin membrane and subjected to vacuum, hence, the force acting on particles at this stage develops at the *boundary*. As desiccation progresses, air gradually invades the specimen. If the fluid phase remains continuous, the medium is in the funicular regime. The average interparticle force $F_{cap}=\pi d^2 u/4$, is computed by invoking Equation 5. Radii r_1 and r_2 are related to the diameter d_{pore} of the largest pore at the air-water interface where the front is currently receding (the effective value for d_{pore} is about the diameter d of the particles surrounding the pore),

$$F_{cap} = \pi \frac{d^2}{d_{pore}} T_s \qquad \text{\textit{Funicular regime}} \qquad (6)$$

At very low moisture content, disconnected fluid remains in the form of menisci at inter-particle contacts; this is the pendular regime. Figure 5 shows a sequence of microphotographs that capture the drying of water in the menisci between two spherical particles. As menisci dry, the negative pressure increases, and the cross section decreases. The contact-level capillary force computed from Laplace's equation at very low moisture content is (asymptotic solution for w%→0; the value of F_{cap} is limited by water cavitation),

$$F_{cap} = \pi d T_s \qquad \text{\textit{Pendular regime (contact-level)}} \qquad (7)$$

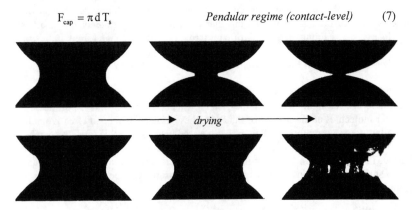

Figure 5.   Evolution of unsaturation – Pendular regime. The lower spherical glass bead (d= 2mm) is being held by the meniscus. Top sequence: de-ionized water. Lower sequence: water has salt in solution, salts eventually precipitate rendering inter-particle cementation (Gathered with D. Fratta).

Note that capillary forces in clays can be very high even at high degree of saturation, because d_{pore} is very small (Equation 6; in the presence of soluble salts and double layers, the total suction combines matric suction and osmotic suction - see Fredlund and Rahardjo 1993).

As the air-water interface gradually recedes during drying, it displaces first along the largest pores it encounters, then, the smaller pores are evacuated. Therefore, the negative pore fluid pressure increases as drying continues, and the evolution of inter-particle forces reflects the pore size distribution of the soil, as seen at the receding front. Pore size distribution is related to grain size distribution, thus, the suction-moisture plot resembles the grain size distribution of the soil (as observed in

Öberg 1997). Because of the stiffening effect of suction on soils, there is also a strong parallelism between the shear wave velocity vs. moisture plot and the grain size distribution plot (Cho and Santamarina, 2001 – continuously drying tests without remixing).

Electrical Forces (Contact-Level)

Can uncemented remolded soils exhibit cohesion? This can be tested by preparing a mud ball, submerging it in water to cancel capillary effects, and observing if the soil crumbles and reaches the angle of repose. This experiment must be carefully conducted to avoid seepage forces, entrapped air, diffusion and osmotic effects. Consider the following simple procedure to avoid these difficulties: a few grams of soil are mixed with some selected solution inside a test tube (diameter much larger than the particle size), the system is allowed to homogenize for 24 hr, a vacuum is applied to extract all the air while shaking the test tube, and the sediment is consolidated by subjecting the test tube to a high g-field in a centrifuge. Finally, the top is sealed with wax to avoid entrapping any air, and the tube is inverted. This is a very simple tension test. Kaolinite specimens prepared following these guidelines have been kept upside down for more than two years and no detachment has been observed (Figure 6). Clearly, the electrical attraction between particles is sufficient to support their buoyant weight.

Figure 6.   Electrical attraction forces are greater than the particle buoyant weight in fine soils (experimental details in the text – gathered with Y.H. Wang).

While the study of contact-level electrical forces started more than a century ago, molecular dynamic simulations and atomic force microscopy studies in the last 20 years have provided unprecedented information. The essence of all phenomena involved is the interplay between geometric compatibility, thermal agitation and Coulomb's electrical force. As a result, various repulsion and attraction forces develop, and the balance between these forces varies with inter-particle distance, rendering a highly non-linear force-distance relation. A brief review follows.

When the inter-particle distance exceeds ~30-40 Å, the response is well described disregarding the molecular nature of water molecules and the atomic nature of charges. Both, van der Waals attraction and electrical repulsion must be considered

(DLVO theory). The van der Waals attraction between two spherical particles of equal diameter d is (Israelachvili 1992),

$$Att = \frac{A_h}{24t^2}d \qquad\qquad van\ der\ Waals\ attraction \qquad (8)$$

where t is the separation between particles, and the Hamaker constant for silica-water-silica is $A_h = 0.64 * 10^{-20}$J. Repulsion results from thermal agitation and kinetic effects (also referred to as osmotic pressure or entropic confinement). The repulsion force between two spherical particles of diameter d is (derived after Israelachvili 1992 – applies to low surface potential and inter-particle distance t>ϑ)

$$\begin{aligned} Rep &= 32\pi RTc_0 d\vartheta e^{-\frac{t}{\vartheta}} \\ &\approx 0.0024d\sqrt{c_0}\ e^{-10^8 t\sqrt{c_0}} \end{aligned} \qquad Electrical\ repulsion\ force \qquad (9)$$

where the gas constant R=8.314 J/(mol.K), T is temperature [°K], c_0 is the ionic concentration of the pore fluid [mole/m^3], and the double layer thickness or Debye-Huckel length in units of [m] is equal to $\vartheta = 9.65 \cdot 10^{-9}(c_0)^{-0.5}$ for monovalent ions. The second simplified equation applies to monovalent ions, T=298°K, distance t in [m] and the force in [N].

At smaller distances, the ionic concentration between the particles exceeds the pore fluid concentration c_0, the osmotic pressure becomes independent of c_0 and it is only a function of the particle surface charge density. In turn, the effective surface charge density and the surface potential gradually decrease as particles come closer together due to ion binding (Delville, 2001). In this range, the electrostatic attraction force may exceed the van der Waals attraction, particularly when di-valent and tri-valent ions are present. Di-valent and tri-valent ions such as Ca^{2+} and Al^{3+} are most effective at shielding the electric field generated by the particle (lowering the surface potential), and if they bind to the particle, they may even render the particle positively charged. High-valence ions also interact among themselves so their positions are correlated, causing an additional attraction force between particles. Because of these effects, the presence of di-valent and tri-valent ions hinders the swelling of clays; these effects are not taken into consideration in the DLVO theory, therefore, this theory applies best to mono-valent ions.

When the inter-particle distance is less than ~10-20Å, the discrete nature of ions and water molecules must be recognized (the size of a water molecule is ~2.8 Å). In this range, the behavior of the particle-fluid-particle system resembles two plates with marbles in between: the molecular structure tends to be crystal-like ordered, friction is understood within the framework of thin film lubrication (Bhushan et al. 1995; Landman et al. 1996; Persson 1998), the water-ion system is organized reflecting the counterions' affinity for water and the density profile oscillates (Skipper et al. 1991; Delville 1995), inter-particle forces vary cyclically with distance with a periodicity of about one molecular diameter ("hydration force" - experimental data first reported in Horn and Israelachvili 1981), swelling progresses by discrete jumps ("hydration swelling"), and ionic mobility and diffusion are reduced (Skipper, 2001). Finally, at very high applied load, mineral to mineral interaction could develop and interpenetration is opposed by Born repulsion at the atomic level.

This framework explains most colloidal phenomena, including coagulation and swelling as a function of ionic concentration, as well as the effect of changes in fluid permittivity (changes in Hamaker constant and ion hydration). Yet, the direct application of these concepts could be misleading. For example, the sedimentation volume observed in test tubes does not decrease monotonically with ionic concentration, but often starts increasing at high concentration (Figure 7): either the form of aggregation of individual particles changes (e.g., from face-to-face "domains" to edge-to-face), or already formed domains flocculate forming open edge-to-face flocculation (Emerson 1959; Bennett and Hulbert 1986). Similar trends can be found in mechanical properties such as viscosity vs. concentration (e.g., van Olphen 1991). The transition concentration depends on clay mineralogy and pH.

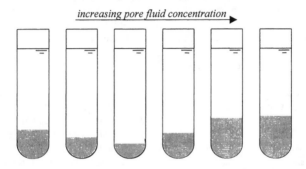

increasing pore fluid concentration ➤

Figure 7.   Electrical forces and fabric. The increase in ionic concentration above a characteristic value may render lower density fabrics (response varies with pH and mineralogy).

Cementation-reactive Force (Contact-Level)

There are many mechanisms leading to cementation. Figure 5 shows the evaporation of water in the meniscus between two particles and the precipitation of salt, forming crystals that bond the particles together. Some agents lithify the soil around particles and at contacts, while other processes change the initial physical-chemical structure (Mitchell 1993; Larsen and Chilingar 1979). Cementation is a natural consequence of aging and the ensuing diagenetic effects in soils. Most natural soils have some degree of inter-particle bonding (e.g., an ingenious device and data for London clay are presented in Bishop and Garga 1969).

Cementation is often accompanied by either shrinkage or swelling, and the ensuing changes in inter-particle skeletal forces. However, the most significant mechanical contribution of cementation is activated when strains are imposed onto the soil. To facilitate comparing this cementation-reactive force to other forces, the tensile force required to break the cement at a contact is computed herein. Consider a homogeneous layer of cementing material of thickness t deposited all around particles of diameter d. The diameter of the cement across the contact is determined by the Pythagorean relation; for small cement thickness, $d^2_{cont}=4d \cdot t$. Then, for a cement with tensile strength σ_{ten}, the maximum tensile force that the contact may withstand is

$$T = \pi \, t \, d \, \sigma_{ten} \tag{10}$$

Even small amounts of cementation may produce significant changes in the behavior of soils if the confinement is relatively low.

Forces: Relative Relevance and Implications

The relative balance between particle forces gives rise to various regimes and phenomena in soils that affect geotechnical engineering practice. A few salient examples follow. Phenomena studied by Prof. C. Ladd or reported in two comprehensive reviews he co-authored are often invoked (Ladd et al. 1977; Jamiolkowski et al. 1985; Ladd 1991 Terzaghi Lecture).

Skeletal -vs- Contact-Level Forces

Skeletal, capillary, and van der Waals forces contribute to the normal compressive contact force (the contribution of particle weight is in the vertical downwards direction and may be compressive if sitting or tensile if the particle is hanging). Their relative contributions for different size spherical particles are depicted in Figure 8 (using previous equations; van der Waals attraction is computed for an inter-particle separation of 30 Å. The skeletal force is shown for $\sigma'=10$ kPa and $\sigma'=1$ MPa). These compressive forces mobilize the electrical repulsion forces and bring particles together until compression and repulsion are balanced. Changing the pore fluid can alter the inter-particle distance at equilibrium; the upper part of the figure shows the strain caused by changing the pore fluid from fresh-water to seawater concentrations (axis on the right - Equation 9 combined with Equations 1 and 8). The following observations can be made:

- Particle weight looses relevance with respect to capillary forces for particles smaller than $d \approx 3$mm (Point 1 in the figure), and with respect to van der Waals attraction for particles smaller than $d \approx 30 \mu m$ (Point 2 in the figure).
- Capillary forces can exceed the contribution of $\sigma'=10$ kPa confinement for particles smaller than $d \approx 20 \mu m$ (Point 3) and the contribution of $\sigma'=1$ MPa for $d<0.2 \mu m$ (Point 4).
- Judging by the strain level, chemical-mechanical coupling gains relevance for micron and sub-micron particles: the smaller the particles or the lower the effective confinement, the greater the effect of changes in pore fluid chemistry.
- Particles are considered "coarse" when skeletal forces due to boundary loads prevail. This is the case for particles larger than $d \approx 20 \mu m$ (Point 3).
- Particles are "fine" when contact-level capillary and electrical forces gain relevance. This is the case when particles are smaller than $d \approx 1-10 \mu m$.

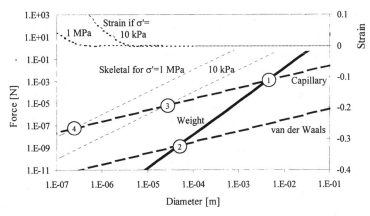

Figure 8.   Skeletal vs. contact-level capillary and electrical forces. The upper part of
the figure shows the strain (axis on right) caused by changing the pore
fluid ionic concentration from fresh-water to seawater conditions. Note
slopes: skeletal 2:1 (Equation 1), weight 3:1 (Equation 2), capillary and
van der Waals 1:1 (Equations 6, 7 and 8).

The contribution of the van der Waals attraction force remains much smaller
than the skeletal force under normal engineering conditions, even for very small
particles. In the absence of cementation (or edge-to-face coordination), the
classification of fine soils as "cohesive soils" is misleading and physically unjustified
(two related views are presented in Santamarina 1997, and in Schofield 1998). Still,
electrical forces determine fabric formation (Figure 7), which in turn affects soil
behavior. A soil that has formed within a high ionic concentration fluid develops a
characteristic fabric; if it is then leached with fresh water while confined, it may
preserve the salient features of its fabric, yet, for a different inter-particle force
condition. Therefore, it is not at its minimum energy configuration (as shown in
Figure 7) and it is unstable. This is the case of sensitive marine clays.

The opposite case is equally important: when the soil fabric is already formed
within a certain pore fluid, leaching a soil with a fluid with higher ionic concentration
and/or valence produces a reduction in the inter-particle repulsion force (Equation 9),
causing: a decrease in volume under a given effective stress condition, an increase in
shear wave velocity, and an increase in hydraulic conductivity. Note that there are
documented exceptions to these trends (an extensive compilation of experimental
studies can be found in Santamarina et al. 2001-b). The response depends on the
history of the test and whether stress or strain-controlled boundary conditions are
imposed. Similar observations also apply to clay swelling (Ladd 1959).

Skeletal Forces -vs- Cementation - Sampling Effects

The stress-strain behavior, the strength and the volume change tendency of soils
can be drastically affected by the degree of cementation (Clough et al. 1981; Lade

and Overton 1989; Airey and Fahey 1991; Reddy and Saxena 1993; Cuccovillo and Coop 1997). Two regions can be identified: the "cementation-controlled" region at low confinement, and the "stress-controlled" region at high confinement. In the cementation-controlled region, the small-strain shear stiffness can increase by an order of magnitude, the strength is cementation controlled, the buckling of chains is hindered (lower initial volume contraction), and the soil tends to brake in blocks (immediately after breaking, the inter block porosity is null, hence shear tends to cause high dilation, even if the cemented soil within the blocks has high void ratio).

The relative relevance of cementation and confinement can be identified by comparing the shear wave velocity in situ V_{So} with the velocity in a remolded specimen $V_{s\text{-remold}}$ that is subjected to the same state of stress. In general, one should suspect cementation if $V_{So} > V_{S\text{-remold}}$ (Note: creep and viscous effects also increase the shear wave velocity; these effects may be altered during sampling as well).

In most cases, natural cementation occurs when the soil is under confinement. When the soil is sampled, the applied confinement is removed, the center-to-center distance between particles increases, and the cement at contacts is put into tension. If the change in contact force $\Delta N = \Delta\sigma'd^2$ due to the stress reduction $\Delta\sigma'$ exceeds the tensile capacity of the cement at contacts (Equation 10), debonding occurs and the soil is permanently damaged or destructured. The cementation thickness can be related to the shear modulus of the soil G_s by considering a modified Hertzian formulation (Fernandez and Santamarina, 2001). Then, the following expression can be derived to predict the magnitude of stress reduction that can cause debonding σ'_{debond}:

$$\Delta\sigma' > \sigma'_{debond} = 1.9\sigma_{ten}\left(\frac{G_s}{G_g}\right)^2 \qquad \textit{condition for de-bonding} \qquad (11)$$

where σ_{ten} is the tensile strength of the bonding agent and G_g is the shear modulus of the mineral that makes the particles. The shear modulus of the soil G_s can be determined from the shear wave velocity V_{so} measured in situ as $G_s = \rho_s V_{so}^2$, where ρ_s is the mass density of the soil mass. The probability of debonding increases if the stress reduction $\Delta\sigma'$ approaches or exceeds the soil capacity $\Delta\sigma'_{debond}$. The upper bound for stress reduction is the in situ state of stress $\Delta\sigma' \approx \sigma_o'$ (details and data in Ladd and Lambe 1963). While G_g and σ_{ten} cannot be readily evaluated, this equation highlights the importance of the in situ shear wave velocity V_{so} in determining whether a soil will experience sampling effects due to stress reduction. Figure 9 summarizes these observations into four regions, and provides a framework for organizing available data and further studies on sampling effects (sampling disturbance is reviewed in Jamiolkowski et al. 1985). Note that it is premature to predict the potential impact of sampling on the bases of the in-situ shear wave velocity V_{So} alone.

De-bonding during sampling affects the behavior of both sands and clays, and it can cause important differences between the soil response measured in the laboratory

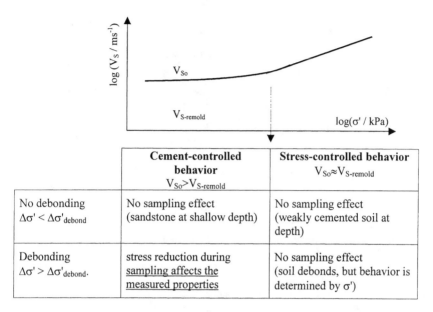

	Cement-controlled behavior $V_{So} > V_{S\text{-remold}}$	Stress-controlled behavior $V_{So} \approx V_{S\text{-remold}}$
No debonding $\Delta\sigma' < \Delta\sigma'_{debond}$	No sampling effect (sandstone at shallow depth)	No sampling effect (weakly cemented soil at depth)
Debonding $\Delta\sigma' > \Delta\sigma'_{debond}$.	stress reduction during sampling affects the measured properties	No sampling effect (soil debonds, but behavior is determined by σ')

Figure 9.   Skeletal forces vs. cementation strength – Sampling and debonding.

and in the field (Tatsuoka and Shibuya 1992; Leroueil, 2001; data by Stokoe published in Stokoe and Santamarina 2000).

In line with the main emphasis of this manuscript, the preceding analysis was done in terms of contact forces and stress reduction. Alternatively, the analysis can be generalized in terms of strains (or particle-level deformation) and compared against the linear and degradation threshold strain of the soil. Stress reduction is the prevailing mechanism in block sampling; with other samplers, insertion and removal of the specimens cause additional stresses, pore pressure changes and volume changes in the soil that must also be taken into consideration.

Drag Force, Weight and Electrical Forces - Fines Migration

The potential for fines migration during seepage depends on the balance between the drag force, the weight of the particle and other resisting contact forces. Figure 10 shows the drag force for different pore flow velocities in comparison with the sum of the weight and the van der Waals attraction (computed for a possible Rep-Att minimum at 30 Å and at 100 Å inter-particle distances). Notice that:

- The migration of particles greater than about 100 μm is unlikely (the required pore flow velocity would render a turbulent regime).
- The migration of particles less than ~10 μm is determined by the electrical forces. In this case, changing the pore fluid chemistry can alter the force balance. For

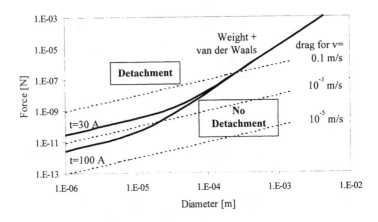

Figure 10. Drag vs. weight and net electrical attraction force.

example, a low concentration front may promote massive particle detachment allowing for particle migration.

• While individual small particles may not be detached, flocks of particles may.

For the conditions considered in the figure, a minimum pore flow velocity $v \approx 10^{-3}$ m/s is needed to cause detachment of any particle. Such pore flow velocity can be attained in sands or in coarse silts at high gradients. Therefore movable particles or "fines migration" is only relevant in the coarser formations and at high gradients, such as near a well.

Particles that are dragged may be flushed out of the soil or may form bridges at pore throats clogging the soil. Flushing and clogging depend on the relative size of the pore throats between skeleton-forming particles d_{large}, the size of the smaller migrating particles d_{small}, their ability to form bridges, and the volumetric concentration of fines in the permeant (Valdes 2002). In general, the required condition for flushing to occur is $d_{large}/d_{small} > 15$-to-30. These microscale considerations provide insight into filter criteria. Whether flushing or clogging develops, the movement of the movable particles renders fluid flow non-linear, causing pressure jumps and changes in effective stress.

Skeletal Force Distribution: Effective Stress Strength (Friction Angle)

Micromechanical analyses and simulations show the relevance of particle coordination, rotational frustration and the buckling of chains on the ability of a soil to mobilize internal shear strength (Figures 1, 2 and 3). Such analyses predict that: (1) the friction angle is highest in plane strain, then in axial extension, and least in axial compression, (2) the difference in peak friction between plane strain and AC increases with inter-particle friction and density due to the enhanced rotational frustration, and (3) the difference among critical state friction angles determined at different b-values is smaller than among peak friction values. All these micromechanical-based predictions are matched with experimental data gathered with sands (Figure 11-a and results in references such as Cornforth 1964; Bolton 1986; Schanz and Vermeer 1996). The evidence for any variation in critical state friction angle with b is less conclusive, yet minor differences may exist. Differences in friction angle in AE and AC are observed in clays as well, as shown in Figure 11.

The effect of the intermediate stress is not captured in Coulomb's failure criterion which predicts equal frictional resistance for all b-values, and it was first considered in the model by Lade and Duncan (1975).

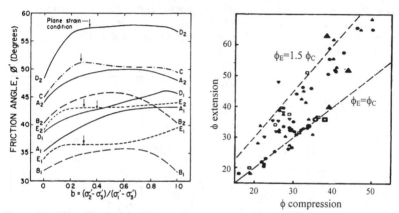

Figure 11. The effect of the intermediate stress on friction angle. (a) Sands - data from different authors compiled by Ladd et al. 1977 – Compare with Figure 3. (b) Clays: friction angle measured in AE b=0 and AC b=1 loading paths - from Mayne and Holtz (1985 - Most specimens are normally consolidated under K_0 conditions).

Skeletal Force Distribution: Undrained Strength (D_r and PI effects)

Ladd (1967) disclosed differences in undrained shear strength measured in different loading paths. As the undrained strength is controlled by the generation of pore pressure, the following microscale mechanisms should be considered:

1. The buckling of particle chains and the consequent transfer of confinement onto the pore fluid pressure. Buckling vulnerability increases in soils that have been anisotropically consolidated (Figure 3), and when subsequent loading reverses the direction of deformation and the lateral stability of columnar chains is altered. Displacement reversal also faces lower skeleton stiffness (Figure 1c). As mentioned earlier, a higher tendency to early volume contraction in AE than AC is observed in micromechanical simulations.

2. Spatial variability in void ratio and the increased instability of chains around large pores (Figure 1b).

3. Cementation (even slight), long-range electrical forces in small particles, and
    menisci at particle contacts in a mixed fluid-phase condition (e.g., soils with oil-
    water mixtures) provide contact stability and hinder buckling.
4. Inherent fabric anisotropy and its effects on skeletal stability and compressibility.

It follows from the first observation that AE loading should be more damaging
than AC loading. The supportive experimental evidence is overwhelming, in sands
(Hanzawa 1980, Vaid and Sivathayalan 1996, Yoshimine et al. 1999, Robertson et
al. 2000, Vaid and Sivathayalan, 2000), silts (Zdravkovic and Jardine 1997), and
clays (Bjerrum 1972, Ladd et al. 1977, Mayne and Holtz 1985, Jamiolkowski et al.
1985). Differences between AC and lateral extension LE, or between AE and lateral
compression LC are less conclusive (e.g., Campanella and Vaid 1972; Parry 1960).
Figure 12 shows data for sands and clays.

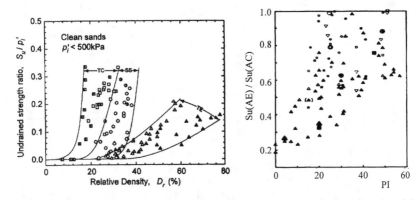

Figure 12. (a) Sands: comparison between undrained strength in triaxial axial
            compression TC, simple shear SS, and triaxial axial extension TE for
            various sands as a function of relative density (Yoshimine et al. 1999). (b)
            Clays: Undrained shear strength anisotropy Su(AE) / Su(AC) as a
            function of the plasticity of the clay; most specimens are normally
            consolidated under K_o conditions. Strength is defined at the maximum
            principal stress difference (Mayne and Holtz 1985).

Clearly, the collapse vulnerability of chains increases with decreasing relative
density of sands, as particle coordination decreases (Figure 12-a). A related factor not
captured in this figure is the effect of spatial variability of void ratio: arches around
large pores are most vulnerable during shear (Figure 1b), hence, large pores tend to
close first, as experimentally observed in soils (Sridharan et al. 1971; Delage and
Lefebvre 1984). The spatial void ratio variability is related to specimen preparation
methods (Castro 1969; Jang and Frost 1998). Hence, the undrained strength in soils is
not only affected by the mean porosity, but by the pore size distribution as well (soil
response for different specimen preparation methods is reviewed in Ishihara 1996).

The effect of plasticity in clays, shown in Figure 12-b, deserves special consideration. The typical role of electrical forces extends from about h=50 to 100 Å. Figure 13 compares a Hertzian-type analysis of the degradation threshold strain γ_{dt} applicable to coarse elastic particles, with an analysis applicable to small rigid particles where the threshold strain is calculated for a limiting thickness h that keeps particles "in touch". The corresponding relations are:

$$\gamma_{dt} = 1.3 \left(\frac{\sigma'}{G_g} \right)^{2/3} \qquad \text{\textit{Coarse grains}} \qquad (12)$$

$$\gamma_{dt} = 1.2 \frac{h}{d} \qquad \text{\textit{Fine grains}} \qquad (13)$$

where d is the particle diameter, G_g is the shear modulus of the mineral that makes the grains and σ' is the applied effective confining stress.

Figure 13. Contact forces and contact deformability - Degradation threshold strain. (a) Large particles: Hertzian deformation. (b) Small particles: electrical interaction.

The distance h for relevant inter-particle electrical interaction can be related to the thickness of the double layer 9. The value of h is in the range of 20 Å to 70 Å. The liquid limit and the plastic index of a soil are proportional to h and 1/d (Muhunthan 1991) therefore, Equation 13 confirms the link between PI and the degradation threshold strain observed by Vucetic and Dobry (1991). Then, the higher the plasticity of the clay, the higher the degradation threshold strain, the less vulnerable force chains are to buckling, and the lower the undrained strength anisotropy in axial extension vs. axial compression (Figure 12) The stabilizing effect of electrical forces can be readily confirmed by saturating soils specimens with non-polar fluids (S. Burns, personal communication).

Strength anisotropy data reflect the combined consequences of inherent fabric anisotropy and stress-induced anisotropy. Inherent fabric anisotropy results from either particle eccentricity and/or the biasing effects of deposition in a gravitational field. Its effect on undrained strength anisotropy can be explicitly studied by rotating the direction of the specimen an angle α with respect to the deposition direction (see data for sands and clays in Ladd et al. 1977; Jamiolkowski et al. 1985; Vaid and Sivathayalan 2000; for drained response in sand: Vaid and Sayao 1995). To facilitate the visualization of depositional anisotropy effects, consider the extreme anisotropic packing of platy particles illustrated in the sketches on the left side of Figure 14: all contact normals and normal contact forces are in the vertical direction, all particle axes and contact shear forces are in the horizontal direction, and all pores are of equal size and geometry. Clearly, particle slenderness enhances contact and force anisotropy. The following responses are expected in the mind experiments proposed in the sketches:

- AC (b=0, α=0) causes minimum pore pressure generation.
- AE or LC (b=1, α=90) causes high initial pore pressure, followed by the development of kinematic constraints and possible dilatancy after "phase transformation".
- Simple shear SS (b>0, α=>0) may not only produce excess pore pressure but also a failure that is aligned with particle orientation so minimum dilatancy may be mobilized.

The structure sketched in Figure 14 could form during the slow deposition of large platy particles, such as mica platelets, so that gravity prevails over electrical forces. However, particles with slenderness as low as ~1.1/1.0 can confer large fabric anisotropy effects to the soil (Rothenburg 1993). This is the case not only in clays but also in most sands; data for Fraser River sand is presented in Figure 14[3].

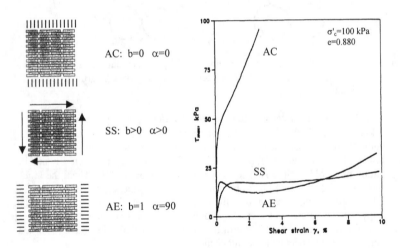

Figure 14. Inherent anisotropy effects on undrained strength. Left: conceptual models. Right: data for sands from Vaid and Sivathayalan (1996).

---

[3] Note: most numerical micromechanical simulations do not include gravity and depositional anisotropy (e.g., simulation in Figure 3). Slender particles accentuate the effects of stress induced anisotropy.

Spatial and Temporal Scales in Particle Forces

The variation in sedimentation volume with concentration shown in Figure 7 and the distribution of inter-particle skeletal forces forming particle chains shown in Figure 1 highlight the presence of multiple internal spatial scales in the medium. These scales add scale-dependent phenomena. For example, the distribution of skeletal forces (which reflects the interplay between chain buckling and rotational frustration) causes an uneven displacement field related to the mobilization of normal and shear forces at particle contacts and the threshold for frictional slippage, $T_{ult}=N\mu$. Individual particles move together by forming wedges that displace relative to each other along inter-wedge planes where the deformation localizes; eventually, the displacement becomes kinematically restricted, columns buckle, wedges break and new inter-wedge planes form. This behavior can be readily verified by assembling a 2D random packing of coins on a flat surface and enforcing the displacement of one boundary (Figure 15 – see Drescher and de Josselin de Jong 1972). Note that domains made of fine particles form conglomerates that can move in wedges as coarse grains; this observation can facilitate explaining the similarities between fine and coarse grained soil response, such as in Figure 11.

Vectors indicate the displacement of disks as the rod shown at the bottom is pushed into the assembly

Figure 15. Localization at the particle level – Wedges. At the particle level, the deformation in granular media is inherently uneven. (Test procedure: pennies on a scanner).

At larger scales, the localization of deformation leads to the development of shear bands, where the buckling of particle columns tends to concentrate (Oda and Kazama 1998). Therefore, the localization of deformation is an inherent characteristic of particulate materials, and it has been observed in dense-dilative soils under drained loading and under undrained loading (if cavitation is reached), loose-contractive soils under undrained loading, lightly cemented soils, unsaturated soils, soils with platy or rod-like particles, and heterogeneous soils (e.g., Vardoulakis 1996; Finno et al. 1997;

Saada et al. 1999; Mokni and Desrues 1998; Cho 2001). The development of global localization affects the interpretation of laboratory data, including results presented in Figures 11 and 12: the strain level at the formation of a shear band is a function of b (Lanier et al.1997, Wang and Lade, 2001), peak strength becomes specimen-size dependent as a function of soil rigidity and brittleness due to the associated "progressive failure", the measured global void ratio at large strain deviates from the critical state void ratio (Desrues et al. 1996), and the interpretation of the critical state friction angle is affected by the inclination of the shear band.

Particle forces also experience time-effects. The following examples apply to the different particle forces addressed above:

- *Normal and shear skeletal forces - Creep.* Two mechanisms are identified: First, material creep within particles near the contact (Kuhn and Mitchell 1993; Rothenburg 1993). Second, frictional slip; in this case the time scale at the particle level is determined by the inertial effects and the stress drop σ, $t=\sqrt{(d^2\rho/\sigma)}$. Supporting evidence is obtained with acoustic emission measurements.

- *Pore fluid pressure - Transient.* The most common time dependent effect in soils is the diffusion of excess pore pressure. While the boundary-level effect is considered in standard practice (this is the typical case in Terzaghi's consolidation), pressure differences also develop at the level of pores, for example, in dual porosity soils. The time scale for the dissipation of these local gradients or pressure diffusion is related to the internal length scale L_{int} and the internal coefficient of consolidation c_{v-int} which depends on the skeletal stiffness and permeability, $t=L_{int}^2/c_{v-int}$.

- *Capillary force.* When an unsaturated soil is subjected to shear, equilibrium in the water-air potentials is not regained immediately. When the water phase is continuous (funicular regime), the time scale is determined by the hydraulic conductivity of the medium and tends to be short. When the water phase is discontinuous and remains only at contacts (pendular regime), pressure homogenization occurs through the vapor pressure and is very slow. Evidence gathered with shear wave velocity is presented in Cho and Santamarina (2001).

- *Electrical forces.* Forcing a relative displacement between particles alters the statistical equilibrium position of counterions around particles, hence, the inter-particle forces. The time for ionic stabilization can be estimated using the ionic diffusion coefficient D, as $t=d^2/D$. Time varying changes in electrical conductivity after remolding a soil specimen was observed by Rinaldi (1998). Additionally, viscous effects take place during the transient.

- *Cementation.* Diagenesis is clearly time-dependent and depends on the rate of chemical reactions and diffusion.

Particle-level time scales and macro-scale time scales in soils can be very different. The steady-state distribution of forces within a particulate medium presumes that equilibrium has been reached at all particles. When equilibrium is not attained at a given particle, the particle displaces and alters the equilibrium of its neighbors. Therefore, while it would appear that the particle-level time scales listed above tend to be short in general, their manifestation at the level of the soil can extend for a long period of time due to the large number of particles that are recursively involved, even if only the particles along main chains are considered.

Time-dependency in particle forces is related to macroscale phenomena observed in soils such as creep, strain rate effects on strength, secondary consolidation, thixotropy, aging, pile capacity, and changes in penetration resistance (Mitchell 1960, Kulhawy and Mayne 1990, Mesri et al. 1990, Schmertmann 1991, Díaz-Rodríguez and Santamarina 1999).

The coexistence of multiple internal spatial and temporal scales in particle forces hints to important potential effects when trying to relate laboratory measurements to field parameters (strength, stiffness, diffusion and conduction parameters). Thus, parameters obtained in the lab must be carefully interpreted in order to select design parameters.

Reassessing Customary Concepts

Many commonly used concepts and accepted soil phenomena gain new clarity when they are re-interpreted at the level of particle forces. The principle of effective stress and the phenomenon of hydraulic fracture in soils are briefly addressed next.

Effective Stress and Modified Effective Stress Expressions

The concept of effective stress plays a pivotal role in understanding and characterizing soil behavior. Several earlier observations are relevant to the concept of effective stress, in particular:

- The hydrostatic pore pressure around a particle provides buoyancy. The intensity of the pore pressure around a particle does not affect the skeletal force transmitted between particles (Figure 4).
- Hydraulic gradients i cause fluid flow. The ensuing drag and velocity gradient forces act on particles and alter the effective stress transmitted by the skeleton.
- Changes in electrical and pendular capillary forces produce changes in volume, stiffness and strength, particularly in the finer soils and at low confinement (Figure 8).

Skeletal forces are defined at the boundaries (e.g., membrane in a triaxial specimen or equipotential lines in seepage) while other forces are determined at the particle or contact levels. Therefore the impact of these forces on soil behavior is different, and mixing both types of forces in a single algebraic expression can lead to incorrect predictions (as observed in Bishop and Blight 1963, for unsaturated soils). For example, some soils collapse upon wetting even though suction decreases and expansion should be expected; others experience a decrease in stiffness with an increase in ionic concentration even though a lower repulsion force is expected.

It follows from this discussion, that the use of modified effective stress expressions to accommodate suction or electrical repulsion and attraction forces should be discontinued (modified effective stress expressions are tabulated in Santamarina et al. 2001-a). Instead, behavior should be re-interpreted taking into consideration the separate and independent contributions of the skeletal force due to boundary loads and the other contact-level forces. This has been recognized in unsaturated media (Fredlund and Morgenstern 1977; Alonso et al. 1990), but it still requires further attention in the context of contact-level electrical forces (developments in constitutive modeling can be found Gajo et al. 2001; Guimaraes et al. 2001; Frijns et al. 1997).

Hydraulic Fracture in Soils

Hydraulic fracturing is intentionally or unintentionally produced in soil masses in a wide range of situations: thermal changes (both in the laboratory and in in situ, such as in thermal ground improvement techniques - Figure 16-a), as an experimental tool to determine the state of stress (similar to applications in intact rocks – reviewed in Ladd et al. 1977), it has been hypothesized as a failure mechanism in the failure of large dams (including Teton dam – exacerbated by arching and stress redistribution), it has been used in the context of deformation control (Mair and Hight 1994; Jafari et al. 2001), it is routinely utilized to increase the hydraulic conductivity (in view of enhanced oil recovery or even in decontamination strategies), and it occurs during grouting (even when compaction grouting is intended).

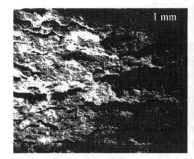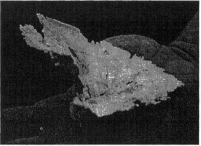

Figure 16.   Hydraulic fracture in soils. It can take place in both fine grained soils (electrical attraction greater than particle weight) and in uncemented coarse grained soils (electrical attraction is irrelevant). (a) Kaolinite specimen mixed with water at the LL, and subjected to microwave radiation for fast heating. (b) Grouting in Ottawa sand: the picture shows the fracture that formed by grouting gypsum after the gypsum hardened and was retrieved - obtained in cooperation with L. Germanovich.

How can hydraulic fracture take place in soils? Current fracture mechanics theories apply to media with tensile strength and fracture propagation involves tensile failure at the tip. Yet, there is virtually no tensile capacity in uncemented soils. For clarity, consider uncemented dry sands where the adhesion force between particles is much smaller than the weight of particles (Figure 16-b; refer to Figure 8). The stress anisotropy in soils cannot exceed the limiting anisotropy determined by friction, say $\sigma_1/\sigma_3 = \tan^2(45+\phi/2)$. Then, it appears that hydraulic fracturing in soils is the result of yield at the tip (stress path similar to AE), probably combined with other phenomena such as hydrodynamic drag and/or cavitation of the pore fluid in the process zone. From this perspective, the application of standard fracture mechanics theory invoking the undrained shear strength of the soil as a form of cohesion violates the fundamental behavior of soils.

Closing remarks

Summary of Main Observations. Discerning soil behavior at the microscale brings enhanced physical meaning and understanding that can be applied to comprehend both available results and new measurements. Such insight guides research as well as the understanding of soil behavior in view of engineering applications. The main observations follow:

- Soils are particulate materials. Therefore, particle forces determine soil response. Particle forces are interrelated with particle characteristics (including size and distribution, slenderness, mineralogy), contact behavior and fabric .
- The behavior of coarse-grained soils is controlled by skeletal forces related to boundary stresses. On the other hand, the finer the particle and the lower the effective confinement, the higher the relevance of contact-level electrical and capillary forces. The transition size from coarse to fine for standard engineering applications is around $d \approx 10 \mu m$.
- Hydrodynamic forces alter the skeletal forces and can displace movable particles. Either fines migration or clogging renders non-linear fluid flow, and affects the pore pressure distribution and the effective stress.
- Cementation, even if small, alters stiffness, strength and volume change tendencies. Two regions can be identified: the low-confinement cementation-controlled region and the high-confinement stress-controlled region.
- While previously suggested modified effective stress expressions can incorporate the various contact-level forces, they are physically incorrect and can lead to inadequate predictions. Therefore, their use should be discontinued.
- The drained strength of a soil reflects the balance between two competing micro-processes: the decrease in inter-particle coordination to reduce rotational frustration (in order to minimize friction), and the buckling of chains (that increases coordination).
- The undrained strength is determined by the (tendency to) volume compressibility of the skeleton, which depends on the vulnerability of load carrying chains.
- The degradation threshold strain in coarse soils increases with the applied load and decreases with the stiffness of the particles. In fine grain soils, it increases with the effective distance of electrical forces and with decreasing particle size, hence, the higher the plastic index the higher the degradation threshold strain.
- There are multiple internal scales inherent to particle forces in soils (both spatial and temporal).
- The re-interpretation of common concepts at the level of particle forces, such as "cohesive soil", "effective stress" and "hydraulic fracture", provides enhanced insight into soil behavior.

Current Capabilities. The soil researcher today has exceptional experimental and numerical tools to study soil behavior, including particle forces. Likewise, today's geotechnical engineer can use robust, versatile and inexpensive numerical capabilities, standard testing methods, and a new generation of testing principles and procedures. These tools must be implemented within the framework of a clear understanding of soil behavior.

Challenging future. There are abundant fascinating research questions in soil behavior, including: the implications of omnipresent strain-localization and multiple internal scales on soil properties and engineering design; the reinterpretation of friction in light of a large number of new studies (coarse and fine soils) and harnessing friction through inherent noise-friction interaction; new characterization-rehabilitation methodologies based on dynamic energy coupling; bio-geo phenomena; methane hydrates (characterization and production); and subsurface imaging (similar to medical diagnosis) combined with the determination of engineering design parameters or within the framework of an advanced observational approach.

Today's availability and easy access to information were unimaginable only 20 years ago. Yet, paradoxically, this great facility to access almost unlimited information appears to enhance the risk of forgetting knowledge. Indeed, we are challenged not only to address new fascinating questions, but also to preserve the great insight and understanding developed by the preceding generations. This should be our commitment today, as we celebrate Prof. C. Ladd's leading example.

Acknowledgements

Prof. Ladd's contributions have profoundly enhanced our understanding of soil behavior and its implementation in engineering design. Many colleagues provided valuable comments and suggestions, including: J. Alvarellos, S. Burns, G. Cascante, G.C. Cho, D. Fratta, D. Frost, L. Germanovich, M.S. Guimaraes, I. Herle, K.A. Klein, J.S. Lee, P. Mayne (his knowledge of the literature is enviable), A.M. Palomino, G. Rix, J.R. Valdes, V. Uchil (for the many runs to the library), Y.H. Wang, and exceptional anonymous reviewers. References acknowledge the insight gained from other authors. Funding was provided by the National Science Foundation.

References

Airey, D.W. and Fahey, M. (1991), Cyclic Response of Calcareous Soil from the North-West Shelf of Australia, Géotechnique, vol. 41, no. 1, pp. 101-121.

Alonso, E.E. Gens A. and Josa, A. (1990), A constitutive Model for Partially Saturated Soils, Géotechnique, vol. 40, no. 3, pp. 405-430.

Bear, J. (1972), Dynamics of Fluids in Porous Media, Dover, New York, 764 pages.

Bennett, R.H. and Hulbert, M.H. (1986), Clay Microstructure, International Human Resources Development Corporation, Boston, 161 pages.

Bhushan, B., Israelachvili, J. N. and Landman, U. (1995), Nanotribology: Friction, wear and lubrication at the atomic scale, Nature, vol. 374, pp. 607-616.

Bishop, A. W. and Blight, G. E. (1963), Some aspects of effective stress in saturated and partly saturated soils, Géotechnique, vol. 13, pp. 177-196.

Bishop, A.W. and Garga, V.K. (1969), Drained Tension Tests on London Clay, Géotechnique, vol. 19, pp. 309-313.

Bjerrum, L. (1972), Embankments on Soft ground, Proc. Earth & Earth-Supported Structures (Purdue Conference), vol. 2, ASCE, Reston, Virginia, pp 1-54.

Bolton, M. D. (1986), The strength and dilatancy of sands, Géotechnique, vol. 36, pp. 65-78.

Campanella R.G. and Vaid, Y.P. (1972), A simple Ko Triaxial Cell, Canadian Geotechnical Journal, vol. 9, pp.249-260.

Castro G. (1969), Liquefaction of Sands, Ph.D. Dissertation, Harvard University, Massachusetts, 112 pages.

Chantawarangul, K. (1993), Numerical Simulation of Three-Dimensional Granular Assemblies, Ph.D. Dissertation, University of Waterloo, Ontario, 219 pages.

Cho, G.C. and Santamarina, J.C. (2001), Unsaturated Particulate Materials - Particle Level Studies, ASCE Geotechnical Journal, vol. 127, no. 1, pp. 84-96.

Cho, G.C. (2001), Unsaturated Soil Stiffness and Post Liquefaction Shear Strength (Chapter III: Omnipresent Localizations), PhD Dissertation, Georgia Institute of Technology, 288 pages.

Clough, G.W., Sitar, N., Bachus, R.C. and Rad, N.S. (1981), Cemented Sands Under Static Loading. Journal of Geotechnical Engineering, ASCE, vol. 107, no. 6, pp. 799-817.

Cornforth, D.H. (1964), Some Experiments on the Influence of Strain Conditions on the Strength of Sand, Géotechnique, vol. 14, no. 2, pp. 143-167.

Cuccovillo, T. and Coop, M.R. (1997), Yielding and Pre-Failure Deformation of Structure Sands, Géotechnique, vol. 47, no. 3, pp. 491-508.

Cundall, P.A. and Strack, O.D.L. (1979), A discrete Numerical Model for Granular Assemblies, Geotechnique, vol. 29, no. 1, pp. 47-65.

Dantu, P. (1968), Etude Statistique des Forces Intergranulaires dans un Milieu Pulverulent, Géotechnique, vol. 18, pp. 50-55.

Delage, P. and Lefebvre, G. (1984), Study of the structure of a sensitive Champlain clay and of its evolution during consolidation, Canadian Geotechnical Journal, vol. 21, pp. 21-35.

Delville, A. (1995), Monte Carlo Simulations of Surface Hydration: An Application to Clay Wetting, J. Phys. Chem. vol. 99, pp. 2033-2037.

Delville, A. (2001), The Influence of Electrostatic Forces on the Stability and the Mechanical Properties of Clay Suspensions, Proc. Workshop on Clay Behaviour: Chemo-Mechanical Coupling, Edts., C. Di Maio, T. Hueckel, and B. Loret, Maratea, Italy. 20 pages.

Deresiewicz, H. (1973), Bodies in contact with applications to granular media, Mindlin and Applied Mechanics, Edited by B. Herrman, Pergamon Press, New York.

Desrues, J., Chambon, R., Mokni, M. and Mazerolle, F. (1996), Void Ratio Evolution Inside Shear Bands in Triaxial Sand Specimens Studied by Computed Tomography, Geotechnique, vol. 46, no. 3, pp. 529-546.

Díaz-Rodríguez, J.A. and Santamarina, J.C. (1999), Thixotropy: The Case of Mexico City Soils, XI Panamerican Conf. on Soil Mech. and Geotech. Eng., Iguazu Falls, Brazil, Vol. 1 pp. 441-448.

Drescher, A. and de Josselin de Jong, G. (1972), Photoelastic Verification of a Mechanical Model for the Flow of a Granular Material, J. Mechanics and Physics of Solids, vol. 20, pp. 337-351.

Duffy, J. and Mindlin, R. D. (1957), Stress-strain relations of a granular medium, Journal of Applied Mechanics, vol. 24, pp. 585-593.

Durelli, A. J. and Wu, D. (1984), Loads between disks in a system of discrete elements, Experimental Mechanics, December, pp. 337-341.

Emerson, W.W. (1959), The structure of Soil Crumbs, J. Soil Science, vol. 10, no. 2, pp. 234.

Fernandez, A. and Santamarina, J.C. (2001), The Effect of Cementation on the Small Strain Parameters of Sands, Canadian Geotechnical Journal, vol. 38, no. 1, pp. 191-199.

Finno, R.J., Harris W.W., Mooney, M.A. and Viggiani, G. (1997), Shear Bands in Plane Strain Compression of Loose Sand, Géotechnique, vol. 47, no. 1, pp. 149-165.

Fredlund, D. G. and Morgenstern, N. R. (1977), Stress state variables for unsaturated soils, Journal of Geotechnical Engineering, ASCE, vol. 103, pp. 447-466.

Fredlund, D. G., and Rahardjo, H. (1993). Soil Mechanics for Saturated Soils. John Wiley and Sons, New York.

Frijns, A.J.H., Huyghe, J.M. and Janssen, J.D. (1997), A Validation of the Quadriphasic Mixture Theory for Intervertebral Disc Tissue, Int. Journal Engineering Science, vol. 35, pp. 1419-1429.

Gajo, A., Loret, B. and Hueckel, T. (2001), Electro-Chemo-Mechanical Coupling in Homoionic and Heteroionic Elastic-Plastic Expansive Clays, Proc. Workshop on Clay Behaviour: Chemo-Mechanical Coupling, Edts., C. Di Maio, T. Hueckel, and B. Loret, Maratea, Italy. 26 pages.

Gherbi, M. Gourves, R. and Oudjehane, F. (1993), Distribution of the Contact Forces Inside a Granular Material, Powders and Grains 93, Prod. 2nd International Conference on Micromechanics of Granular Media, Brimingham, Elsevier, Amsterdam, pp. 167-171.

Graf, W.H. (1984), Hydraulics of Sediment Transport, Water Resources Publications, Highlands Ranch, Colorado, 513 pages.

Guimaraes, L., Gens, A, Sanchez, M. and Olivella, S. (2001), A Chemo-Mechanical Model for Unsaturated Expansive Clays, Proc. Workshop on Clay Behaviour: Chemo-Mechanical Coupling, Edts., C. Di Maio, T. Hueckel, and B. Loret, Maratea, Italy. 28 pages.

Hanzawa, H. (1980), Undrained Strength and Stability Analysis for a Quick Sand, Soils and Foundations, vol. 20, no. 2, pp. 17-29.

Horn, R.G. and Israelachvili, J. (1981), Direct Measurement of Structural Forces Between Two Surfaces in a Non-Polar Liquid, J. Chemical Physics, vol. 75, no. 3, pp. 1400-1411.

Ingles, O.G. (1962), Bonding Forces in Soils - Part 3, Proc. Of the First Conference of the Australian Road Research Board, vol. 1, pp. 1025-1047.

Ishihara, K. (1996), Soil Behaviour in Earthquake Geotechnics, Calderon Press, Oxford, 350 pages.

Israelachvili, J. (1992), Intermolecular and Surface Forces, Academic Press, New York, 450 pages.

Jafari, M.R., Au, S.K., Soga, K., Bolton, M.D. and Komiya, K. (2001), Fundamental Laboratory Investigation of Compensation Grouting in Clay, Cambridge University, 15 pages.

Jaeger, H. M., Nagel, S. R. and Behringer, R. (1996), The physics of granular materials, Physics Today, April, pp. 32-38.

Jamiolkowski, M., Ladd, C.C., Germaine, J.T. and Lancellotta, R. (1985), New Developments in Field and Laboratory Testing of Soils, Proc. 11th ICSMFE, San Francisco, vol. 1, pp 67-153.

Jang, D.J. and Frost, J.D. (1998), Sand Structure Differences Resulting from Specimen Preparation Procedures, Prod. ASCE Conf. Geotechnical Earthquake Engineering and Soil Dynamics, Seattle, vol. 1, pp. 234-245.

Kulhawy, F.H. and Mayne, P.W. (1990), Manual on Estimating Soil Properties for Foundation Design, EPRI EL-6800, Electric Power Research Institute, Palo Alto.

Kuhn, M. R. and Mitchell, J.K. (1993), New Perspectives on Soil Creep, ASCE J. Geotechnical Engineering, 119(3): 507-524.

Ladd, C.C. (1959), Mechanisms of Swelling by Compacted Clays, Highway Research Board Bulletin 245, pp. 10-26.

Ladd, C.C. and Lambe, T.W. (1963), The strength of Undisturbed Clay Determined from Undrained Tests, NRC-ASTM Symposium on Laboratory Shear Testing of Soils, Ottawa, ASTM STP 361, pp. 342-371.

Ladd, C.C. (1967), Discussion, Proc. Geotechnical Conference on Shear Strength Properties of Natural Soils and Rocks, NGI, Oslo, vol. 2, pp. 112-115.

Ladd, C.C., Foott, R., Ishihara, K., Schlosser, F. and Poulos, H.G. (1977), Stress-Deformation and Strength, Proc. 9th ICSMFE, vol. 2, Tokyo, pp. 421-494.

Ladd, C.C. (1991), Stability Evaluation during Staged Construction, J. Geotechnical Engineering, vol. 117, no. 4, pp. 540-615.

Lade, P.V. and Duncan, J.M. (1975), Elasto-plastic Stress-Strain Theory for Cohesionless Soil, ASCE Journal of Geotechnical Engineering, vol. 101, no. 10, pp. 1037-1053.

Lade, P.V. and Overton, D. D. (1989), Cementation effects in Frictional Materials, ASCE Journal of Geotechnical Engineering. 115, pp. 1373-1387.

Landman, U., Luedtke, W.D. and Gao, J.P. (1996), Atomic-Scale Issues in Tribology: Interfacial Junctions and Nano-Elastohydrodynamics. Langmuir, vol. 12, pp. 4514-4528.

Lanier, J., Viggiani, G. and Desrues, J. (1997), Discussion to the paper by J. Chu, R. Lo and I.K.Lee entitled Strain Softening and Shear band Formation of Sand in Multi-axial Testing, Geotechnique, vol. 47, no. 5 pp. 1073-1077.

Larsen, G. and Chilingar, G.V. (1979), Diagenesis in Sediments and Sedimentary Rocks, Elsevier Scientific Publishing Company, New York, N.Y.

Leroueil, M.S. (2001) Natural Slopes and Cuts: Movement and Failure Mechanisms, Géotechnique, vol. 51, no. 3, pp. 195-244

Mair, R.J. and Hight, D.W. (1994), Compensation Grouting, World Tunneling, November, pp. 361-367.

Mayne, P.W. and Holtz, R.D. (1985), Effect of principal stress rotation on clay strength, Proc. 11th ICSMFE, San Francisco, vol. 2, pp 579-582.

Mesri, G., Feng, T.W. and Benak, J.M. (1990), Postdensification Penetration Resistance of Clean Sands, vol. 116, no. 7, pp. 1095-1115.

Mitchell, J.K. (1993), Fundamentals of Soil Behavior, J.Wiley, New York, 437 pages.

Mitchell, J.K. (1960), Fundamentals Aspects of Thixotropy in Soils, Journal of the Soil Mechanics and Foundations Division, ASCE, vol. 86, no. 3, pp. 19-52.

Mokni, M. and Desrues, J. (1998), Strain Localization Measurements in Undrained Plane-Strain Biaxial Tests on Hostun RF Sand, Mechanics of Cohesive-Frictional Materials, vol. 4, pp. 419-441.

Muhunthan, B. (1991), Liquid Limit and Surface Area of Clays, Géotechnique, vol. 41, pp. 135-138.

Öberg, A.L. (1997), Matrix Suction in Silt and Sand Slopes, Ph.D. Dissertation, Chalmers University of Technology, Sweden, 160 pages.

Oda, M., Nemat-Nasser, S. and Konishi, J. (1985), Stress-Induced Anisotropy in Granular Masses, Soil and Foundation, vol. 25, pp. 85-97.

Oda, M. (1972), The mechanism of fabric changes during compressional deformation of sand, Soils and Foundations, vol. 12, no. 2, pp. 1-18.

Oda, M. and Kazama, H. (1998), Microstructure of Shear Bands and its Relation to the Mechanism of Dilatancy and Failure of Dense Granular Soils, Géotechnique, vol. 48, no. 4, pp. 465-481.

Parry, R.H.G. (1960), Triaxial Compression and Extension Tests on Remoulded Saturated Clay, Géotechnique, vol. 10, pp. 166-180.

Persson, B. N. J. (1998), Sliding Friction - Physical Principles and Applications, Springer, New York, 462p.

Radjai, F., Wolf, D.E., Jean, M. and Moreau, J.J. (1998), Bimodal Character of Stress Transmission in Granular Packings, Physical review Letters, vol. 80., no. 1, pp. 61-64.

Reddy, K.R. and Saxena, S.K. (1993), Effects of Cementation on Stress-Strain and Strength Characteristics of Sands, Soils and Foundations, vol. 33, no. 4, pp. 121-134.

Rinaldi, V.A. (1998), Changes in electrical conductivity after remolding a clayey soil, personal communication. A similar effect was observed in Mexico City soils and is documented in Diaz-Rodriguez and Santamarina, 1999.

Robertson, P.K. et al. (2000), The CANLEX Project: Summary and Conclusions, Can. Geotechnical J., vol. 37, pp. 563-591.

Rothenburg, L. (1993), Effects of particle shape and creep in contacts on micromechanical behavior of simulated sintered granular media, Mechanics of Granular Materials and Powder Systems, vol. 37, pp. 133-142.

Rothenburg, L. and Bathurst, R. J. (1989), Analytical study of induced anisotropy in idealized granular material, Géotechnique, vol. 49, pp. 601-614.

Saada, A.S., Liang, L., Figueroa, J.L. and Cope, C.T. (1999), Bifurcation and Shar Band Propagation in Sands, Géotechnique, vol. 49, no. 3, pp. 367-385.

Santamarina, J.C., Klein, K. and Fam, M. (2001-a), Soils and Waves, J. Wiley, Chichester, UK, 488 pages.

Santamarina, J.C, Klein, K. Palomino, A. and Guimaraes, M (2001-b), Micro-Scale Aspects Of Chemical-Mechanical Coupling - Interparticle Forces And Fabric; Edts., C. Di Maio, T. Hueckel, and B. Loret, Maratea, Italy. 26 pages.

Santamarina, J.C. (1997), 'Cohesive Soil:' A Dangerous Oxymoron, The Electronic J. Geotechnical Eng. - Magazine, August (URL: under Electronic Publications in http://www.ce.gatech.edu/~carlos).

Schanz, T. and Vermeer, P.A. (1996), Angles of friction and dilatancy in sand, Géotechnique, vol. 46, no. 1 , pp. 145-151.

Schmertmann, J.H. (1991), The Mechanical Aging of Soils, ASCE J. Geotechnical Engineering, vol. 117, no. 9, pp.1288-1330.

Schofield, A. (1998), Don't Use the C-Word, Ground Engineering, August, pp. 29-32.

Skipper, N.T. (2001), Influence of Pore-Liquid Composition on Clay Behaviour: Molecular Dynamics Simulations of Nano-Structure, Proc. Workshop on Clay Behaviour: Chemo-Mechanical Coupling, Edts., C. Di Maio, T. Hueckel, and B. Loret, Maratea, Italy. 18 pages.

Skipper, N.T., Refson, K. and McConnell, J.D.C. (1991), Computer Simulation of Interlayer Water in 2:1 Clays, J. Chemical Physics 94, no. 11, pp. 7434-7445.

Sridharan, A., Altschaeffle, A. G. and Diamond, S. (1971), Pore size distribution studies, Journal of the Soil Mechanics and Foundations Division, Proceedings of the ASCE, vol. 97, SM 5, pp. 771-787.

Stokoe, K.H. and J.C. Santamarina (2000), Seismic-Wave-Based Testing in Geotechnical Engineering, GeoEng 2000, Melbourne, Australia, November, pp. 1490-1536

Tatsuoka, F. and Shibuya, S. (1992), Deformation Characteristics of Soils and Rocks from Field and laboratory Tests, The University of Tokyo, vol. 37, no. 1, 136 pages.

Terzaghi, K (1920), Old Earth Pressure Theories and New Test Results, Engineering News-Record, vol. 85, n. 14, pp. 632-637.

Thornton, C. (2000), Numerical Simulations of Deviatoric Shear Deformation of Granular Media, Geotechnique, vol. 50, pp. 43-53.

Vaid, Y.P. and Sayao, A. (1995), Proportional Loading Behavior in Sand Under Multiaxial Stresses, Soils and Foundations, vol. 35, no.3, pp. 23-29.

Vaid, Y.P. and Sivathayalan, S. (1996), Static and Cyclic Liquefaction Potential of Fraser Delta Sand in Simple Shear and Triaxial Tests, Canadian Geotechnical J., vol. 33, pp. 281-289.

Vaid, Y.P. and Sivathayalan, S. (2000), Fundamental Factors Affecting Liquefaction Susceptibility of Sands, Can. Geotechnical J., vol. 37, pp. 592-606.

Valdes, J.R. (2002), Fines Migration and Formation Damage, PhD Dissertation, Georgia Institute of Technology.

van Olphen, H. (1991), An Introduction to Clay Colloid Chemistry, Krieger Publishing Cy, Florida, pp. 318.

Vardoulakis, I. (1996), Deformation of Water-Saturated Sand: I. Uniform Undrained Deformation and Shear Banding, Géotechnique, vol. 46, no. 3, pp. 441-456.

Vucetic, M. and Dobry, R. (1991), Effect of soil plasticity on cyclic response, Journal of Geotechnical Engineering, ASCE, vol. 117, pp. 89-107.

Wang, Q. and Lade, P.V. (2001), Shear Banding in True Triaxial Tests and Its Effect on Failure of Sand, in print.

Yoshimine, M., Robertson, P.K. and Wride, C.E. (1999), Undrained Shear Strength of Clean Sands to Trigger Flow Liquefaction, Canadian Geotechnical Journal, vol. 36, pp. 891-906.

Zdravkovic, L. and Jardine, R.J. (1997), Some Anisotropic Stiffness Characteristics of a Silt Under General Stress Conditions, Géotechnique, vol. 47, no. 3, pp. 407-437.

Coupled Chemical And Liquid Fluxes In Earthen Materials

Harold W. Olsen[1]

Abstract

The potential importance of coupled chemical and liquid fluxes in earthen materials has been recognized and investigated in the geologic and soil science disciplines for several decades. The mechanisms involved cause a family of coupled transport phenomena that are commonly referred to as osmotic or membrane phenomena. Recently these phenomena have attracted the interest of geotechnical researchers because they appear to provide mechanisms for enhancing the capability of clay barriers to impede the migration of contaminants. To facilitate geotechnical interest in these phenomena, this paper presents a brief review of the pioneering field studies of membrane phenomena in geologic systems, a summary of the pertinent coupled flow theory developed to date, and a synthesis of the experimental evidence obtained in laboratory studies during the last five decades.

Introduction

The coupling or interactions between chemical and liquid fluxes in earthen materials cause a family of coupled transport phenomena that are commonly referred to as membrane or osmotic phenomena in the geologic and soil science literature. These phenomena arise in materials that selectively restrict the passage of dissolved species in an aqueous solution.

The possibility that earthen materials could exhibit membrane behavior was first suggested near the beginning of the 20th century (Becker, 1893; Briggs, 1902), and again in the nineteen thirties and forties (Russell, 1933; De Sitter, 1947). In the middle of the 20th century and for about three decades thereafter, geologists and geochemists investigated the extent to which these phenomena could explain anomalous groundwater pressures and chemistries in deep sedimentary basins. Concurrently, soil scientists were investigating the significance of these phenomena for movements of moisture and dissolved chemical species in agricultural soils.

---

[1] Research Professor, Colorado School of Mines, Golden, CO 80401

Recently these membrane phenomena have attracted the interest of geotechnical researchers because they appear to provide mechanisms for enhancing the capability of clay barriers to impede the migration of contaminants.   During the last two decades, geotechnical engineers have become increasingly involved in analyzing the migration and fate of dissolved chemical species through hazardous waste containment barriers and groundwater systems.   To date, this need has been met primarily with the advection-dispersion equation, which is based on Darcy's Law and Fick's Law for describing the transport of fluid and dissolved solutes in response to hydraulic and solute concentration gradients.   Recent studies suggest that the coupling of chemical and hydraulic fluxes in clays provides mechanisms for enhancing the capability of clay barriers beyond what can be anticipated from the classical advective-dispersion equation (Keijzer, 2000; Shackelford et al, 2001; Malusis et al, 2001, 2002).

To facilitate geotechnical interest in membrane phenomena, this paper presents a brief review of the pioneering field studies of membrane phenomena in geologic systems, a summary of the pertinent coupled flow theory developed to date, and a synthesis of the experimental evidence obtained in laboratory studies during the last five decades concerning the extent to which clay soils act as membranes.

Background

Coupled Transport Phenomena.   In general, coupled transport phenomena in soils include the fluxes of fluid, dissolved chemical species, electric charge, and heat in response to hydraulic, chemical, electrical, and thermal gradients.   However, the scope of this paper is limited to the fluxes of liquid and electrolytes, such as NaCl, through earthen materials in response to chemical and hydraulic gradients.   These phenomena include (1) groundwater flow in response to chemical gradients, otherwise known as chemico-osmosis; (2) chemico-osmotic pressure that can be generated by chemico-osmosis; and (3) the filtration of dissolved solutes as groundwater migrates through an earthen material, commonly known as ultrafiltration or reverse osmosis.   These phenomena are also known as membrane phenomena because membranes selectively restrict the passage of solutes relative to water. Materials that partially restrict the passage of solutes are called leaky membranes. Materials that completely restrict the passage of solutes are called ideal membranes.

Figure 1 illustrates these phenomena for a leaky membrane that separates aqueous solutions of high and low solute concentrations.   In Figure 1a, the chemical gradient drives water flow towards the high concentrated solution (chemico-osmosis), and also drives dissolved solutes towards the low concentrated solution (diffusion).   In Figure 1b, liquid flow ceases when the induced hydraulic head (chemico-osmotic pressure) causes the chemically driven liquid flow to cease.   In Figure 1c, an externally applied pressure drives both water and dissolved solutes through the clay specimen.   Because the soil selectively restricts the transport of solutes, the concentration of the inflow solution increases. This process is commonly referred to as ultrafiltration or reverse osmosis. The magnitudes of chemico-osmosis, osmotic pressure, and ultrafiltration vary with the degree to which the solute flux is restricted.   The maximum magnitudes of these membrane processes occur in ideal membranes where both diffusion and advection vanish.

Figure 2 illustrates the fundamental cause of these phenomena, namely, the difference between the concentrations of dissolved cations and anions in soil pores caused by overlapping electric double layers extending from the soil pore walls. The left side of Figure 2 illustrates one extreme where only dissolved cations are present, and electroneutrality is provided by the fixed negative charges in the pore walls. In this case, the electric fields prevent any anions from entering the soil, and electroneutrality prevents cations from entering also. Materials that prevent the flux of any solutes are ideal membranes. The right side of Figure 2 illustrates a soil with larger pores wherein some anions are present, but fewer in number than cations. In this case, the pressure that can be generated by chemico-osmosis, and the degree of salt restriction during ultrafiltration is smaller. Such materials are leaky membranes. Finally, when the influences of double layers on the relative abundance of cations and anions in soil pores are negligible, differences in the populations of mobile cations and anions in soil pores become negligible.

Figure 3 illustrates that very large osmotic pressures, compared with hydraulic pressures associated with insitu hydraulic gradients, can develop across ideal membranes. This figure also suggests that very significant osmotic pressures may develop across leaky membranes. Evidence presented subsequently will show that the ideality, or membrane efficiency, of soils is relatively high for highly compacted clays with high surface area and exchange capacity like smectite, relatively low for loosely compacted clays of low surface area and exchange capacity like kaolinite, and negligible for silts and sands.

Pioneering Field Studies. Since the late fifties, numerous field studies have examined whether abnormal distributions of pore pressures and chemical compositions in sedimentary basins could be attributed to membrane or osmotic behavior of clays and clay shales. Chemico-osmosis between aquifers separated by tight (impermeable) shales was proposed to explain abnormal highs and lows in the piezometric surfaces of the San Juan Basin regions of Utah, Colorado, and New Mexico (Hanshaw, 1962; Hanshaw and Hill, 1969). The same mechanism was used to interpret high hydraulic heads in the San Juan Basin, New Mexico and Colorado, and Wheeler Ridge anticline, San Joaquin Valley California (Berry, 1960; and Berry and Hanshaw, 1960), and also in the Triassic Dumbarton Basin, South Carolina (Marine, 1974; Marine and Fritz, 1981). It was also proposed to be a cause of high pore pressures that facilitate overthrust faulting (Hanshaw and Zen, 1965). Ultrafiltration was proposed to explain the compositions of brines in the Imperial Valley of California (Berry, 1966, 1967), the composition of sedimentary formation fluids in Western Canada (Hitchon et al, 1971), and the geo-chemistry of formation fluids in the Kettleman Dome area of California (Kharaka and Berry, 1974). Graf (1982) and Neuzil (1986) have reviewed these and other studies in detail.

However, attempts to evaluate the significance of membrane behavior in these pioneering field studies led to the emergence of alternative plausible mechanisms for abnormal fluid pressures and concentrated brines (Graf, 1982; Neuzil, 1986; Raven et al., 1992). The difficulty in evaluating the role of chemico-osmosis in the generation of high in situ fluid pressures is illustrated in the paper by Raven et al. (1992) where 10 mechanisms of supernormal pressure generation suggested in previous studies were evaluated in the context of the observed hydrogeologic conditions associated with measurements of the super-normal pressures of Paleozoic Rocks in Southern Ontario (see Table 1). In this case, Raven argues that the most reasonable explanation for the observed supernormal pressures is the upward migration and accumulation of gas generated from deeper distant sources in the Michigan and Appalachian basins or underlying basement rocks.

The best-documented and least ambiguous example of abnormal pressure attributed to chemico-osmosis is that illustrated in Figure 4 and documented by Marine (1974) and Marine and Fritz (1981). Neuzil (1995) summarizes this example as follows. "Hydraulic heads approximately 130 m above hydrostatic occur in Triassic sediments, coinciding with salinities of 12,000 to 19,000 mg/l. The surrounding normally pressured crystalline rocks and overlying coastal plain sediments have salinities less than 6000 mg/l. As Figure 4 shows, there is good correspondence between high salinity and high fluid pressures." This claim of osmotic pressuring is noteworthy because Marine and Fritz (1981) carefully considered and ruled out alternative causes of the pressures closely similar to those considered by Raven (1992). They also demonstrated with laboratory experimental and theoretical studies that osmosis could be the cause of the high pressures (Marine and Fritz, 1981; Fritz and Marine, 1983).

Recently, Neuzil (2000) reported the first direct in situ evidence for pore pressures generated by chemical gradients. This evidence was obtained in a nine-year experiment conducted in Cretaceous-age Pierre Shale in central South Dakota. At the test site the shale is saturated, has a permeability of 10^{-11} to 10^{-12} cm/s, and is 70-80% clay, of which about 80% is mixed-layer smectite-illite. The shale pore water contains ~3.5 g/l NaCl, with minor amounts of other solutes.

Four boreholes were drilled in the shale, 15 m apart, as illustrated in Figure 5a. The spacing between boreholes was chosen to ensure that they did not interact during the test. The boreholes were capped to prevent rainfall or runoff from entering. Breather holes were provided to accommodate water level changes. Estimated evaporative losses from barometric pumping through the breather holes during the 9 year test were ~1 cm of water. The test was started by adding waters with different total dissolved solids (TDS) to the boreholes. Fluid squeezed from cores guided the test design, with one borehole receiving a near duplicate of the expelled fluid (designated borehole DUP), two receiving water with ten times the solute concentration (high TDS boreholes HC1 and HC2), and one receiving deionized water (borehole DI).

For nine years, water levels were monitored with an electrical sounding cable, and samples of borehole waters were collected for analysis. The water level and TDS changes are plotted in Figures 5b and 5c. Water levels are plotted as differences from borehole DUP to isolate osmotic responses. The figure shows that water levels in HC1 And HC2 rose compared to DUP as osmosis drove water from the shale into the boreholes. The water level in DI initially declined slightly, as osmosis drew water into the shale, and then slowly approached DUP after the TDS difference between DI and DUP became small.

Neuzil (2000) concludes his report on this experiment as follows. "The conditions necessary for significant osmotic pressures appear to be common in the subsurface. . . . . . . Anomalous pressures are frequently encountered in sedimentary basins . . . . but have been attributed to continuing and otherwise undetected dynamic processes that include diagenesis, oil generation, and tectonic deformation. It now appears that in many cases the pressure could also be derived osmotically from pre-existing chemical potential differences in the pore water."

Theoretical Framework

Advection-Dispersion and Advection-Diffusion Equations.   The advection-dispersion equation describes the flux of solute mass through a unit total cross-sectional area of soil in terms of advective, J_s^a, and dispersive, J_s^d, components as shown in equation 1.

$$J_s = J_s^a + J_s^d \tag{1}$$

The advective solute flux equals the product of the solute concentration, c_s, and the liquid flux through a unit total cross-sectional area of the soil, J_v. The dispersive mass flux reduces to the diffusive mass flux, J_s^d, in low permeability soils where mechanical dispersion is negligible (Shackelford, 1988). Accordingly,

$$J_s = c_s J_v + J_s^d \tag{2}$$

Darcy's Law gives the liquid flux, $J_v = k\nabla(-h)$, where k is the hydraulic conductivity and h is the hydraulic head. Fick's law gives the diffusive flux, $J_s^d = -D^* n\nabla(-c_S)$, where D^* is the effective diffusion coefficient and n is the porosity (Shackelford and Daniel, 1991). Substituting these relationships into equation 1 yields the advection-diffusion equation

$$J_s = c_s k\nabla(-h) + D^* n\nabla(-c_s) \tag{3}$$

Generalized Advection-Diffusion Equation for Low Permeability Soils. Darcy's Law and Fick's Law are special cases of a more general coupled flow theory provided by irreversible thermodynamics for low permeability materials that was initially developed for biophysical membrane phenomena in the mid-20th century. Katchalsky and Curran (1965) present this theory in a useful form for the physical model in Figure 6 that consists of two homogeneous solutions (designated by compartments 1 and 2) separated by a homogeneous membrane. The fundamental processes in the system are the mass fluxes of water, cations, and anions through the membrane in response to differences in the chemical potential of water and the electrochemical potentials of the cations and anions between the solutions in compartments 1 and 2.

Greenberg (1971), Greenberg et al (1973), Mitchell (1991) and Yeung and Mitchell (1993) extended this theory for soil engineering applications by expressing the driving forces in terms of gradients within rather than differences across the membrane. Otherwise the fundamental thermodynamic and chemical relationships involved are the same as those used by Katchalsky and Curran (1965). However, one difference in terminology is noted because it leads to apparent differences in relationships involving the flux of solute. Katchalsky and Curran (1965) use the term J_s^d to represent the velocity of the solute, whereas Greenberg, Mitchell, and Yeung (1973) use the same term to represent the mass flux of the solute.

Yeung and Mitchell (1993) express the fundamental basis of the theory in terms of the dissipation function, Φ, which describes the energy dissipated by the mass fluxes of water, J_w, cations, J_c, and anions, J_a, in response to the gradients for the chemical potential of water, $\nabla(-\mu_w)$, and the electrochemical potentials of the cations, $\nabla(-\mu_c^{ec})$, and anions, $\nabla(-\mu_a^{ec})$, as follows.

$$\Phi = J_w\nabla(-\mu_w) + J_c\nabla(-\mu_c^{ec}) + J_a\nabla(-\mu_a^{ec}) \tag{4}$$

Yeung and Mitchell (1993) further derived the following version of the dissipation function in terms of more experimentally accessible quantities, including the fluxes of liquid volume, J_v, and electric current, J_I, and the gradients of pressure, $\nabla(-P)$, and electric potential, $\nabla(-\psi)$.

$$\Phi = J_v\nabla(-P) + J_I\nabla(-\psi) + J_c^d\nabla(-\mu_c^c) + J_a^d\nabla(-\mu_a^c) \tag{5}$$

Mitchell (1991, p.307) points out that the cationic and anionic components of the electrolyte flow were kept separate in equation 5 "because of the need to consider different chemical constituents in the particular study; namely, the potential for application of electrokinetic barriers for waste containment."

For the model in Figure 6, and in the absence of externally applied electrical gradients, the fluxes of cations and anions and their driving forces are coupled by electroneutrality such that they can be expressed in terms of the flux of electrolyte solute, J_s, and the gradient of its chemical potential, $\nabla(-\mu_s^c)$ (Olsen et al, 2000). Accordingly, the dissipation function can be expressed with three terms rather than four, as follows.

$$\Phi = J_v\nabla(-P) + J_I\nabla(-\psi) + J_s^d\nabla(-\mu_s^c) \tag{6}$$

or

$$\Phi = J_v\nabla(-P) + J_I\nabla(-\psi) + \frac{J_s^d}{c_s}\nabla\pi \tag{7}$$

where J_I is the electrical charge flux, ψ is the electric field potential, P is pressure, μ_s is the chemical potential of the electrolyte solute, c_s is the solute concentration, and $\nabla \pi = c_s \nabla \mu_s^c$ equals the osmotic pressure gradient in the membrane.

Based on equation 7, the governing equations relating the fluxes and forces can be written as

$$J_v = L_{11}\nabla(-P) + L_{12}\nabla(-\psi) + L_{13}\nabla(-\pi) \tag{8}$$

$$J_I = L_{21}\nabla(-P) + L_{22}\nabla(-\psi) + L_{23}\nabla(-\pi) \tag{9}$$

$$\frac{J_s^d}{c_s} = L_{31}\nabla(-P) + L_{32}\nabla(-\psi) + L_{33}\nabla(-\pi) \tag{10}$$

where the L coefficients are phenomenological coefficients that govern the fluxes of liquid, electric charge, and a dissolved electrolyte in terms of parameters that can either be measured or controlled. The coefficients relating liquid flux to pressure, electric charge flux to electric potential, and solute flux to osmotic pressure are commonly referred to as direct coefficients because they characterize the hydraulic, electrical, and solute diffusion processes that are generally described by Darcy's, Ohm's, and Fick's laws. The other coefficents characterize the interactions or coupling between these processes: L_{12} and L_{21} express the coupling between hydraulic and electrical processes; L_{13} and L_{31} express the coupling between hydraulic and solute diffusion processes, and L_{23} and L_{32} express the coupling between electrical and diffusion processes. In addition, each pair of coupling coefficients is equal, as follows.

$$L_{12} = L_{21} \tag{11}$$

$$L_{13} = L_{31} \tag{12}$$

$$L_{23} = L_{32} \tag{13}$$

In Figure 6, electroneutrality requires that there be no net electric charge flux through the membrane, i.e., $J_I = 0$. Accordingly, Equation 9 can be rewritten and substituted into equations 8 and 10 as follows.

$$\nabla(-\psi) = \frac{J_I}{L_{22}} - \frac{L_{21}}{L_{22}}\nabla(-P) - \frac{L_{23}}{L_{22}}\nabla(-\pi) \tag{14}$$

$$J_v = \left[L_{11} - \frac{L_{12}L_{21}}{L_{22}}\right]\nabla(-P) + \left[L_{13} - \frac{L_{12}L_{23}}{L_{22}}\right]\nabla(-\pi) \tag{15}$$

$$\frac{J_s^d}{c_s} = \left[L_{31} - \frac{L_{32}L_{21}}{L_{22}}\right]\nabla(-P) + \left[L_{33} - \frac{L_{32}L_{23}}{L_{22}}\right]\nabla(-\pi) \tag{16}$$

Note that each coefficient group in equations 15 and 16 consists of two terms, the former being the direct and coupled hydraulic and diffusion coefficients, and the latter consisting of the direct and coupled electrical coefficients. Because there are no variables within the coefficient groups, equations 15 and 16 can be simplified by expressing the coefficient groups in terms of the following direct and coupled hydraulic and diffusion coefficents that include embedded electrical coefficients:

$$L_P = \left[L_{11} - \frac{L_{12}L_{21}}{L_{22}} \right] \tag{17}$$

$$L_{PD} = \left[L_{13} - \frac{L_{12}L_{23}}{L_{22}} \right] \tag{18}$$

$$L_{DP} = \left[L_{31} - \frac{L_{32}L_{21}}{L_{22}} \right] \tag{19}$$

$$L_D = \left[L_{33} - \frac{L_{32}L_{23}}{L_{22}} \right] \tag{20}$$

Combining equations 11-13 and 17-20 with equations 15 and 16 gives

$$J_v = L_P \nabla(-P) + L_{PD} \nabla(-\pi) \tag{21}$$

$$\frac{J_s^d}{c_s} = L_{DP} \nabla(-P) + L_D \nabla(-\pi) \tag{22}$$

$$L_{PD} = L_{DP} \tag{23}$$

Combining equations 21 and 22 with equation 2 yields the following generalized form of the advection-diffusion equation.

$$\frac{J_s}{c_s} = J_v + \frac{J_s^d}{c_s} = \{L_P + L_{DP}\}\nabla(-P) + \{L_{PD} + L_D\}\nabla(-\pi) \tag{24}$$

When coupling is negligible ($L_{DP}=L_{PD}=0$), equation 24 reduces to the following form of the advection-diffusion equation.

$$J_s = c_s L_P \nabla(-P) + c_s L_D \nabla(-\pi) \tag{25}$$

Equations 3 and 25 both describe the solute mass flux in terms of advective and diffusive components. However, they express the chemical and hydraulic driving forces differently. The hydraulic driving force is expressed in terms of hydraulic head in equation 3, and in terms of pressure in equation 25. The chemical driving force is expressed in terms of solute concentration in equation 3, and in terms of osmotic pressure in equation 25. These differences can be reconciled with the following two relationships.

$$\nabla(-P) = \rho g \nabla(-h) \tag{26}$$

$$\nabla(-\pi) = RT\nabla(-c_S) \tag{27}$$

Chemico-Osmotic Efficiency. A convenient parameter for quantifying the extent to which materials act as membranes was first introduced by Staverman (1952). Katchalsky and Curran (1965) explain the underlying concepts as follows. First, an ideal membrane prevents the passage of any solute. For this condition, equation 24 shows that both terms in parentheses must be zero for all values of $\nabla(-P)$ and $\Delta\nabla(-\pi)$.

$$L_P = -L_{DP}; \quad L_{PD} = -L_D; \quad and \quad L_{PD} = L_{DP}$$
$$\therefore L_P = -L_{DP} = -L_{PD} = L_D \tag{28}$$

Second, liquid flow ceases in an ideal membrane when the hydraulic pressure gradient equals and opposes the theoretical osmotic pressure gradient. For this condition ($J_v = 0$), equation 21 provides the following relationship for the ratio of the hydraulic and osmotic pressure gradients.

$$\frac{\nabla(-P)}{\nabla(-\pi)}\bigg|_{J_V=0} = -\frac{L_{PD}}{L_P} = 1 \tag{29}$$

In cases when a material partially restricts the passage of solute,

$$\frac{\nabla(-P)}{\nabla(-\pi)}\bigg|_{J_1=0} = -\frac{L_{PD}}{L_P} < 1 \tag{30}$$

When a material exhibits no membrane behavior,

$$\frac{\nabla(-P)}{\nabla(-\pi)}\bigg|_{J_V=0} = -\frac{L_{PD}}{L_P} = 0 \tag{31}$$

because $L_{PD} = L_{PD}/L_P = 0$. Since the ratio $-L_{PD}/L_P$ varies from 0 to 1, it provides a convenient measure of the efficiency of a membrane, ω.

$$\omega = \frac{\nabla(-P)}{\nabla(-\pi)}\bigg|_{J_1=0} = -\frac{L_{PD}}{L_P} \tag{32}$$

When multiplied by 100, this ratio can also be expressed in percent.

Staverman (1952) first denoted this ratio by the symbol, σ, and called it the reflection coefficient. Subsequently, in the geologic and soil science literature this ratio bas been commonly denoted by the symbol σ and called the membrane or osmotic efficiency. More recently in the geotechnical literature this ratio is denoted by the symbol ω and called the osmotic efficiency coefficient (Mitchell, 1991 and 1993), the osmotic selectivity coefficient (Yeung and Mitchell, 1993); and the chemico-osmotic efficiency coefficient (Malusis et al, 2002). In this paper, the ratio is denoted by ω and called the chemico-osmotic efficiency coefficient.

One way to express the significance of the chemico-osmotic efficiency coefficient in the generalized advection-diffusion equation is the following. Consider a version of the generalized advection-diffusion equation that can be obtained by solving equation 21 for $\nabla(-P)$ and using this equation to eliminate $\nabla(-P)$ from equation 24.

$$J_s = c_s(1 + \frac{L_{DP}}{L_P})J_v + \beta\nabla(-\pi) \tag{33}$$

where $\quad \beta = \dfrac{c_s(L_P L_D - L_{DP} L_{PD})}{L_P}$

When equation 32 is substituted into equation 33, we obtain

$$J_s = c_s(1 - \omega)J_v + \beta\nabla(-\pi) \tag{34}$$

where $\quad \beta = c_s(L_D - \omega^2 L_P)$

Equation 34 shows how the chemico-osmotic efficiency coefficient, ω, influences both the advective and diffusive components of the solute flux. Both the advective and diffusive components decrease with increasing values of ω, and become zero for an ideal membrane (ω=1). When ω become negligible, equation 34 reduces to the following form of the advection-diffusion equation

$$J_s = c_s J_v + c_s L_D \nabla(-\pi) \tag{35}$$

Chemico-Osmotic Efficiency Measurement Methods. The most direct laboratory method that has been used for measuring the chemico-osmotic efficiency (reflection) coefficient is to measure the hydraulic pressure gradient induced by a chemical gradient in cylindrical, volume controlled, and saturated specimens in which liquid flux through the specimen is prevented (J_v=0). With this measurement, the chemico-osmotic efficiency is calculated from equation 32.

The chemico-osmotic efficiency coefficient has also been obtained from separate hydraulic and chemical osmotic flow measurements (Kemper and Rollins, 1966) as follows. By substituting equation 32 into equation 21,

$$J_v = L_P \nabla(-P) - \omega L_P \nabla(-\pi) \tag{36}$$

For a hydraulic flow measurement when $\nabla(-\pi) = 0$, the flow, J_{VH}, equals

$$J_{VH} = L_P \nabla(-P) \tag{37}$$

Similarly for a chemico-osmotic flow measurement when $\nabla(-P) = 0$, the flow, J_{VO}, equals

$$J_{VO} = -\omega L_P \nabla(-\pi) \tag{38}$$

Dividing equation 38 by equation 37,

$$\omega = \frac{-J_{VO}[\nabla(-P)]}{J_{VH}[\nabla(-\pi)]} \tag{39}$$

In most ultrafiltration studies, membrane efficiency has been described empirically in terms of either the Filtration Ratio (FR) of inflow (c_{SO}) and outflow (c_{SE}) solute concentrations (McKelvey and Milne, 1962; Kharaka and Berry, 1973; Kharaka and Smalley, 1976),

$$FR = \frac{c_{SO}}{c_{SE}} \tag{40}$$

or the Filtration Efficiency (FE) which can be expressed either as a ratio (Milne and McKelvey, 1964; Benzil and Graf, 1984; Hayden and Graf, 1986)

$$FE = \frac{c_{SO} - c_{SE}}{c_{SO}} \tag{41}$$

or as a percent when multiplied by 100.

In one ultrafiltration study, Fritz and Marine (1983) propose a method for calculating the chemico-osmotic efficiency coefficient, ω, from ultrafiltration data. This method is based on the assumption that the $\beta\nabla(-\pi)$ term in Equation 34 is negligible. Therefore, $J_s \cong c_s(1-\omega)J_V$ and since $c_s = \dfrac{(c_{SO} + c_{SE})}{2}$ and $J_s = c_{SE}J_v$, then

$$\omega = \frac{c_{SO} - c_{SE}}{c_{SO} + c_{SE}} \tag{42}$$

Experimental Evidence

General. The sections below illustrate, compare, and synthesize the evidence obtained in: (1) initial studies aimed at verifying the occurrence of membrane phenomena in earthen materials; (2) geologically oriented studies that were undertaken to clarify the factors governing membrane efficiency of low permeability clays and clay shales in relatively deep sedimentary environments; and (3) studies conducted by soil scientists to provide a basis for evaluating the significance of membrane phenomena in agricultural soils.

Occurrence of membrane phenomena in earthen materials. The earliest studies designed to demonstrate the occurrence of membrane phenomena in earth materials are illustrated in Figures 7-9. Figure 7 shows data Kemper (1960) obtained in a pressure membrane apparatus to demonstrate ultrafiltration in a Na-saturated bentonite mixed with 0.01 N (0.01 molar) NaCl and permeated with the same solution under a pressure of approximately 150 psi (1034 kPa). The data show the concentration of the outflow is 32% less than that of the inflow solution. The data also illustrate the concentrations of the inflow and outflow solutions increase with time, due to the filtration and accumulation of NaCl in the inflow solution. Kemper obtained similar data on two additional samples of the same clay that were mixed and permeated with 0.10 N (0.10 molar) NaCl and 0.94 N (0.94 molar) NaCl solutions. With these higher NaCl concentrations, the filtration efficiency of these clays decreased to 10% and 4%, respectively.

Shortly thereafter, Kemper (1961) demonstrated chemico-osmotic pressures across clay specimens in response to a 5.4 N (5.4 molar) NaCl solution above, and distilled water below each specimen. His experimental system and the data he obtained are illustrated in Figure 8. The test specimens were consolidated in the test cell under a pressure of about 5000 psi (34.5 MPa). The materials tested were Wyoming bentonite, ground Pierre shale, and a fine fraction of Pierre shale prepared homo-ionic to sodium, and having cation exchange capacities of 94, 18, and 30.5 meq per 100 g. respectively. For each material the data in Figure 8 show (a) the initial transient pressure rise followed by a relatively sustained pressure in response to the imposed chemical gradient, and (b) the subsequent pressure decay in response to the removal of the chemical gradient. The figure clearly shows the osmotic pressure varies substantially with the composition and/or the cation exchange capacity of the materials.

Figure 9 shows Young and Low's (1965) experimental system and the data they obtained to demonstrate chemico-osmotic flow of water in two siltstone samples composed of 8 to 12 percent of clay-size material of which 40 to 50 percent was illite, 30 to 40 percent kaolinite, and the remainder was montmorillonite or mixed layer clays. The samples were ¼- to ¾-inch (0.64 to 1.91-cm) discs cut from 3 ¼-inch (8.26-cm) diameter cores from the Lower Cretaceous Viking Formation in central Alberta, Canada. The data for the upper curves were obtained with distilled water above, and 1N (1 molar) NaCl below, the specimens. The data for the lower curves were obtained with 2N (2 molar) NaCl above, and 1N (1 molar) NaCl below the specimens. These data demonstrate the occurrence of chemico-osmosis. They further show that flow rate differs by nearly a factor of two for the two salinity difference situations, even though the theoretical osmotic pressure (about 700 psi or 4.83 MPa) is about the same in the two situations. Chemico-osmosis is greater for the case with the lower average solute concentration.

Geologically Oriented Studies. These studies were conducted on clay samples mounted between porous discs and pistons in cylindrical test cells. Most of these cells were equipped with flushing systems in one or both pistons for controlling the chemistry of the solutions near the boundaries of the test specimens. Some of the chemico-osmotic efficiency data were obtained with hydraulic conductivity and osmotic pressure measurements using equation 32. Other data were obtained from hydraulic and osmotic flow measurements using equation 39. Additional data were obtained from ultrafiltration experiments designed to measure membrane efficiency in terms of the filtration efficiency as defined in equation 41. A basis for comparing the results from these different approaches was obtained by expressing the filtration efficiency in terms of the chemico-osmotic efficiency using equation 42.

Table 2 summarizes the materials tested and the factors that increase chemico-osmotic efficiency. They are increasing surface area and/or ion exchange capacity of the soil, decreasing valence and hydrated radius of the solute cations, increasing consolidation and preconsolidation pressures, increasing dispersion of the soil fabric, and increasing temperature. The effects of these factors on chemico-osmotic efficiency are illustrated in Figures 10 through 13.

In Figure 10a, four data sets show increasing membrane efficiency with increasing consolidation pressure (Kharaka and Berry, 1973; Olsen, 1969, 1972; and Kemper and Maasland, 1964). The kaolinite data set further shows a very small decrease in membrane efficiency for decreasing consolidation pressure. Hence membrane efficiency varies with both the current consolidation pressure and the maximum past or preconsolidation pressure on the sample.

The data in Figure 10a also indicate effects of soil type and solute concentration. The data for bentonite and illite with seawater (Kharaka and Berry, 1973) show much greater membrane efficiency for bentonite. The data for bentonite also show much greater membrane efficiency for the sample used by Kemper and Maasland (1964) having a much lower solute concentration in its pore fluid. The membrane efficiency for the kaolinite data (Olsen, 1969,1972) is much lower than that for the bentonite having low solute concentration, and slightly higher than that for the bentonite whose pore fluid is seawater. The data for the bentonite with the low solute concentration show much higher chemico-osmotic efficiencies at low consolidation pressures than either the kaolinite with a lower solute concentration or the bentonite with the much higher solute concentration.

Figure 10b compares chemico-osmotic efficiency versus inflow solute concentration data from three studies on bentonites that were consolidated at four porosities ranging from 0.19 to 0.7. The data show chemico-osmotic efficiency increasing with decreasing porosity from nearly zero to nearly 100%. Kemper (1960) reported the data for the sample having the highest porosity (0.7). Milne et al (1964) reported the data for the sample having the lowest porosity (0.19). Fritz and Marine (1983) reported the data for the samples having porosities of 0.59 and 0.41.

Figure 11a illustrates some of the data reported by Kharaka and Smalley (1976) that show the chemico-osmotic efficiency varies substantially for different solute species. The figure shows the efficiency is greater for monovalent than divalent cations, and increases with decreasing hydrated radius (number in parentheses in angstroms, Å) within the monovalent and divalent cation groups. Figure 11a also shows that chemico-osmotic efficiencies for some of the ions decrease, and for others increase, with increasing magnitudes of hydraulic pressure driving flow through the test specimen. Figure 11b shows the latter trend for chemico-osmotic efficiency in terms of the total dissolved solutes, using data reported by Kharaka and Smalley (1976) and Milne et al (1964).

Figure 12a shows the chemico-osmotic efficiency of bentonite samples varies with fabric orientation ratio, which is a relative and semi-quantitative measure of particle alignment perpendicular to the axial stress used to consolidate the test specimens. Benzil and Graf (1984) devised this parameter based on x-ray diffraction measurements to depict differences in the fabrics of specimens that were initially prepared in flocculated and dispersed states. In a subsequent study conducted with the same experimental system, Haydon and Graf (1986) show (Figure 12b) that the chemico-osmotic efficiency of a bentonite sample varies with temperature, and that the variation with temperature differs for the sodium and calcium species in the pore fluid.

Soil Science Oriented Studies. These studies were conducted by Kemper and Rollins (1966) and Kemper and Quirk (1972) on unconsolidated and volume controlled samples that were prepared in pastes of known water content. Each sample was prepared by placing soil paste in a retainer ring 2-mm thick and 63 mm in diameter. Porous stainless steel plates were placed on each side of the ring and the assembly was then mounted in the center of the test cell between two solution chambers, as illustrated in Figure 13a. The cell was equipped with stirring blocks for maintaining uniform solute concentrations in each solution chamber. Chemico-osmotic flow was generated first by introducing and maintaining different solute concentrations in the solution reservoirs, and measuring the rates of movement of the menisci in the calibrated horizontal capillary tubes, which were maintained at the same elevation. The chamber solutions were then equalized by mixing, a hydraulic head difference was applied in the two chambers, and the flow rate was again measured from the rate of movement of the menisci in the capillary tubes. With these measurements, the chemico-osmotic efficiency was calculated using equation 39.

The principal results from these studies were obtained on bentonite, illite, and kaolinite samples that were initially prepared with sodium on the exchange sites. Data were initially obtained (Kemper and Rollins, 1966) on bentonite samples having porosities of 0.8, 0.84, and 0.91. Each sample was sequentially permeated with solutions containing a range of concentrations of NaCl, Na_2SO_4, $CaCl_2$, and $CaSO_4$. Subsequently, additional data were obtained (Kemper and Quirk, 1972) on bentonite, illite, and kaolinite. Figures 13b-14 show the variation of chemico-osmotic efficiencies with soil type, solute type, solute concentration, and porosity are consistent with those in the ultrafiltration data that were summarized in Table 2, namely, that chemico-osmotic efficiency increases with increasing surface area and/or ion exchange capacity of the soil, decreasing porosity, and decreasing cation valence and hydrated radius. Figure 13b further illustrates that the chemico-osmotic efficiency can decrease to small negative values at high solute concentrations.

Comparisons Among the Results from all of the Studies. Figures 15 and 16 illustrate both similarities and differences among the results from the previously cited studies in terms of porosity and hydraulic conductivity, respectively. Not all of the data sets in these figures are the same because many of the data sets did not include both porosity and hydraulic conductivity measurements.

In Figure 15, the data are grouped in three categories. The data from all the ultrafiltration tests on bentonites are shown by open triangles. These data are widely scattered, in part, because they were obtained with solutions having a wide range of solute concentrations. The data shown in solid circles were obtained from osmotic pressure measurements on kaolinite (Olsen, 1969, 1972). They follow a well-defined relation because they were obtained over a wide range of consolidation pressures on one sample having a constant pore fluid solute concentration. The data shown in solid squares are those from the unconsolidated and volume controlled samples of Wyoming bentonite prepared with sodium on the exchange sites and tested with NaCl and Na_2SO_4 solutions (Kemper and Rollins, 1966 and Kemper and Quirk, 1972). The wide variation in chemico-osmotic efficiency is due to the wide range of solute concentrations employed in the chamber solutions.

Overall, these data show chemico-osmotic efficiencies ranging from nearly zero to about 90% over porosities ranging from about 0.20 to 0.91. The data on kaolinite that were obtained over a wide range of consolidation loads appear to be a lower bound for most of the ultrafiltration data. A corresponding upper bound is suggested by the data on unconsolidated samples of bentonite.

Figures 16a and 16b compare several of the data sets in terms of hydraulic conductivity. Unfortunately, the data set on unconsolidated samples of bentonite could not be included because the hydraulic conductivity values were not reported.

Figure 16a compares the data sets from ultrafiltration tests (McKelvey and Milne, 1962; Milne et al, 1964; Kharaka and Berry, 1973; Kharaka and Smalley, 1976; Fritz and Marine, 1983; Benzel and Graf, 1984; Haydon and Graf, 1986). These data sets generally show increasing chemico-osmotic efficiencies with decreasing solution concentrations with a few exceptions. For example, the two data sets having the lowest hydraulic conductivities show higher chemico-osmotic efficiency for the bentonite having a NaCl solution concentration of 0.5 moles/liter than the bentonite having a seawater concentration of 0.109 moles/liter.

Figure 16b compares the data in Figure 16a on bentonites obtained from ultrafiltration tests with the data obtained on kaolinite using osmotic pressure tests (Olsen, 1969, 1972). Generally, appreciable chemico-osmotic efficiencies correlate with much higher hydraulic conductivities in kaolinite with very dilute NaCl pore fluid (0.001 moles/liter) compared with the bentonite samples having dilute to highly concentrated sodium rich pore fluids with solute concentrations ranging from 0.011 to >5 moles per liter.

Discussion

The qualitative trends in each of the data sets are generally consistent with the trends summarized in Table 2. Moreover, most of these trends are consistent with the mechanism illustrated in Figure 2 in that the factors causing increasing chemico-osmotic efficiency are either decreasing pore sizes or increasing thickness of double layers within the pores. This includes the trends involving consolidation pressure, surface area and/or ion exchange capacity, decreasing solute concentration, temperature, and increasing dispersion of soil. For the trends involving different solute species and decreasing ultrafiltration flow pressure, Kharaka and Berry (1973) show the above mechanism needs to be expanded to include consideration of four factors that fall in two groups. The first group includes the ionic charge and hydrated radii of ions that affect the selective adsorption of species and control their concentrations within soil pores. The second group includes the interactions of species with exchange sites and the forces exerted on the species by flowing water and streaming potentials that control the relative velocity of species within soil pores.

The differing magnitudes and trends for different solute species in Figure 11 focuses attention on one limitation of the existing theory. That is, the theory only considers solutions with one non-charged species or one single electrolyte. A more general coupled flow theory is needed that takes into account multi-constituent pore fluids and ion exchange (Olsen et al, 2000).

In addition, there are differences between data sets in Figures 15 and 16b that do not appear to be consistent with the aforementioned trends. First, in Figure 15, the data in solid squares (reported by Kemper and Rollins, 1966 and Kemper and Quirk, 1972) show very low to very high chemico-osmotic efficiencies in unconsolidated pastes of Na bentonite having high porosities ranging from about 0.75 to 0.90. In contrast, the data represented by open triangles that were obtained from ultrafiltration tests on heavily consolidated samples of bentonite, some of which were initially Na saturated, show high chemico-osmotic efficiencies at much lower porosities, mostly ranging from about 0.1 to 0.5. Second, Figure 16b shows chemico-osmotic efficiencies in kaolinite are appreciable in samples having hydraulic conductivities on the order of 10^{-8} to 10^{-9} cm/s. In contrast, comparable chemico-osmotic efficiencies obtained from ultrafiltration tests on bentonite occur in samples having hydraulic conductivities that are lower by two orders of magnitude.

It is not known whether the source(s) of these differences arise from limitations in the theory or the experimental methods or both. Nevertheless, one limitation in both the theory and the ultrafiltration test systems became apparent during this review, namely, the lack of sufficient consideration for the filtration and accumulation of solute at the specimen boundary where solution is being forced into the clay specimen.

All of the previous investigators recognized the need to remove the filtered solutes at the inflow boundary of a test specimen during ultrafiltration by flushing the boundary with fresh inflow solution. However, only the most recent ultrafiltration study (Fritz and Marine, 1983) recognized that the solute concentration in their flushing system was not the same as that at the specimen boundary, because a porous disc separated the flushing system from the specimen boundary. Fritz and Marine (1983) developed a model for the exponential rise of solute concentration from the flushing system to the specimen boundary, based on advection towards and diffusion away from the specimen boundary. They called this model the "Concentration Polarization Layer, CPL," and used it to estimate the solute concentrations at the specimen boundaries where solutions were being forced into the test specimens. Their estimates indicate that solute concentrations at their specimen boundaries may have been several times higher than in their flushing system.

Because the CPL was not considered for most of the ultrafiltration data in Figures 15 and 16, both the chemico-osmotic efficiencies and hydraulic conductivities appear to be underestimated. Chemico-osmotic efficiencies should be larger because they were calculated from Equation 42 using the inflow solute concentrations in the flushing systems. If the higher solute concentrations at the inflow specimen boundary had been known, larger chemico-osmotic efficiencies would have been obtained. Hydraulic conductivities should also be higher because the CPL increases the solute concentration gradient that causes chemico-osmosis in the direction opposite to the externally imposed hydraulic flow.

Summary And Conclusions

Coupled chemical and liquid fluxes in earthen materials have long been recognized as membrane phenomena in the geologic and soil science disciplines, and have recently attracted the interest of geotechnical researchers because they appear to provide mechanisms for enhancing the capability of clay barriers to impede the migration of contaminants. This paper presents a brief review of the pioneering field studies of membrane phenomena in geologic systems, a summary of the pertinent coupled flow theory developed to date, and a comparison and synthesis of the laboratory experimental evidence obtained during the last five decades concerning the extent to which clay soils act as membranes.

The significance of the pioneering field studies has been clouded by the emergence of alternative plausible mechanisms for abnormal groundwater pressures and chemistries in geologic systems. Recently, the first direct in situ evidence for pore pressures generated by chemical gradients was obtained from a nine-year experiment in the Pierre Shale. This study concludes, "the conditions necessary for significant osmotic pressures appear to be common in the subsurface."

The scope of existing coupled flow theories for analyzing the migration of groundwater and its dissolved species through earthen materials is limited in that it does not consider the different behaviors of dissimilar species. A more general coupled flow theory is needed that takes into account multi-constituent pore fluids and ion exchange.

The trends in the various data sets are generally consistent in that they show the chemico-osmotic efficiency increasing with consolidation and preconsolidation pressures, decreasing ultrafiltration flow pressure, increasing surface area and/or ion exchange capacity, decreasing solute concentration, decreasing valence and hydrated radii of solute cations, increasing temperature, and increasing dispersion of soil fabric.

However, substantial quantitative differences occur between the data sets obtained from ultrafiltration, osmotic pressure, and hydraulic and chemical conductivity tests. Although the causes of these differences are not known, a significant contributing factor appears to be the accumulation of excess solute due to filtration at the boundary of a test specimen where solution is being forced into the specimen during ultrafiltration tests. This factor was explicitly recognized, and called the "concentration polarization layer, CPL," in the most recent of the ultrafiltration studies reviewed.

Acknowledgements

Support for this study was provided by the U. S. National Science Foundation under Grant numbers CMS-9616855 and CMS-9713442. This support is greatly appreciated.

References

Becker, G. F. (1893). "Quicksilver ore deposits." *Mineral resources of the United States for 1892*, United States Geological Survey, Washington, DC, USA, 138-168.

Benzel, W. M. and Graf, D. L. (1984). "Studies of smectite membrane behavior: Importance of layer thickness and fabric in experiments at 20°C." *Geochimica et Cosmochimica Acta*, 48, 1769-1778.

Berry, F. A. F. (1960). "Geologic field evidence suggesting membrane properties of shales." (abs) *American Association of Petroleum Geologists Bulletin*, 44, 953-954.

Berry, F. A. F. and Hanshaw, B. B. (1960). "Geologic field evidence suggesting membrane properties of shales." (abs) *International Geologic Congress*, Session 21, Copenhagen, 209.

Berry, F. A. F. (1966). "Proposed origin of subsurface thermal brines, Imperial Valley California." (abs) *American Association Petroleum Geologists Bulletin*, 50, 655-645.

Berry, F. A. F. (1967). "Role of membrane hyperfiltration on origin of thermal brines, Imperial Valley, California." (abs) *American Association Petroleum Geologists Bulletin*, 51, 454-455.

Briggs, L. J. (1902). "Filtration of suspended clay from solutions." *United States Bureau of Soils Bulletin* 19, 31-49.

De Sitter, L. U. (1947). "Diagenesis of oil-field brines." *Bulletin of the American Association of Petroleum Geologists*, 31, 2030-2040.

Fritz, S. J. and Marine, I. W. (1983). "Experimental support for a predictive osmotic model of clay membranes." *Geochimica et Cosmochimica Acta*, 47(8), 1515-1522.

Graf, D. L. (1982). "Chemical osmosis, reverse chemical osmosis, and the origin of subsurface brines." *Geochimica et Cosmochimica Acta*, 46, 1431-1448.

Greenberg, J. (1971). *Diffusional flow of salt and water in soils.* PhD. dissertation, University of California at Berkeley.

Greenberg, J. A., Mitchell, J. K., and Witherspoon, P. A. (1973). "Coupled salt and water flows in a groundwater basin." *Journal of Geophysical Research*, 78(27), 6341-6353.

Hanshaw, B. B. (1962). *Membrane properties of compacted clays*, Ph.D. dissertation, Harvard University, Cambridge, Mass, 113 p.

Hanshaw, B. B. and E-An Zen, E. (1965). "Osmotic equilibrium and overthrust faulting." *Geological Society of America Bulletin*, 76, 1379-1386.

Hanshaw, B. B. and Hill, G. A. (1969). "Geochemistry and hydrodynamics of the Paradox Basin region, Utah, Colorado, and New Mexico." *Chemical Geology*,

4, 263-294.

Haydon, P. R. and Graf, D. L. (1986). "Studies of smectite membrane behavior: Temperature dependence, 20-180°C." *Geochimica et Cosmochimica Acta*, 50, 115-121.

Hitchon, B., Gillings G. K., and Klovan H. E. (1971). "Geochemistry and origin of formation waters in the western Canada sedimentary basin III. Factors controlling chemical composition." *Geochimica et. Cosmochimica. Acta*, 33, 1321-1349.

Katchalsky, A. and Curran, P. F. (1965). *Non-Equilibrium Thermodynamics in Biophysics*, Harvard University Press, Cambridge, Mass., 248 p.

Keijzer, T. J. S. (2000). *Chemical osmosis in natural clayey materials,* Ph. D. Thesis, Mededelingen van de Faculteit Aardwetenschappen, Universiteit Utrecht No. 196. 161p.

Kemper, W. D. (1960). "Water and ion movement in thin films as influenced by the electrostatic charge and diffuse layer of cations associated with clay mineral surfaces." *Soil Science Society of American Proceedings*, 24, 10-16.

Kemper, W. D. (1961). "Movement of waters as affected by free energy and pressure gradients, II, experimental analysis of porous systems in which free energy and pressure gradients act in opposite directions." *Soil Science Society of America Proceedings*, 25, 260-265.

Kemper, W. D. and Maasland, D. E. L. (1964). "Reduction in salt content of solution passing through thin films adjacent to charged surfaces." *Soil Science Society of America Proceedings*, 28, 318-323.

Kemper, W. D. and Rollins, J. B. (1966). "Osmotic efficiency coefficients across compacted clays." *Soil Science Society of America Proceedings*, 30(5), 529-540.

Kemper, W. D. and Quirk, J. P. (1972). "Ion mobilities and electric charge of external clay surfaces inferred from potential differences and osmotic flow." *Soil Science Society of America Proceedings*, 36, 426-433.

Kharaka, Y. K. and Berry, F. A. F. (1973). "Simultaneous flow of water and solutes through geologic membranes I, Experimental Investigation." *Geochimica et Cosmochimica Acta*, 37, 2577-2603.

Kharaka, Y. K. and Berry, F.A.F. (1974). "The influence of geological membranes on the geochemistry of subsurface waters from Miocene sediments at Kettleman North Dome in California." *Water Resources Research*, 10, 313-327.

Kharaka, Y. K. and Smalley, W. C. (1976). "Flow of water and solutes through compacted clays." *American Association of Petroleum Geologists Bulletin*, 60, 973-980.

McKelvey, J. G. and Milne, I. H. (1962). "The flow of salt solutions through compacted clay." *Proceedings of the 9th National Conference on Clays and Clay Minerals*, Ada Swineford, ed., 246-259.

Malusis, M. A., Shackelford, C. D., and Olsen, H. W. (2001). "A laboratory apparatus to measure chemico-osmotic efficiency coefficients for clay soils." *Geotechnical Testing Journal*, 24(3), 229-242.

Malusis, M. A., Shackelford, C. D., and Olsen, H. W. (2002). "Chemico-osmotic efficiency of a geosynthetic clay liner." *Journal of Geotechnical and Geoenvironmental Engineering*, 128(2), 97-106.

Marine, I. W. (1974). "Geohydrology of buried Triassic Basin at Savannah River Plant, South Carolina." *American Association of Petroleum Geologists Bulletin*, 58, 1825-1837.

Marine, I. W. and Fritz, S. J. (1981). "Osmotic model to explain anomalous hydraulic heads." *Water Resources Research*, 17, 73-82.

Milne, I. H., McKelvey, H. G. and Trump, R. P. (1964). "Semi-permeability of bentonite membranes to brines." *American Association of Petroleum Geologists Bulletin*, 48, 103-105.

Mitchell, J. K. (1991). "Conduction phenomena: from theory to geotechnical practice." *Geotechnique* 41(3) 299-340.

Neuzil, C. E. (1986). "Groundwater Flow in Low-Permeability Environments." *Water Resources Research*, 22(8), 1163-1195.

Neuzil, C. E. (1995). "Abnormal pressures as hydrodynamic phenomena." *American Journal of Science*, 295, 742-786.

Neuzil, C. E. (2000). "Osmotic generation of 'anomalous' fluid pressures in geological environments." *Nature*, 403, 182-184.

Olsen, H. W. (1969). "Simultaneous fluxes of liquid and charge in saturated kaolinite." *Soil Science Society of America, Proceedings*, 33, 338-344.

Olsen, H. W. (1972). "Liquid movement through kaolinite under hydraulic, electric, and osmotic gradients." *American Association of Petroleum Geologists Bulletin*, 56(10), 2022-2028.

Olsen, H. W., Gui, S., Lu, N. (2000). "A critical review of coupled flow theories for clay barriers." *Journal of the Transportation Research Board*, Transportation Research Record 1714, Paper No. 00-0913, p. 57-64.

Raven, K. G., Novakowski, K. S., Yager, R. M., and Heystee, R. J. (1992). "Supernormal fluid pressures in sedimentary rocks of southern Ontario - western New York State." *Canadian Geotechnical Journal*, 29(1), 80-93.

Russell, W. L. (1933). "Subsurface concentration of chloride brines." *Bulletin of American Association of Petroleum Geologists*, 17, 1213-1228.

Shackelford, C. D. (1988). "Diffusion as a transport process in fine-grained barrier materials." *Geotechnical News*, 6(2), 24-27,

Shackelford, C. D. and Daniel, D. E. (1991). "Diffusion is saturated soil: I. Background." *Journal of Geotechnical Engineering*, ASCE, 117(3), 467-484.

Shackelford, C. D., Malusis, M. A., and Olsen, H. W. (2001). "Clay membrane barriers for waste containment." *Geotechnical News*, BiTech Publ. Ltd., Richmond, BC, Canada, 19(2), 39-43.

Staverman, A. J. (1952). "Non-equilibrium thermodynamics of membrane processes." *Transactions of the Faraday Society*, 48, 176-185.

Yeung, A. T. and Mitchell, J. K. (1993). "Coupled fluid, electrical, and chemical flows in soil." *Geotechnique*, 43(1) 121-134.

Young, A. and Low, P. F. (1965). "Osmosis in argillaceous rocks." *American Association of Petroleum Geologists Bulletin*, 49, 1004-1007.

Table 1   Possible Abnormal Pressure Mechanisms

Raven et al (1992)	Marine (1974)
Regional groundwater flow and variations in formation fluid density	Aquifer head
Sediment loading	Rapid loading and compaction
Uplift and erosion	Fossil head and Igneous water
Thermal effects	Temperature increase
Mineral diagenesis	Phase or chemical changes and Mineral precipitation
Osmosis	Osmotic-membrane phenomena
Tectonic compression	Tectonic compression
Gas generation and Gas migration	Gas infiltration

Table 2 – Factors That Increase Chemico-Osmotic Efficiency

Item	Factor	Soil Type	References
1	Increasing consolidation and preconsolidation pressures	Bentonite, kaolinite	Kemper and Maasland (1964) Olsen (1969, 1972) Kharaka and Berry (1973)
2	Decreasing ultrafiltration flow pressure	Bentonite	Milne et al (1964) Kharaka and Berry (1973)
3	Increasing surface area and/or ion exchange capacity	Bentonite, kaolinite, and illite	Kemper and Maasland (1964) Kharaka and Berry (1973) Kharaka and Smalley (1976)
4	Decreasing valence and hydrated radius of solute cations	Bentonite	Haydon and Graf (1986) Kharaka and Berry (1973)
5	Decreasing solute concentration	Bentonite	Kemper (1960) McKelvey and Milne (1962) Milne et al (1964) Fritz and Marine (1983)
6	Increasing temperature	Bentonite, montmorillonite, illite	Kharaka and Berry (1973) Kharaka and Smalley (1976) Haydon and Graf (1986)
7	Increasing dispersion of soil fabric	Bentonite	Benzel and Graf (1984)

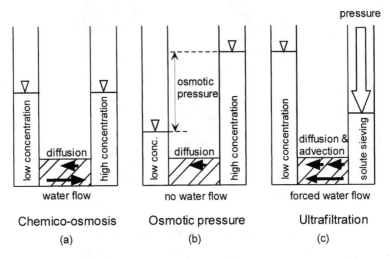

Figure 1.   Schematic illustration of chemico-osmosis (a), osmotic pressure (b), and
             ultrafiltration (c) through a leaky membrane (after Keijzer, 2000).

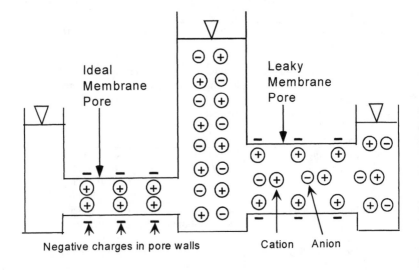

Figure 2   Distributions of cations and anions in ideal and leaky membrane pores.

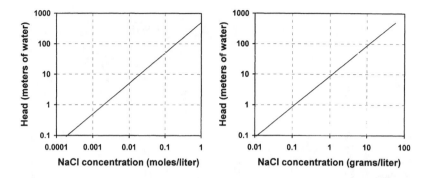

Figure 3   Theoretical magnitudes of osmotic pressure across an ideal membrane between distilled water and a NaCl solution at 20°C, based on the Van't Hoff equation.

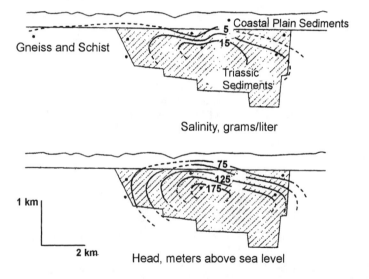

Figure 4   Northwest to southeast cross sections through the Dunbarton Basin, South Carolina and Georgia, showing abnormally high pressures thought to be generated by osmosis. Water table and ground surface elevations at 58 and 100 m above sea level respectively. The cross sections were constructed by Neuzil (1995) using data published by Marine and Fritz (1981).

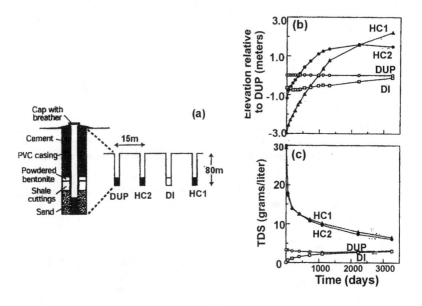

Figure 5   (a) Pierre Shale boreholes used for the osmosis test. (b) Borehole water
level changes with time relative to borehole DUP. (c) Borehole solute
concentration (in terms of TDS) changes with time. (After Neuzil, 2000)

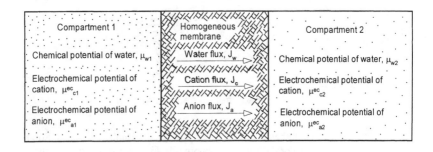

Figure 6   Model for generalized advection-diffusion equation

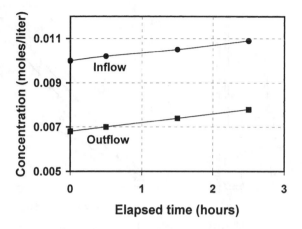

Figure 7   Kemper's (1960) initial ultrafiltration data obtained on Na-saturated bentonite mixed and permeated with 0.01 molar NaCl in a pressure membrane cell.

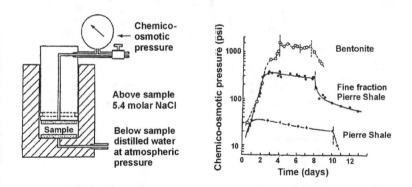

Figure 8   Kemper's (1961) initial osmotic pressure apparatus and data.
(1 psi = 6.895 kPa)

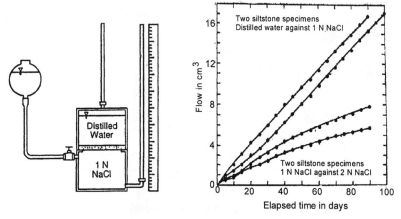

Figure 9   Young and Low's (1965) apparatus and chemico-osmosis data. Note that
            1 N and 2 N NaCl are equivalent to 1 molar and 2 molar NaCl
            respectively

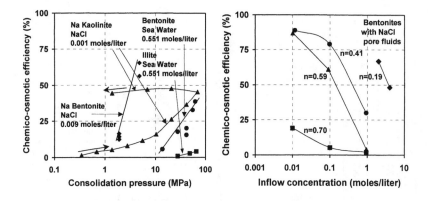

Figure 10   Chemico-osmotic efficiency versus (a) consolidation pressure (data from
            Kemper and Maasland, 1964; Kharaka and Berry, 1973; and Olsen, 1969,
            1972) and (b) inflow concentration and porosity data from Kemper, 1960;
            Milne et al, 1964; and Fritz and Marine, 1984).

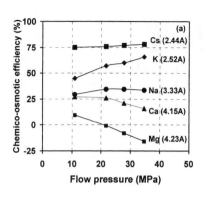 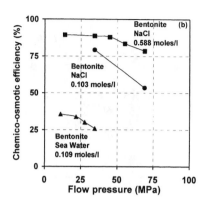

Figure 11 Chemico-osmotic efficiency versus solute composition and ultrafiltration flow pressure. (a) Data from Kharaka and Smalley (1976). (b) Data from Kharaka and Smalley (1976) and Milne et al (1964).

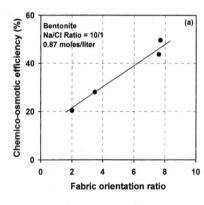 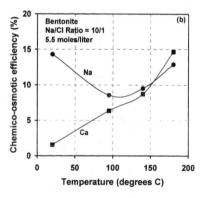

Figure 12. Chemico-osmotic efficiency versus (a) fabric orientation ratio (data from Benzil and Graf, 1984), and (b) temperature (data from Haydon and Graf, 1986).

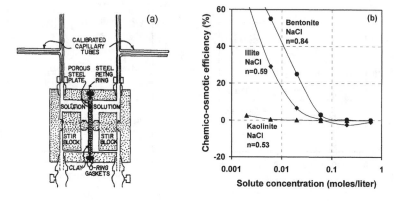

Figure 13. (a) Kemper's test cell for hydraulic and osmotic conductivity measurements. (b) Chemico-osmotic efficiency versus solute concentration and soil type. (Data from Kemper and Rollins, 1966 and Kemper and Quirk, 1972).

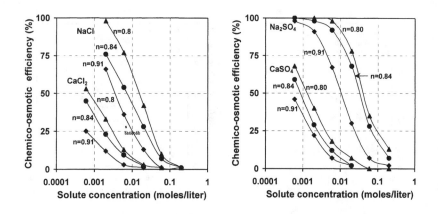

Figure 14 Chemico-osmotic efficiency of bentonite versus solute concentration, solute type, and porosity. (Data from Kemper and Rollins, 1966).

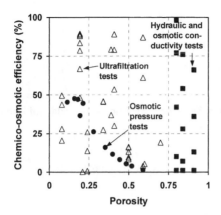

Figure 15 Synthesis of data on chemico-osmotic efficiency versus porosity. Open
         triangles show ultrafiltration data on bentonites. Solid circles show
         osmotic pressure data on kaolinite. Solid circles show hydraulic and
         osmotic flow data on bentonites. See text for data sources.

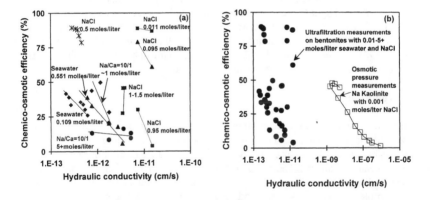

Figure 16 (a) Synthesis of ultrafiltration data on chemico-osmotic efficiency versus
         hydraulic conductivity. (b) Comparison of ultrafiltration data and osmotic
         pressure data versus hydraulic conductivity. See text for data sources.

Sampling Effects in Soft Clay: An Update on Ladd and Lambe (1963)

D. W. Hight
Geotechnical Consulting Group

Abstract

The paper presents a review of the effects of sampling on the subsequent behaviour of soft clay. It takes for its starting point the classic paper by Ladd and Lambe (1963) on this subject. The effects of perfect sampling, stress relief and of tube sampling are considered. Particular emphasis is placed on the effects of desaturation and gas exsolution, and of destructuring as a result of tube sampling strains. Evaluation of sample quality is discussed and serves as a link to a companion paper to this Symposium by O'Neill and Hight (2002). The benefits of sampling soft clays with a sampling tube of large diameter, thin wall, sharp cutting edge and no inside clearance, and of immediate extrusion and trimming of the sample, all as recommended by LaRochelle et al (1981), are emphasised. The dangers of improving only sample quality in design methods which rely on empiricism are highlighted.

Introduction

Both our engineering characterisation of soils for design and construction and our scientific understanding of natural soil behaviour depend on obtaining samples from the ground in which disturbance has been minimised, and the level of disturbance and its effects can be evaluated. This paper presents an update on the effects of sampling in soft clays, taking as its starting point the classic work of Ladd and Lambe (1963). Evaluation of sample quality is discussed and reference made to a companion paper to the Symposium by O'Neill and Hight (2002).

Soil sampling herein concerns the recovery of an element of soil in its natural state from the ground to the laboratory for mechanical testing, to measure properties such as strength, stiffness and compressibility. In situ, the soil is under total vertical and horizontal boundary stresses of σ_{vo} and σ_{ho} (taking the case of level ground) and with an initial pore pressure u_o. At some stage in the sampling process, the boundary stresses have to be removed before they can be reimposed in the laboratory. The stress and strain paths that are followed and the stage at which the boundary stresses are fully removed depend on the method of sampling.

Ladd and Lambe (1963) depicted the process of tube sampling in a stress path plot (Figure1) which illustrated the changes in effective stress that occur during drilling (forming the borehole), tube sampling, extrusion of the sample from the tube, cavitation and water content redistribution, and trimming and mounting the specimen in the test apparatus. They also introduced the concept of perfect sampling as the effects of the inevitable removal during sampling of the deviator stress, $(\sigma_{vo}-\sigma_{ho})$; the effective stress remaining after perfect sampling was denoted by $\bar{\sigma}_{ps}$.

Figure 1 encapsulated the effects of sampling of soft clays in terms of a reduction in effective stress. In fact Ladd and Lambe suggested 'that the measurement of the residual effective stress $\bar{\sigma}_r$ of laboratory specimens should become a standard test for jobs requiring a rational interpretation of lab strength data'. They proposed that the ratio of $\bar{\sigma}_r/\bar{\sigma}_{ps}$ be used as a measure of the amount of additional disturbance caused by actual sampling. Figure 1 did not consider the effects of mechanical damage to structure because at that time these effects were not fully appreciated, ironically because of the very effects of sampling that were being studied. Nevertheless, Ladd and Lambe were careful to restrict their comments to 'normally consolidated, moderately sensitive, plastic clays which do not possess significant natural cementation'.

In this paper we will consider in detail the effects of perfect sampling, total stress relief, tube sampling, and preparation for testing. Clays having natural cementation, i.e. structure, will be included.

Perfect Sampling

Perfect sampling effects have been studied by, among others, Skempton and Sowa (1963), as well as Ladd and Lambe (1963). The examination here considers the effects of removing the in situ deviatoric stress $(\sigma_{vo}-\sigma_{ho})$ from soils having different stress histories and plasticity, and of loading them without reconsolidation in different directions.

Figure 2 shows the effective stress paths followed when the deviator stress is removed from a reconstituted low plasticity clay, the Lower Cromer Till, having OCRs of 1, 1.25, 1.5, 2, 4 and 7. There is, inevitably, a change in effective stress in the sample, except for OCR 4. The magnitude and sign of the change varies with stress history. Reductions in mean effective stress, p', occur in lightly overconsolidated samples (OCRs 1, 1.25, 1.5 and 2), the size of the reduction becoming less as OCR increases. In heavy overconsolidated clays there is a small increase in p'.

The unloading path followed by normally consolidated or lightly overconsolidated soils during perfect sampling depends on the soil's plasticity, as illustrated in Figure 3. The reduction in p' reduces as plasticity increases.

The behaviour after perfect sampling is illustrated in Figure 4 for the case of unconsolidated undrained (UU) triaxial testing. Effective stress paths during compression and extension after perfect sampling are shown for Lower Cromer Till having OCRs of 1, 2 and 4 (PSC1, PSC2, PSC4, PSE1, PSE2, PSE4) and may be compared with the corresponding stress paths that would be followed by the soil in situ, without sampling (ISC1, ISC2, ISC4, ISE1, ISE2, ISE4). It can be seen that the effects depend on a combination of loading direction and OCR. The behaviour of normally and lightly overconsolidated clays is modified in subsequent triaxial compression, but not in subsequent triaxial extension - the extension path is simply a continuation of the perfect sampling path. The pattern reverses for heavily overconsolidated clays and, for OCR of 4, there is no effect of perfect sampling, because for this soil Ko is equal to 1.0 at an OCR of 4.

If we accept the concept of mobile kinematic yield surfaces that surround a soil at its current stress state and which are dragged around as the stress state changes (see, for example, Jardine, 1992), then we can also draw some conclusions on the effects of perfect sampling on subsequent stiffness. Reference to Figure 5 shows that the location of the stress state relative to the kinematic yield surface changes as a result of perfect sampling. A normally consolidated soil after perfect sampling will be stiffer in compression and less stiff in extension than the in situ soil, because its stress path reverses across the small strain zone in compression before yielding, but immediately crosses the kinematic yield surface when loaded in extension . The reverse is true for a heavily overconsolidated soil.

It follows simply then, from a consideration of perfect sampling, that one of the effects of imperfect sampling is likely to be a change in mean effective stress and the magnitude and sign of this change is likely to depend on the stress history and plasticity of the clay. The subsequent behaviour will depend on whether or not there is reconsolidation of the sample. Following conventional practice of unconsolidated undrained triaxial testing, there will be a change in stiffness and strength, which will depend on loading direction and stress history. UU testing following even perfect sampling modifies a soil's apparent anisotropy, by an amount that depends on the soil's stress history.

Total Stress Relief

The discussion on perfect sampling above considered only the effects of removing the deviatoric component of the total stress. At some stage in the sampling process the remaining component of the total stress, i.e. the isotropic component, has to be removed. It is often assumed that if the soil is saturated, i.e. has a pore pressure parameter, B, of one if it is a soft clay, it is not affected by the removal and reinstatement of isotropic total stress. In fact, there are several potential effects of removing the isotropic total stress which are now considered, and some of which are summarised in Figure 6. Deviation of B below one is, of course one such effect, which becomes more important with increasing depth from which the soil is retrieved. The effects considered here and illustrated in Figure 6 are desaturation by air entry, gas exsolution and cavitation of the pore water.

Desaturation by Air Entry and Sustainable Suction. At some stage in bringing a soil element from the ground to the laboratory, as well as removing the boundary stresses, the boundary of the sample is exposed to the air. Under undrained conditions, the removal of the total boundary stresses will result in the pore pressures becoming negative. As the total stress is removed, the tendency for the soil to expand is resisted by the development of surface tension forces at the air-soil boundary; these resist the negative pore pressure pulling the pore water into the soil. The pore water is in tension when the pore water pressure is less than atmospheric pressure.

The ability to sustain a suction, i.e. to retain a positive effective stress, at this stage, is essential to the process of sampling. If the imposed suction cannot be sustained, the soil will lose coherence and water will drain from the sample.

Whether or not the imposed suction can be sustained depends on the magnitude of the surface tension forces than can develop at the air-soil boundary. These forces depend, in turn, on the radii of the menisci at the air-soil boundary. As the radii decrease, their ability to resist the stress pulling the menisci into the soil increases. The radii of the menisci decrease as the pore size and particle size decrease. Therefore, the suction that can be sustained increases as pore size decreases. This relationship between maximum sustainable suction before desaturation by air invasion is shown in Figure 7(a), based on a simple capillary model. A similar trend can be inferred from the data presented by Lane and Washburn (1946) which related the height of saturated capillary rise (equivalent to sustainable suction) to permeability (see Figure 7(b)).

Silts and fine sands can sustain only low suctions. In coarse sands, gravitational forces are sufficient to pull the menisci into the soil and to cause the water to drain. Alternative methods, such as freezing, are required to sample sands and to cater for their inability to sustain suctions and so remain coherent.

In layered or laminated sands or silts and clays, there is a mismatch between the suctions that can be sustained (see Figure 8). When the imposed suction exceeds the sustainable suction in the sand or silt, it desaturates, giving up its water to the surrounding clay which then swells.

The suction that is imposed depends on the effective stresses in the sample before the isotropic stress is finally removed - this may be quite different to the in situ effective stress.

Gas Exsolution. As illustrated in Figure 6(b), whether or not gases will exsolve from the pore water depends on the magnitude of the reduction in pore pressure, Δu, and on the gas concentration in solution. Gases can exsolve while pore pressures are positive and represent a different process to that of cavitation, which occurs when pore pressures are negative. The pressure at which gas exsolves will also depend on the type of gas and its solubility.

When gas exsolves, it is either trapped within the soil sample (undrained gas exsolution) or it breaks through and drains from the sample (drained gas exsolution). Gas is trapped (occluded) if the permeability of the soil is low and it can sustain a suction which is high enough to resist the break out pressure of the gas bubble. In this case, the water content of the sample remains constant, the sample volume may increase, as the gas expands, and its density reduce. In a soil sample in which exsolved gas is trapped, diffusion will take place with time, with gas moving towards the larger pores and to the top of a vertically stored sample.

Gas drains from the sample if the soil is of high permeability. In this case, the water content may reduce but the dry density may stay constant.

The temperature of the soil will also change as a result of sampling, influencing the amount of gas that may come out of solution and modifying permeability and other viscous effects.

Cavitation. Water has a finite tensile strength which varies with the amount of impurities and dissolved gases within it. There is a limit, therefore, to the negative pore pressures that can develop after sampling, before there is cavitation. This limit is likely to reduce as pore size increases, because the likelihood of there being imperfections or impurities in the water within those pores increases. A limit to the suction that can develop before there is cavitation and its link with pore size is qualitatively shown in Figure 6(a). Bishop et al (1975) have shown that very high negative pore pressures, in excess of 19MPa, can be sustained in heavily overconsolidated clays before there is cavitation.

Desaturation by cavitation is an internal process as opposed to desaturation by air entry which is a process initiating at the boundary of the sample.

Tube Sampling Strains

Strain Path Method. Although the effects of perfect sampling and stress relief are inevitable, they are not sufficient to define the full effects of sampling in practice. While block sampling is the nearest we can come to perfect sampling, its stress path is different and additional damage cannot always be avoided. Still the most common method of obtaining a soil sample is by forming a borehole and pushing a tube into the ground, involving the stages shown in Figure 9. Disturbances can occur at each of the stages shown in Figure 9, but we will consider in detail in this section only those caused by penetration of the sampling tube.

There is little doubt that the most significant advance in the last two decades in understanding disturbance by undrained tube sampling has been the introduction of the

Strain Path Method by Professor Baligh and his co-workers (Baligh, 1985; Baligh, Azzouz and Chin, 1987) and its application to the deep penetration of a sampling tube into the ground. By superimposing sinks into a uniform inviscid fluid flow, Baligh was able to model the shape of a sampling tube, with outside diameter B and wall thickness t (Figure 10(a)), and to derive the deformation pattern of the fluid, which represents the undrained soil, as it flows into and around the sampling tube (Figure 10(b), which shows the deformation pattern for B/t = 20). The sampling tube, referred to as the Simple Sampler, has a rounded tip and inside clearance. From the predicted deformation pattern, the strain pattern could be defined, leading to the two classic figures shown here as Figures 11 and 12. Figure 11 shows the complete pattern of strains around a Simple Sampler having B/t = 40, being penetrated into the ground. The diagram is split into four parts, each showing half of the sample and one of the following strains: radial, tangential, shear and vertical. Note the high shear strains that are predicted to occur at the wall of the tube as shown in the bottom left quadrant. The pattern of straining demonstrates that conditions of triaxial compression and extension apply along the centre-line of the tube ($\varepsilon_{\theta\theta} = \varepsilon_{rr}$; $\varepsilon_{rz} = 0$). Figure 12 shows the strain history experienced by a soil element positioned along the centre-line of the sampler as it is approached by the sampler, enters the tube and arrives at its final resting place in the tube.

Baligh et al's findings apply to undrained tube penetration in which there is no side shear on the tube walls, i.e. similar to tube sampling in sensitive clay, and they demonstrate the following:

(i)   the soil captured by a sampling tube experiences a complex sequence of strains, which varies with position in the tube;

(ii)   on the centre-line and in a sampling tube having inside clearance, the soil experiences a sequence of triaxial compression strains followed by triaxial extension strains, followed by unloading from extension;

(iii)   at the periphery, there is a zone of intense shearing, the thickness of which is controlled more by the tube wall thickness, t, than the tube diameter, B;

(iv)   in the Simple Sampler, the magnitude of the maximum compression strain reduces as the ratio B/t increases.

Because the strains vary across the diameter and along the length of the sample, in a clay undergoing undrained tube penetration pore pressures will also vary. There is, therefore, an important time effect as pore pressures will equilibrate within the sample and water content redistribution will take place, as recognised by Ladd and Lambe (1963) (DE in Figure 1).

Recent Developments with the Strain Path Method. The precise shape of the Simple Sampler, with its rounded tip and inside clearance, is not typical of a high quality thin wall sampling tube, which may have a sharp cutting edge and, possibly, no inside clearance. Clayton (pers. comm.), using the Strain Path Method, has compared the centre-line strain paths followed using three sampling tubes having no inside clearance and the geometries which are summarised in Figure 13. These represent:(a) the conventional UK piston sampling tube, which has a 30° cutting edge on a 100mm diameter aluminium alloy tube, (b) a modified stainless steel tube, which we introduced in the UK, which has a 5° cutting edge and 100mm diameter, and (c) the Laval sampling tube, with a 5° cutting edge and a 200mm diameter, . The predicted strain paths are shown in Figure 14 and illustrate that, with no inside clearance, there are no extension strains as the sample enters and moves up inside the tube. The maximum strains on the centre-line are strongly influenced by the sharpness of the cutting edge, being approximately 0.6% for the 5° edge on the Laval tube and the modified tube and more than 1% for the conventional UK tube with its 30° edge. A comparison of the performance of these different samplers is presented subsequently.

Clayton et al (1998) have extended Baligh et al's work further to examine the effects of different details of sampling tube geometry, including inside clearance ratio (ICR), outside cutting edge angle (OCA), inside cutting edge angle (ICA) and area ratio (AR). Their predictions are summarised in Figure 15 and emphasise the sensitivity of imposed strain levels to details of tube geometry.

Effects of Tube Sampling Strains

Centre Line Strains. The effects of the predicted tube sampling strains can be examined by appeal to models of soil behaviour and seeking confirmation from laboratory and field experiments (Hight, 1993). The problem can be simplified by considering the separate effects of the strains on the centre-line of the tube sample (referred to by Baligh et al as causing 'minimum disturbance') and those around the periphery, and subsequently combining these effects.

The model of soil behaviour that is used is illustrated in Figure 16. It assumes the existence of a local bounding surface, the location of which, relative to the current stress state, is determined by stress history and by the level of structure, i.e. the effects of ageing and cementing. With increasing levels of structure, the local bounding surface expands. As a corollary, the surface shrinks when the structure is damaged - destructuring - by an amount which depends on the level of damage.

For brevity, we will consider here the effects of the centre-line tube sampling strains, using the results of laboratory tests described by Clayton et al (1992) on high quality samples of the natural Bothkennar Clay. The test sequence is illustrated in Figure 17 and involved the following stages:

(i)   install a specimen of 100mm diameter by 200mm height in the triaxial cell and measure the initial effective stress, p_i' (point A),

(ii)   reconsolidate the specimen anisotropically to the estimated in situ effective stresses (path ABD),

(iii)   subject the sample to undrained triaxial compression to a maximum axial strain of $+\epsilon_a$, followed by extension to an axial strain of $-\epsilon_a$, followed by unloading to zero axial strain (path DEFG),

(iv)   remove the remaining deviator stress (path GH) and measure the mean effective stress, p', at H, i.e. the residual effective stress, σ_r, of Ladd and Lambe (1963),

(v)   reconsolidate the specimen for a second time to in situ effective stresses (path HABD),

(vi)   shear undrained in triaxial compression (path DJ).

In the work on Bothkennar Clay, the magnitude of the strains, $\pm\epsilon_a$, imposed in the cycle of triaxial compression-extension-unloading was varied in different tests. Thus, strains of ±0.5% were imposed in one test, before subsequent reconsolidation and shearing to failure in triaxial compression; in other tests, the strains were ±1% and ±2%. The increasing levels of imposed strains can be considered to represent reducing quality of sampling tube, e.g. reducing B/t or reducing cutting edge sharpness. The results of the experiments are summarised in Figures 18 to 20 and demonstrate the following:

(i)   there is a reduction in mean effective stress, p', in this soft clay as a result of imposing centre-line tube sampling strains,

(ii)   the magnitude of Δp' increases with the magnitude of the imposed strains (Figure 20),

(iii)   there is destructuring of the clay, manifest as a reduction in peak strength and in stiffness (Figure 19) after reconsolidation and as a shrinking of the soil's bounding surface,

(iv)   destructuring is progressive, increasing with increasing levels of imposed strains, i.e. with reducing quality of sampling tube,

(v)   the volumetric strains during reconsolidation to in situ stresses after simulated sampling increase with imposed strain level (Figure 20).

Strains at Periphery. The predictions of Baligh et al (1987) are that there is intense shearing in a narrow zone adjacent to the wall of the sampling tube, where shear strains may exceed 100% (Figure 11). This prediction is, of course, consistent with our observations of distortion around the perimeter of a tube sample. The effects of these strains when sampling clay are to create a zone of destructured or remoulded soil in which excess pore pressures are generated which are:

(i)   positive in normally consolidated or lightly overconsolidated clays, or

(ii)   negative in heavily overconsolidated clays.

The situation is summarised in Figure 21 for the two types of clay. With time, there is pore pressure equilibration and water content redistribution. In the soft clay, the outer zone consolidates and the water content reduces; the central part swells and the water content increases. In the heavily overconsolidated clay, the outer zone swells and increases in water content and the central part consolidates and reduces in water content.

Overall, the effects of the peripheral strains superimposed on those of the centre-line strains are to reduce further the effective stress in the soft clay, but possibly to increase the effective stress in the stiff heavily overconsolidated clay.

The combination of the centre-line and peripheral strains introduces an important sample diameter effect. Baligh et al have shown that the thickness of the distorted peripheral zone is controlled more by wall thickness, t, than tube diameter, B. Therefore, the proportion of the volume of the whole sample occupied by the distorted zone reduces as the sample diameter increases and the effects of the peripheral distortion and the pore pressure change to which they lead become less.

There is, incidentally, a sample length effect when one takes into account the presence of additionally disturbed zones at the top of the sample, resulting from yield at the base of the borehole, and at the bottom of the sample, resulting from suction forces which may develop on withdrawing the sampling tube from the ground. In short samples, these two effects may even overlap. Clearly, the larger the diameter and the longer the sampling tube, within reason, the better.

Field Evidence for the Effects of Tube Sampling Strains in Soft Clays

The predicted effects of the combined tube sampling strains on samples of soft clay are:

(i)   a reduction in mean effective stress,

(ii)   damage to structure, manifest as shrinking of the soil's bounding surface, and

(iii)   an increase in water content in the centre of the sample.

Field evidence supporting these predicted effects is presented in Figures 22,23,24 and 25. Evidence from the field for reductions in mean effective stress as a result of tube sampling of lightly overconsolidated clays, and for the dependence of that reduction on sample tube geometry, is presented in Figure 22. Measurements of p_i' were made either through suction measurements on unconfined specimens using the Imperial College suction probe (Ridley and Burland, 1993), or by measuring equilibrium pore water pressures in triaxial specimens after setting them up in the triaxial cell and applying a cell pressure. Values of p_i' in the modified sampling tube are similar to those in the Laval sampling tube and much higher than in the conventional sampling tube, all as would be predicted on the basis of the previous discussions.

Figure 23 presents data from Tanaka (2000) on the measurement of residual effective stress in samples of lightly overconsolidated clays having a wide range of plasticities, but all taken with the Japanese standard piston sampler. The dependence of residual effective stress on soil plasticity, predicted on the basis of perfect sampling, is confirmed.

Figure 24 shows the bounding surfaces found for Saint Louis clay, from Eastern Canada, in block samples, Laval samples and 50mm diameter piston samples. Tests on the block and Laval samples define the same bounding surface, which sits well outside the equivalent surface defined on the basis of the poorer quality piston samples.

Figure 25 shows water contents measured across the diameter of tube samples of soft clay after water content redistribution has taken place and which are higher in the centre.

Consequences of Tube Sampling Disturbance

Acknowledging that there are two key effects of tube sampling strains on soft clays, a reduction in p' and shrinking of the soil's bounding surface, it is possible to anticipate the consequences of these effects in testing tube samples.

The behaviour in unconsolidated undrained, UU, tests is illustrated in Figure 26(a) and compared to the behaviour of the intact in situ soil. Two cases are shown for sampled soil, one representing a high quality sample and one a poor quality sample, the latter having a lower value of p' and a lower bounding surface, i.e. having suffered greater destructuring. Strengths and stiffness are seriously underestimated in triaxial compression of tube samples, but less so in triaxial extension. (The reduction in p' and the importance of loading direction were anticipated on the basis of perfect sampling.)

Proof of these effects, as far as sample quality are concerned, is presented in Figure 27. This compares the effective stress paths measured in UU tests on samples of Bothkennar Clay taken with the Sherbrooke block sampler, the conventional UK piston sampling tube, with its 30° cutting edge, and the modified stainless steel tube, with its 5° cutting edge, referred to earlier. On the basis of the Strain Path predictions presented in Figure 14, we would expect the modified tube to perform better than the conventional tube because of the lower imposed strains. This was found to be the case, lower strengths being measured with the conventional tube. In fact, the modified tube performed as well as the Sherbrooke sampler in this instance. The differences between samplers of different quality were confirmed as being the result of larger reductions in p' and greater destructuring in the poor quality samplers.

The behaviour in CAU (or CK_oU) tests in which samples of different quality are reconsolidated to in situ stresses prior to shear is illustrated in Figure 26(b). The effects of differences in p' are now eliminated and any differences are due to different levels of destructuring and to the opposing effects of additional damage during reconsolidation and of reductions in water content during reconsolidation. In Figure 26(b) it is assumed that the effects of changes of water content are small compared to the effects of damage to structure. Confirmation of these effects is presented in Figure 28, using data from CAU tests on poor quality samples of Bothkennar Clay (taken with the conventional tube with its 30^0 cutting edge) and high quality samples taken with the Laval and Sherbrooke samplers.

The behaviour in oedometer tests on samples of different quality is illustrated in Figure 26(c). In this test, the effective stress paths are not controlled but depend on factors such as p_i', the fit of the specimen in the oedometer ring and the amount of destructuring. It is possible for an oedometer test on a disturbed sample to indicate a similar apparent yield stress as an oedometer test on an undisturbed sample. This is simply the result of the stress paths intercepting the local bounding surfaces at different locations. A comparison of oedometer tests on samples of different quality is presented in Figure 29; the contrast between samples of different quality is less than in UU triaxial compression tests.

Effects of Gas Exsolution

The effects of gas exsolution as a result of sampling are likely to depend, firstly, on whether the process was undrained (gas occluded) or drained (gas continuous). Damage to structure is more likely in the former case, when the gas is trapped and expands the soil skeleton. The damage may become sufficient to rupture the soil and allow the gas to escape. There is also likely to be an effect on the mean effective stress in the sample.

The second key factor is whether or not the sample is resaturated prior to testing, and how it is resaturated. In tests on samples which are not resaturated, the presence of occluded gas will increase the overall compressibility of the soil; it will also reduce permeability and coefficient of consolidation, c_v. Resaturation (or measurement of the pore pressure parameter B) involving an application of confining stress at constant water content will cause closure or collapse of bubbles, with the possibility of additional damage to the soil structure. The volume change during the initial application of confining stress may not be measured.

The effects of gas exsolution on the results of oedometer tests are idealised in Figure 30(a). The higher initial compressibility with increasing gas content is the result of the occluded gas in the unsaturated samples, as well as destructuring. Once the gas bubbles have collapsed, the compression curves become parallel. An example of the effects of gas exsolution on the subsequent response in conventional oedometer tests is shown in Figure 31 using data from the Nile Delta Clays (Hight et al, 2000).

Destructuring by gas exsolution can be expected to be different to destructuring by tube sampling strains. Tube sampling strains are imposed externally along the full length of the sample - destructuring occurs along the whole sample. Assuming a uniform distribution of gas in solution and a uniform reduction in pore pressure, it is probable that gas will exsolve first in the largest pores, expanding and damaging the bridges around these pores. The onset of yield would then occur earlier than in an undamaged sample. This points to the importance of pore size distribution in examining the potential for gas exsolution and its effects.

If the presumption above is right, gas exsolution damage occurs locally, is restricted to soil around the largest pores and will be variable. It causes internal swelling, shear strains and water content redistribution. Its significance is again dependent on the initial level of structure.

Destructuring by gas exsolution (and subsequent resaturation and consolidation) produces a characteristic change to the effective stress path (illustrated in Figure 30(b)).

There is a discontinuity in the effective stress path, i.e. a horizontal section, when excess pore pressures are generated with no increase in shear stress. This corresponds to collapse of the largest pores, damaged by gas exsolution and reconsolidation. The extent of the discontinuity increases with the initial degree of gas saturation and, therefore, with the amount of damage. Confirmation of these effects of damage by gas exsolution in triaxial compression is presented in Figure32, using data from the study by Lunne at al (2001).

Damage during extrusion is more likely in a sample in which there has been gas exsolution than in one that has remained saturated.

Other Sources of Disturbance

In Figure 1 Ladd and Lambe (1963) considered other sources of disturbance that might change the effective stress in samples of soft clay. A more complete list of sources of disturbance and the change in initial effective stress, p_i', to which they may lead is presented in Figure 33. Some of these disturbances may also lead to destructuring. An example is shown in Figure 34 of the destructuring that can occur during the preparation of a specimen for testing, a source of disturbance considered by Ladd and Lambe. The effective stress paths and stress strain curves are compared for CAU tests on specimens prepared by trimming with a wire saw and those prepared by trimming a specimen while simultaneously pushing in a 38mm diameter sampling tube.

Methods of storage, handling, specimen preparation, and reconsolidation paths, all become more critical when high quality sampling is introduced and structural effects resulting from cementing and ageing are retained.

Natural Versus Induced Variability

By virtue of the various depositional environments in which they may be formed and the post-depositional processes to which they may then be subjected, soils are naturally variable. An example of the natural variability of the Bothkennar Clay is presented in Figure 35(a). This clay is remarkably consistent in terms of composition, which does not vary with depth. However, in high quality samples, variability can be seen because of different levels of structure - cementing - resulting from the different environments of deposition and post-depositional processes.

The tube sampling process and its effects are not systematic. In soft clay, the reduction in effective stress will vary because of the factors summarised in Figure 33. In UU testing, p' is not controlled and variability in strength and stiffness is inevitably introduced.

Since destructuring is progressive, as demonstrated above, the level of damage will also vary, depending on the initial level of structure and on a similar set of factors: sampler quality, borehole formation, transport, handling, specimen preparation, reconsolidation path, etc. In UU testing, different amounts of damage simply add to the scatter, while in CAU testing they may well be the main source of the scatter.

Figure 35(b) shows the result of a series of CAU triaxial compression tests on samples of Bothkennar Clay from the same depth, but which have been subjected to the different sources of damage that are indicated: drying during storage, strains imposed during specimen preparation, damage during transport - these are additional to the different levels of damage associated with samplers of reducing quality: Sherbrooke, Laval, conventional piston tube (30° edge). The variability shown in Figure 35(b) is not natural, but has been induced entirely by the different sources of disturbance during sampling.

Assessing Sample Quality

Various methods for assessing sample quality, i.e. the level of sample disturbance, have been proposed and these are introduced below. In any method it is essential to recognise the different forms of damage that can occur and to distinguish between local damage, that may be caused by gas exsolution, and general damage, that may be caused by tube sampling strains.

Fabric Inspection. Although of major importance for assessing some of the potential effects of sampling, in particular the risk of there having been water content redistribution between adjacent sand and clay layers, fabric inspection is not sufficient to determine the likely level of disturbance. Only gross distortion, for example in the distorted peripheral zone, can be seen, whereas only relatively small strains cause yield and damage to a bonded structure. In addition, strain histories in the central zone of the sample involve unloading (see Figures 11, 12 and 14) so that the maximum imposed strains cannot be deduced.

Measurement of Initial Effective Stress, p_i'. Measurement of initial effective stress, p_i', (or residual effective stress, $\bar{\sigma}_r$) in samples taken from the ground and set up in the laboratory, was suggested by Ladd and Lambe (1963) as a means of assessing sample quality by comparing p_i' with the value of p' after perfect sampling, p_{ps}'. Data presented in Figure 20 shows how the increase in pore pressure, and, therefore, the reduction in p_i', increases with increasing levels of imposed tube sampling strains, i.e. increasing disturbance. Data presented in Figure 22 shows how measurements of p_i' indicate a difference in quality between Laval and conventional UK piston samples. However, the measurement of p_i' alone is not sufficient, as it cannot indicate the amount of destructuring that has occurred. Indeed, as illustrated in Figure 36, higher quality samples can show a greater reduction in effective stress than poorer quality samples when water content redistribution takes place inside the tube. The poorer quality sample shows greater expansibility, because of the destructuring that has taken place, and requires a smaller reduction in effective stress to accommodate the available water. It is worth noting that Ladd and Lambe said 'it is probably unwarranted to assume with the present state of knowledge that all the detrimental effects of sampling can be expressed simply as a reduction in effective stress'.

Measurement of Strains During Reconsolidation. In reconsolidating samples to in situ stresses, the strains will depend on both the reduction in effective stress that has occurred as a result of sampling and the amount of destructuring. In this respect, they are a useful comparative measure of quality between samples, as demonstrated by the data on volume strains during reconsolidation of Bothkennar Clay, presented in Figure 20. The absolute value of the strains will depend on the reconsolidation path followed and the soil compressibility. To take account of the latter, Lunne et al (1997) have proposed expressing the volume strains in terms of $\Delta e/e_o$, where Δe is the change in void ratio and e_o the initial void ratio. Lunne at al's approach has been applied successfully in assessing the quality of tube samples of the Nile Delta Clays (Hight et al, 2000).

Comparison of Field and Laboratory Measurements of Shear Wave Velocity/Dynamic Shear Modulus. The basis for adopting comparisons of shear wave velocity, V_s (or small-strain shear modulus, G_{max}), in the laboratory and field, as a measure of mechanical disturbance is presented by O'Neill and Hight (2002) who give examples of its application in practice.

Improving Sample Quality

With our present understanding of tube sampling strains and the factors determining these, we are in a position to improve our design of sampling tubes and to improve sample quality. Similarly, with our improved understanding of the effects of sample disturbance, we are better able to assess sample quality, and this is discussed by O'Neill and Hight (2002) in their paper to this Symposium.

It is clearly beneficial to operate with a sampling tube of large diameter, thin wall, sharp cutting edge and, at least in soft clays, with no inside clearance. To reduce the impact of end effects, associated with disturbance at the base of the borehole and with suction effects when withdrawing the sample, it is also preferable to operate with a long sample. It is also important to work with a mud-supported borehole in soft clays and to ensure a clean and level base to the borehole. To avoid the water content redistribution that occurs in tube samples, there are clearly benefits in immediate extrusion on site, removal of the peripheral zone and sealing and protection of the extruded sample. All these features of a sampler and details of sampling technique are embodied in the approach described by LaRochelle at al (1981).

There are inherent dangers in improving sample quality, however, if this is to be done in conjunction with the use of empirical or semi-empirical design procedures which have been developed on the basis of poorer quality samples and which ignore anisotropy of soil properties. Consider, for example, the case of an embankment constructed on the Bothkennar soft clay. Figure 37 shows the following undrained strength profiles at the Bothkennar soft clay site:

(i)   the profile based on UU triaxial compression (UUTC) tests on conventional piston samples,

(ii)   the profile based on UUTC tests on Laval samples,

(iii)   the profiles based on CAU tests in triaxial compression, triaxial extension and simple shear on Laval samples - these profiles have been adjusted to allow for rate effects and disturbance and represent the best estimates of these particular strengths.

Note how the improved sample quality (Laval versus conventional piston) leads to a higher strength profile, based on UUTC tests, and how the UUTC profile for piston samples corresponds approximately to the best estimate of simple shear strength, and the UUTC profile for Laval samples corresponds approximately to the best estimate of in situ shear strength in triaxial compression.

The strength available around the critical potential failure surface abc beneath an embankment constructed on this clay will vary, as illustrated in Figure 38, by an amount reflecting the level of anisotropy in the soil. The average mobilised strength is indicated by AA. Experience has shown that AA is often close to the average strength measured in simple shear, which in this case is similar to the strength measured in UU tests on poor quality samples, so that, when UUTC data is combined with a conventional factor of safety, a safe design results. An improvement in sample quality will lead to a higher UUTC strength profile (see Figure 37), and if the same design procedure is adopted, i.e. using UUTC strengths to represent an average mobilised strength, an unsafe design may result, unless the factor of safety is modified.

Concluding Remarks

Tube sampling of soft clays can result in a reduction in mean effective stress in the sample, mechanical damage to the soil's structure, modification to its anisotropy and an increase in water content in the central core of the sample, if it is not extruded and the periphery trimmed immediately. The effects depend on the geometry of the sampling tube, the soil plasticity and stress history - sample disturbance reduces with increasing OCR and soil plasticity - and on the initial level of structure. Sampling effects are not systematic and damage to structure is progressive; as a result the apparent variability in a soil deposit is often increased by sampling effects. The effects of stress relief alone can cause damage as a result of gas exsolution or cavitation, which tend to cause local damage, as opposed to the general damage caused by tube sampling strains.

Sampling effects are most apparent in UU testing, in which the effects of both the reduction in mean effective stress and mechanical damage are manifest. CAU testing eliminates the effects of reductions in mean effective stress but is still affected by the mechanical damage during tube sampling and by additional damage and water content changes that occur during reconsolidation.

References

Baligh, M. M. (1985). "Strain path method." *J. Geotech. Engng Div.*, ASCE, Vol. 111, GT9, 1108-1136.

Baligh, M. M., Azzouz, A. S. and Chin, C. T. (1987). "Disturbance due to ideal tube sampling." *J. Geotech. Engng Div.*, ASCE, Vol. 113, GT7, 739-757.

Bishop, A. W., Kumapley, N. K. and El-Ruwayih, A. (1975). "The influence of pore-water tension on the strength of clay." *Phil. Trans. Royl. Soc.*, 278, 511-554.

Clayton, C. R. I., Hight, D. W. and Hopper, R. J. (1992). "Progressive destructuring of Bothkennar Clay: implications for sampling and reconsolidation procedures." *Géotechnique*, 42, 2, 219-239.

Clayton, C. R. I., Siddique, A. and Hopper, R. J. (1998). " Effects of sampler design on tube sampling disturbance - numerical and analytical investigations." *Géotechnique*, 48, 6, 847-867.

Gens, A. (1982). "Stress-strain and strength characteristics of a low plasticity clay." *PhD thesis*. University of London.

Hight, D. W. (1986). "Laboratory testing: assessing BS5930." *Proc. 20th Regional Meeting of Engineering Group of Geological Society. University of Surrey.* (Presented in 1984). Engineering Geology Special Publication No. 2, 43-52.

Hight, D. W., Böese, R., Butcher, A. P., Clayton, C. R. I. and Smith, P. R. (1992a). "Disturbance of the Bothkennar clay prior to laboratory testing." *Géotechnique*, 42, 2, 199-217.

Hight, D. W., Bond, A. B. and Legge, J. D.(1992b). "Characterization of the Bothkennar clay: an overview." *Géotechnique*, 42, 2, 303-347.

Hight, D. W. (1993). "A review of sampling effects in clays and sands." *Proc. SUT Conf: Offshore Site Investigation and Foundation Behaviour*, 115-146.

Hight, D. W. (1998). "Soil characterisation: the importance of structure and anisotropy." *38th Rankine Lecture.* To be published in Géotechnique.

Hight, D. W., Hamza, M. M., El Sayed, A. S. (2000). "Engineering characterisation of the Nile Delta clays." *International Symp. on Coastal Geotechnical Engineering in Practice (IS-Yokohama 2000)*. Preprints, 81-94.

Jardine, R. J. (1992). "Some observations on the kinematic nature of soil stiffness." *Soils and Foundations*, 32, 2, 111-124.

La Rochelle, P., Sarrailh, J., Tavenas, F., Roy, M. and Leroueil, S. (1981). "Causes of sampling disturbance and design of a new sampler for sensitive soils", *Can. Geotech. J.*, 18, 1, 52-66.

Ladd, C. C. and Lambe, T. W. (1963). "The strength of undisturbed clay determined from undrained tests." *Symp. on Laboratory Shear Testing of Soils*, ASTM, STP 361, 342-371.

Lane, K. S. and Washburn, D. E. (1946). "Capillarity tests by capillarimeter and soil filled tubes." *Proc. Highway Research Board.*

Lunne, T., Berre, T. and Strandvik, S. (1997). "Sample disturbance effects in soft low plastic Norwegian clay." *Symp. on Recent Developments in Soil and Pavement Mechanics, Rio de Janiero,* 81-102.

Lunne, T., Berre, T., Strandvik, S., Andersen, K .H. and Tjelta, T .I. (2001). "Deepwater sample disturbance due to stress relief." *Proceedings of OTRC 2001 International Conference on Geotechnical, and Geophysical Properties of Deepwater Sediments*, 64-102.

O'Neill, D. A. and Hight, D. W. (2002). "Sample quality diagnostics (SQUAD) using field measurements of shear wave velocity." *This volume.*

Ridley, A. M. and Burland, J. B. (1993). "A new instrument for the measurement of soil moisture suction." *Géotechnique*, 43, 2, 321-324.

Skempton, A. W. and Sowa, V. A. (1963). "The behaviour of saturated clays during sampling and testing." *Géotechnique*, 13, 4, 269-290.

Tanaka, H. (2000). "Re-examination of established relations between index properties and soil parameters." *Keynote Address. International Symp. on Coastal Geotechnical Engineering in Practice (IS-Yokohama 2000)*. Preprints, 2-24

Tavenas, F. and Leroueil, S. (1987). "Laboratory and in situ stress-strain-time behaviour of soft clays: a state-of-the-art", *International Symposium on Geotechnical Engineering of Soft Soils, Mexico City*, Vol 2, 1-46.

Vaughan, P. R., Chandler, R. J., Apted, J. P., Maguire, W. M. and Sandroni, S. S. (1993). "Sampling disturbance with particular reference to its effect on stiff clays." *Proc. Wroth Memorial Symp., Oxford*, 685-708.

Notation

σ_a, σ_a' axial total and effective stress in cylindrical triaxial test

σ_r, σ_r' radial total and effective stress in cylindrical triaxial test

σ_{ac}' value of σ_a' for consolidation

$\sigma_{vo}, \sigma_{vo}'$ in situ vertical total and effective stress

$\sigma_{ho}, \sigma_{ho}'$ in situ horizontal total and effective stress

u_o in situ pore water pressure

u_c pore pressure in central core of sample

u_p pore pressure at periphery of sample

$\bar{\sigma}_r$ residual effective stress (Ladd & Lambe, 1963)

$\bar{\sigma}_{ps}$ mean effective stress after perfect sampling (Ladd & Lambe, 1963)

p_t' initial mean effective stress in a sample

p_o' initial mean effective stress during consolidation

p_{ts}' mean effective stress after tube sampling

$t \quad = \quad (\sigma_a - \sigma_r)/2$

$s' \quad = \quad (\sigma_a' + \sigma_r')/2$

E_u secant modulus in undrained shear

w_o initial water content

Δw_s change in water content as a result of sampling

w_c water content in central core of sample

w_p water content at periphery of sample

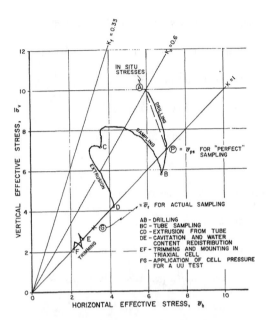

Figure 1 Hypothetical stress path for a normally consolidated clay
element during tube sampling
(Ladd and Lambe,1963)

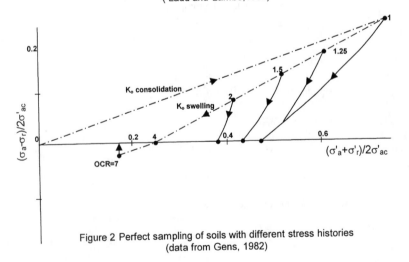

Figure 2 Perfect sampling of soils with different stress histories
(data from Gens, 1982)

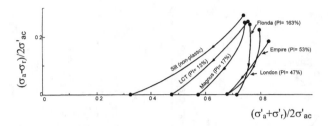

Figure 3 The effect of plasticity on perfect sampling stress paths for normally consolidated soils

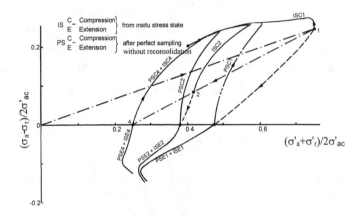

Figure 4 Response in undrained triaxial compression and extension before and after perfect sampling (data from Gens, 1982)

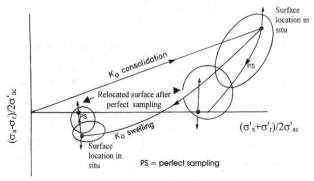

Figure 5 Relocation of kinematic yield surfaces as a result of perfect sampling

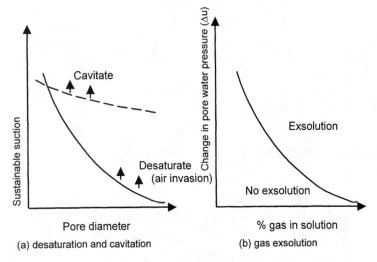

(a) desaturation and cavitation   (b) gas exsolution

Figure 6 Potential effects of undrained stress relief

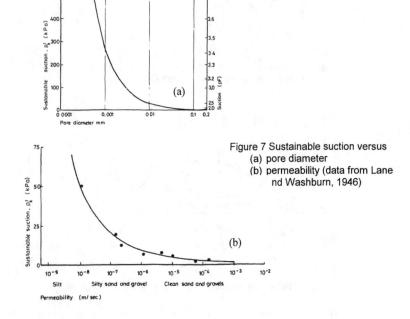

Figure 7 Sustainable suction versus
(a) pore diameter
(b) permeability (data from Lane nd Washburn, 1946)

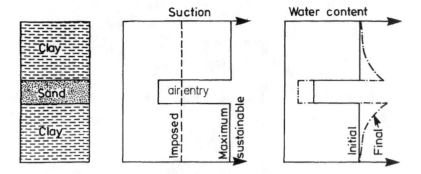

Figure 8 Effects of desaturation in laminated clay

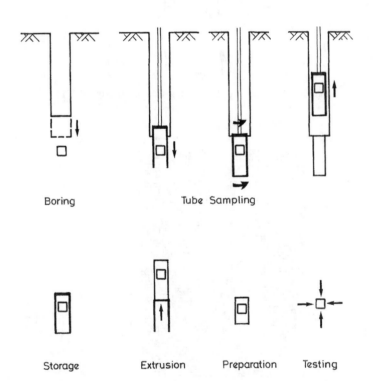

Figure 9 Stages involved in the tube sampling and testing of soils

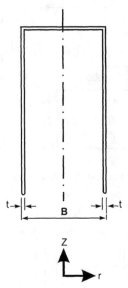

Figure 10(a) Geometry of the Simple Sampler (Baligh, 1985)

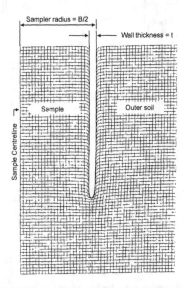

Figure 10(b) Predicted deformation pattern during undrained Simple Sampler
penetration in saturated clays (B/t=20) (Baligh, 1985)

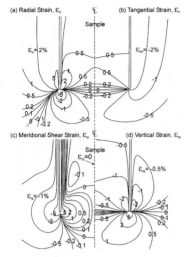

Figure 11 Complete description of strains caused by undrained penetration of
Simple Sampler (B/t=40) (Baligh et al, 1987)

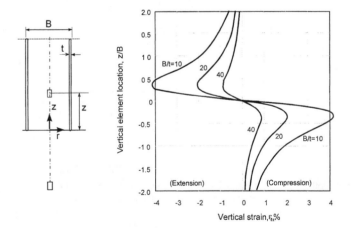

Figure 12 Strain path for an element on the centre-line of the advancing Simple
Sampler (Baligh, 1985)

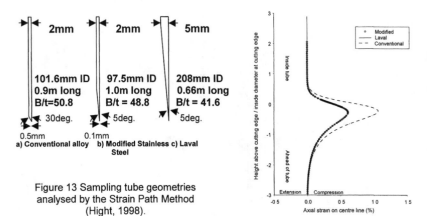

Figure 13 Sampling tube geometries analysed by the Strain Path Method (Hight, 1998).

Figure 14 Predicted strain paths for Laval, conventional and modified sampling tubes (Clayton, pers comm.)

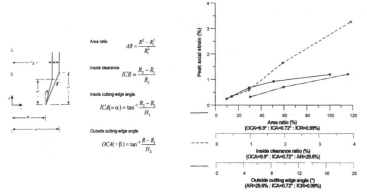

Figure 15 Influence of sampling tube geometry on maximum centre-line sampling strains predicted by Strain Path Method (based on Clayton et al, 1998)

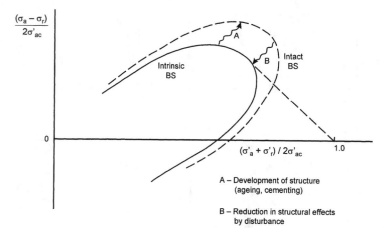

A – Development of structure
(ageing, cementing)

B – Reduction in structural effects
by disturbance

Figure 16 Model of soil behaviour

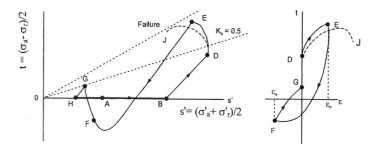

A - Initial Effective Stress   DEFG - Undrained Strain Path   HABD - Second Reconsolidation
D - Estimated In-situ Stresses
ABD - Initial Reconsolidation   GH - Deviatoric Stress Relief   DJ - Undrained Shearing in Compression

Figure 17 Test sequence to simulate centre-line tube sampling effects on
Bothkennar Clay (Clayton et al, 1992).

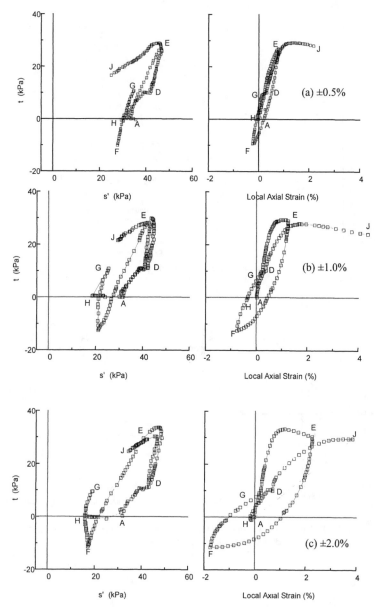

Figure 18 Response of Bothkennar Clay to centre-line tube sampling strains
(Clayton et al, 1992)

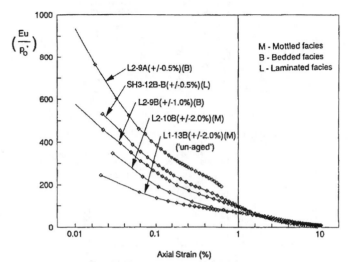

Figure 19 Progressive reduction in stiffness with increasing levels of tube
sampling strains (Clayton et al, 1992)

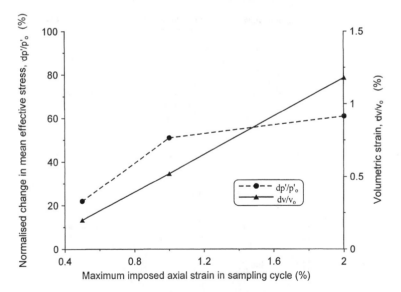

Figure 20 Changes in mean effective stress and volumetric strains during
reconsolidation caused by different levels of tube sampling strains (based on
Clayton et al, 1992)

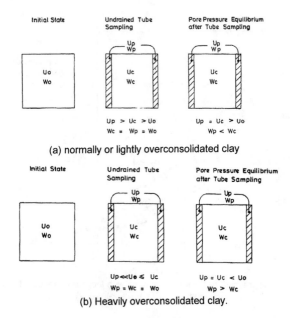

(a) normally or lightly overconsolidated clay

(b) Heavily overconsolidated clay.

Figure 21 Simplified view of pore pressure and water content changes
after tube sampling.

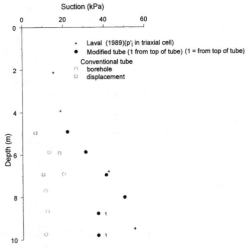

Figure 22 Suction versus depth, Bothkennar Clay (Hight, 1998)

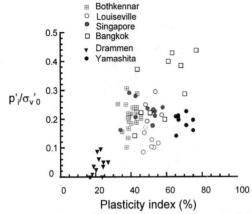

Figure 23 Dependence of residual effective stress on plasticity index
(Tanaka, 2000)

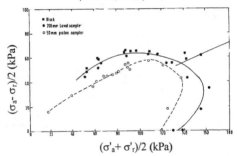

Figure 24 Shrinking of the bounding surface for Saint Louis Clay as a result of
disturbance during sampling (Tavenas and Leroueil, 1987)

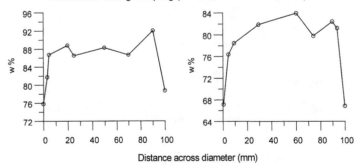

Figure 25 Variations in water content across soft clay sample diameter
(Vaughan et al, 1993)

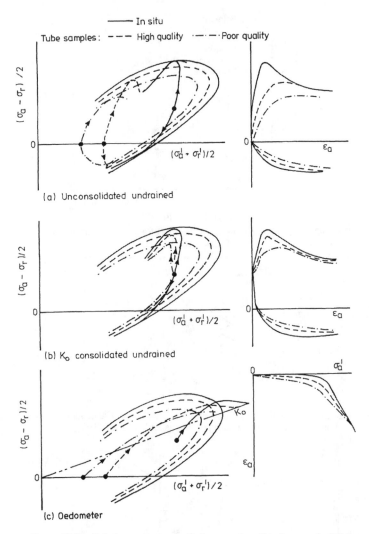

Figure 26 Predicted consequence of tube sampling disturbance of a lightly
overconsolidated clay (Hight, 1993).

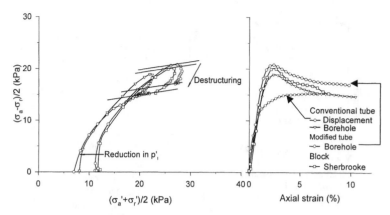

Figure 27 UU triaxial compression tests on Bothkennar Clay, 4.7-4.9m
(Hight,1998)

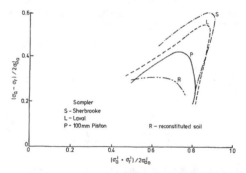

Figure 28 CAU triaxial compression tests on Bothkennar Clay
(based on Hight et al,1992a)

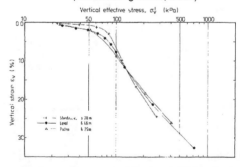

Figure 29 Oedometer tests on Bothkennar Clay (Hight et al,1992a)

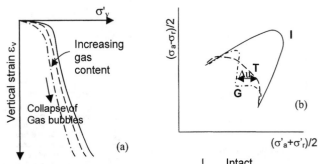

Figure 30 Effects of gas exsolution on
(a) compressibility
(b) undrained effective stress path

I   Intact
T   Damage due to tube
    sampling strains
G   Damage due to gas
    exsolution

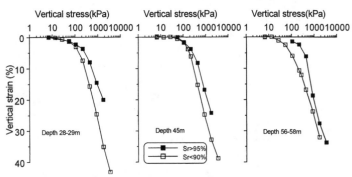

Figure 31 Effects of gas exsolution on compressibility of Nile Delta Clay
(Hight et al, 2001)

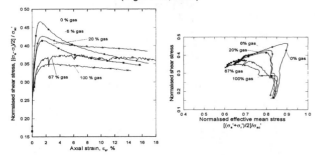

Figure 32 CAUC tests on Lierstranda Clay samples after gas exsolution,
without tube sampling strains (Lunne et al,2001)

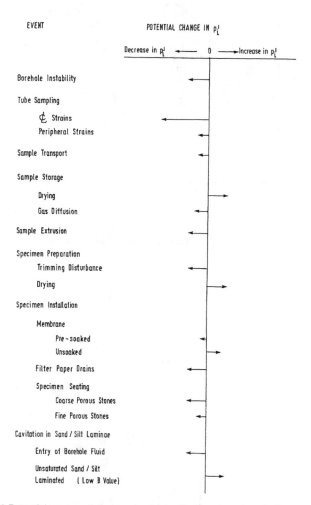

Figure 33 Potential causes of changes in mean effective stress in soft clay
(Hight,1986)

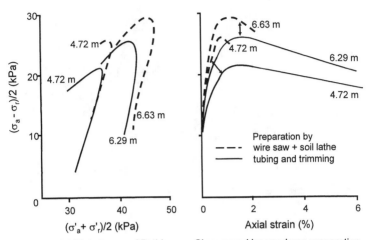

Figure 34 Disturbance of Bothkennar Clay caused by specimen preparation
(based on Hight et al,1992a)

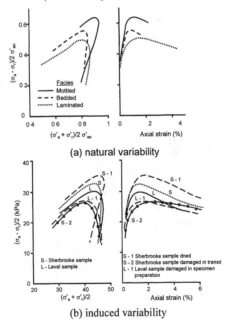

(a) natural variability

(b) induced variability

Figure 35 Natural versus induced variability in Bothkennar Clay
(Hight et al,1992b)

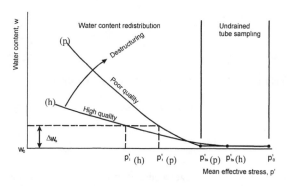

Figure 36 Shortcomings in the use of p_i' in assessing sample quality

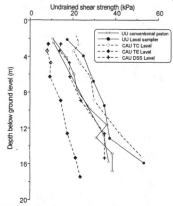

Figure 37 Undrained strength profiles at Bothkennar
(based on Hight et al, 1992b)

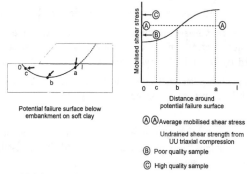

Figure 38 Inherent dangers in improving sample quality while neglecting
anisotropy

Primary Compression and Secondary Compression

Gholamreza Mesri[1]

Abstract

One-dimensional compression of a saturated soil, subjected to an increment of total vertical stress, consists of a primary compression that takes place during the increase in effective vertical stress and a secondary compression that follows at constant effective vertical stress. A soil structure change commences as soon as the increase in effective vertical stress begins, and it continues with time to establish internal equilibrium under the new external condition. At any instant during consolidation, soil compression in response to the effective vertical stress increase is characterized by the compressibility parameter $(\partial e / \partial \sigma'_v)_t$, and the continued compression with time is characterized by the compressibility parameter $(\partial e / \partial t)_{\sigma'_v}$. As soon as primary consolidation begins, both $(\partial e / \partial \sigma'_v)_t$ and $(\partial e / \partial t)_{\sigma'_v}$ contribute to compression; however, only $(\partial e / \partial t)_{\sigma'_v}$ contributes to compression during secondary consolidation when $d\sigma'_v / dt = 0$. In spite of the much longer duration of primary consolidation in the field as compared to that in laboratory oedometer tests, laboratory and field observations and analyses suggest an EOP compression independent of the duration of primary consolidation. The explanation is that for any soil $(\partial e / \partial \sigma'_v)_t$ and $(\partial e / \partial t)_{\sigma'_v}$ are not constants; they are interrelated and both depend on total rate of compression as influenced by drainage boundary conditions. The C_α / C_c law of compressibility reliably defines the contribution of $(\partial e / \partial t)_{\sigma'_v}$ during secondary consolidation, both after loading and associated increase in effective vertical stress as well as following the removal of a surcharge and associated decrease in effective vertical stress. The contribution of $(\partial e / \partial t)_{\sigma'_v}$, introduced by Donald W. Taylor, during both primary consolidation and secondary consolidation, has been well established. Charles C. Ladd has played a leading part in this significant progress in geotechnical engineering.

---

[1]Ralph B. Peck Professor of Civil Engineering, University of Illinois at Urbana-Champaign, 205 North Mathews Avenue, Urbana, Illinois 61801-2352; phone 217-333-6934; g-mesri@uiuc.edu

Introduction

The first attempt to revise the theory of consolidation after Karl Terzaghi (1923) was made by Donald Taylor and Wilfred Merchant (1940) at the Massachusetts Institute of Technology. The motivation for a revised consolidation theory came from (Taylor 1942):

1.   "The most evident difference between the Terzaghi theory and the action of samples in the laboratory consolidation test is the occurrence of secondary compression in the tests."

2.   "Of fully as much importance, but not so evident from conventional test data, are the differences occurring during primary compression."

Taylor and Merchant (1940) realized that during the process of consolidation, void ratio of an element of soil is a function of effective vertical stress and time:

$$e = f(\sigma_v', t) \tag{1}$$

Therefore, the rate of compression in terms of rate of change of void ratio is:

$$\frac{de}{dt} = \left(\frac{\partial e}{\partial \sigma_v'}\right)_t \frac{d\sigma_v'}{dt} + \left(\frac{\partial e}{\partial t}\right)_{\sigma_v'} \tag{2}$$

where $\left(\partial e / \partial \sigma_v'\right)_t$ is compressibility of soil structure, at any time, in response to an increase in effective vertical stress, and $\left(\partial e / \partial t\right)_{\sigma_v'}$ is compressibility of soil structure, at any effective vertical stress, in response to passage of time. Therefore, total compression rate de/dt consists of two parts. The first part of compression rate, $\left(\partial e / \partial \sigma_v'\right)_t \, d\sigma_v' / dt$, is produced by the rate of increase in effective vertical stress $d\sigma_v' / dt$, and the second part of the compression rate, $\left(\partial e / \partial t\right)_{\sigma_v'}$, follows with time.

The effective vertical stress increase necessitates a particle rearrangement and changes in interparticle forces that continue with time to achieve internal equilibrium.

Equation 2 is here employed to define primary compression and secondary compression, and also to elaborate on Taylor's (1942) observations 1 and 2. This paper describes the current state of knowledge on primary compression and secondary compression of soils.

Primary Compression and Secondary Compression

Equation 2 is integrated to obtain total compression in a period of elapsed time t:

$$\int_0^t de = \int_0^t \left[\left(\frac{\partial e}{\partial \sigma_v'}\right)_t \frac{d\sigma_v'}{dt} + \left(\frac{\partial e}{\partial t}\right)_{\sigma_v'} \right] dt \tag{3}$$

Alternatively, Eq. 3 can be written as:

$$\int_{o}^{t} de = \int_{o}^{t_p} \left[\left(\frac{\partial e}{\partial \sigma'_v} \right)_t \frac{d\sigma'_v}{dt} + \left(\frac{\partial e}{\partial t} \right)_{\sigma'_v} \right] dt + \int_{t_p}^{t} \left(\frac{\partial e}{\partial t} \right)_{\sigma'_v} dt \tag{4}$$

A comparison of Eqs. 3 and 4 indicates that beyond the time t_p, $d\sigma'_v / dt = 0$. In other words, t_p represents the period of consolidation during which effective vertical stress increases. Beyond t_p effective vertical stress remains constant. The period of consolidation during which effective vertical stress increases is called primary consolidation stage, and total compression that takes place during primary consolidation stage is called primary compression. The endless period that follows primary consolidation stage is called secondary consolidation stage, and total compression that takes place during secondary consolidation stage is called secondary compression.

The first integral on the right side of Eq. 4 is primary compression, and the second integral on the right side is secondary compression. Equation 4 clearly shows that during primary consolidation stage both $(\partial e / \partial \sigma'_v)_t$ and $(\partial e / \partial t)_{\sigma'_v}$ contribute to primary compression. The second integral on the right side of Eq. 4 suggests that as long as $(\partial e / \partial t)_{\sigma'_v}$ does not become zero, secondary compression will continue indefinitely. However, it is very important to realize that $(\partial e / \partial \sigma'_v)_t$ and $(\partial e / \partial t)_{\sigma'_v}$ are not constant soil properties. Specifically, $(\partial e / \partial \sigma'_v)_t$ and $(\partial e / \partial t)_{\sigma'_v}$ do not remain constant during primary and secondary consolidation stages, and values of $(\partial e / \partial t)_{\sigma'_v}$ during primary consolidation and secondary consolidation are not necessarily the same.

Taylor's (1942) observation 1 was referring to the second integral on the right side of Eq. 4, whereas his observation 2 was concerned with manifestations of the first integral on the right side in contrast to the Terzaghi consolidation theory which assumed $(\partial e / \partial \sigma'_v)_t$ to be a constant and $(\partial e / \partial t)_{\sigma'_v}$ to be zero.

Primary Compression

Because it has not been readily possible to evaluate $(\partial e / \partial \sigma'_v)_t$ and $(\partial e / \partial t)_{\sigma'_v}$ during primary consolidation stage, Eq. 3 is rarely used to compute total compression and associated settlement. Total one-dimensional compression is divided into primary compression and secondary compression as in Eq. 4, and primary compression is determined directly, most often using an oedometer test rather than from the first integral on the right side of Eq. 4. Therefore,

$$\int_{o}^{t_p} \left[\left(\frac{\partial e}{\partial \sigma'_v} \right)_t \frac{d\sigma'_v}{dt} + \left(\frac{\partial e}{\partial t} \right)_{\sigma'_v} \right] dt = [\Delta e]_p \tag{5}$$

where $[\Delta e]_p$ is determined from the end-of-primary (EOP) e versus σ'_v relationship of 20 mm thick undisturbed oedometer specimens.

A fundamental question has been whether or not the $[\Delta e]_p$ from a thin laboratory soil specimen should be directly used in settlement analysis of a thick soil layer in the field. The question was first posed explicitly by Leonards (1972). When a pressure increment is applied, the excess porewater pressure dissipates in a short time (minutes or hours) in the laboratory, whereas porewater pressure dissipates in a long time (years or decades) in the field. "The question is, at the time the excess pore pressure has dissipated ... will the clay layer in the field have compressed much more than that measured in the laboratory test ...?" If the answer to the question were positive, then Leonards (1972) reasoned "... there should at least be some cases where the observed consolidation settlement ... is substantially greater than that predicted. To date, no such cases have come to the writer's attention."

Ladd et al. (1977) stated the question in terms of Fig. 1 which "shows two extremes for the effect of sample thickness," concluding that "Little definitive data exists to show which of the two hypothesis is more nearly correct for the majority of cohesive soils (though the General Reporter is obviously biased toward curve A)." In addition to Ladd (1973), Mesri and Rokhsar (1974) was listed on Fig. 1 as supporting the hypothesis A because the Mesri and Rokhsar theory of consolidation predicted end-of-primary (EOP) settlements independent of the duration of primary consolidation.

Because hypothesis B was mainly based on theoretical analyses of consolidation using separate springs and dashpots (Taylor 1942, Barden 1969), Mesri (1977) pointed out that "If a model of soil structure is assumed in which separate and independent elements represent effective stress- and time-compressibilities, then the simplistic conclusion inevitably follows that the longer the duration of primary consolidation, the lower the end-of-primary void ratio. However, this conclusion need not be true, if in fact during primary consolidation, the incremental compressibilities with respect to effective stress and with respect to time are interrelated."

Mesri and Choi (1985b) reviewed existing data on soil thickness effect on magnitude of EOP compression, and presented new data on EOP void ratio versus effective stress relationship of thin and thick undisturbed specimens of three natural clay deposits. Mesri and Choi (1985b) concluded that EOP e versus σ'_v relationship of each sublayer of a natural soil deposit is unique and independent of the duration of primary consolidation. This conclusion has been both accepted (Jamiolkowski et al. 1985, Mesri et al. 1994, Leonards and Deschamps 1995) and has been challenged (Leroueil 1988, 1994). Additional data and interpretation on uniqueness of EOP e versus σ'_v was presented by Mesri et al. (1995). Leroueil (1995) attempted to explain in terms of soil disturbance, temperature and structuring phenomena the success of EOP e versus σ'_v from laboratory tests to predict field behavior. The last factor, i.e., structuring, is significant because it recognizes the interrelationship between compressibility with time and compressibility with respect to effective stress.

At this writing, empirical data that support an EOP e versus σ'_v independent of the duration of primary consolidation include the following.

1.   Primary compression of thin and thick clay specimens

Mesri and Choi (1985b) described the experimental problems associated with comparing, for a single pressure increment, compression versus time curves of a thin and a thick specimen. The most reliable approach for examining the relationship between magnitude of EOP compression and duration of primary consolidation is to compare EOP void ratio versus effective vertical pressure relationships of identical thin and thick specimens. This objective was achieved by determining and comparing EOP e versus log σ'_v curves of a 125 mm thick specimen and a 500 mm thick specimen which was obtained by connecting in series four 125 mm thick specimens. This is illustrated in Fig. 2 by the initial and final water content profiles in a 500 mm thick specimen of St. Hilaire clay. Five identical 125 mm thick specimens were cut side by side, i.e., at the same elevation, from a 200 mm diameter or a 250 mm diameter undisturbed sample taken by Laval or Sherbrooke samplers, respectively (La Rochelle et al. 1981, Lefebvre and Poulin 1979). One 125 mm thick specimen was tested as the thin specimen, and the remaining four specimens were connected in series and tested as the thick specimen. Note in Fig. 2 that within each 125 mm sublayer, there is a natural variation in water content with depth; however, the change in water content from the initial to the final condition in a consolidation test that lasted for over 2 years, is quite uniform along the height of the 500 mm thick specimen. For the 125 mm thick specimen drainage was allowed from the top and porewater pressure was measured at the bottom, whereas for the 500 mm thick specimen drainage was allowed from the top of the first specimen in the series and porewater pressure was measured at the bottom of all four 125 mm sublayers. In the last two series of tests, in addition to the measurements of porewater pressure and total compression (volume of water leaving the specimen), axial compression of each 125 mm sublayer was measured by a deformation dial. Therefore, it was possible to plot isochrons of excess porewater pressure and isochrons of axial compression, as shown in Fig. 3.

A main requirement of the investigation reported by Mesri and Choi (1985b) and Mesri et al. (1995) was to test undisturbed specimens of natural soil deposits, and for the thick specimen to use a maximum drainage distance of 500 mm. These requirements together could not be met for one-dimensional oedometer tests. In order to minimize ring friction, a 500 mm thick oedometer specimen must have a diameter of at least 1500 mm. It is not readily possible to obtain undisturbed samples of this size from natural soil deposits. Therefore, the consolidation tests were performed on 63.5 mm diameter cylindrical specimens which were subjected to increments of equal all-around pressure. The compression was three-dimensional, whereas drainage was one-dimensional in the axial direction. This method of loading avoided such experimental problems as ring friction and variation of lateral pressure during consolidation, and it was possible to maintain a constant imposed total stress condition over a period of years.

Five series of consolidation tests have been completed using undisturbed specimens of five soft clay deposits (Table 1, Choi 1982, Feng 1991). The EOP e versus log σ'_v curves of 125 mm thick and 500 mm thick specimens are compared in Fig. 4. Note that the values of duration of primary consolidation t_p for 125 mm thick specimens are in the range of 7 to 25 days whereas for the 500 mm thick specimens, they are in the range of 54 to 354 days. These correspond to typical values of t_p for pressure increments in the compression range. The EOP e versus log σ'_v data in Fig. 4 suggest that for the range of layer thickness examined here, for all practical purposes, EOP compression is independent of duration of primary consolidation.

2.   Preconsolidation pressure mobilized in the field compared to σ'_p from
      laboratory tests

Preconsolidation pressure is one of the most important characteristics of soft clay and silt deposits. Because it defines the boundary between stiff and soft deformation response of soil to loading, it plays a major role in settlement analysis. As a significant yield stress, σ'_p from laboratory oedometer tests has been used to normalize undrained shear strength measured by in situ and laboratory tests and mobilized in full-scale loading in the field (e.g., Ladd and Foott 1974, Terzaghi et al. 1996). Therefore, σ'_p from oedometer tests also plays a major role in evaluation of undrained shear strength for stability analysis. If EOP e versus σ'_v relationship depended on the duration of primary consolidation, then preconsolidation pressure mobilized in the field would be different from σ'_p determined from laboratory oedometer test. Specifically, hypothesis B predicts a preconsolidation pressure mobilized in the field less than the EOP σ'_p determined from laboratory oedometer test. However porewater pressure measurements during embankment construction on soft clay deposits suggest that in fact preconsolidation pressure mobilized in the field is similar to EOP σ'_p determined from 20 mm thick laboratory specimens.

It was first demonstrated by Hoeg et al. (1969) and later confirmed by D'Appolonia et al. (1971), Sällfors (1975), Dascal and Tournier (1975), Morin et al. (1983), and Crooks et al. (1984) that $\Delta u / \Delta \sigma_v$, where Δu is the porewater pressure increase measured at a certain depth in a soft clay deposit and $\Delta \sigma_v$ is the increase in total vertical stress at the same location, is initially small and much less than unity; however, it abruptly increases and becomes equal to one after local yielding of clay structure takes place. Small compressibility of soil structure in the recompression range is responsible for measured values of $\Delta u / \Delta \sigma_v$ less than unity as it (a) contributes to rapid dissipation of excess porewater pressure and (b) requires a rather incompressible porewater pressure measuring systems for accurate sensing of the porewater pressure response of clay.

The porewater pressure measurements, together with values of pre-construction effective vertical pressure, σ'_{vo}, and the increase in total vertical stress, $\Delta \sigma_v$, have been used to compute the preconsolidation pressure that is mobilized in the field during construction of embankments or the filling of storage facilities (Sällfors 1975, Leroueil et al. 1978). According to Sällfors (1975), "the breaks in the pore pressure curves probably reflect the change in modulus of the clay at the time when the preconsolidation pressure is reached". The preconsolidation pressure mobilized in the field is determined as the sum of σ'_{vo} and $\Delta \sigma'_v$ when the break in the porewater pressure curve occurs. These data are listed in Table 2 and are plotted in Fig. 5 against corresponding EOP σ'_p from laboratory oedometer tests on 20 mm thick specimens. When only σ'_p values from the so-called conventional 24-hour interpretation of the incremental loading oedometer test were available (e.g., Leroueil et al. 1983), they were increased by a factor of 1.10 to obtain EOP σ'_p. Data in Fig. 5 show that the preconsolidation pressures mobilized in 5 to 20 m-thick soft clay layers in the field are similar to EOP σ'_p determined from 20 mm thick oedometer specimens. The agreement between the field and laboratory preconsolidation pressures suggests an EOP compression independent of the duration of primary consolidation.

3.   Settlement analysis based on EOP e versus σ'_v from laboratory oedometer tests

The most conclusive approach for examining the possible uniqueness of EOP e versus σ'_v relationship is a comparison, for full-scale field loading situations, of surface and subsurface settlement observations with predictions based on oedometer EOP e versus log σ'_v data (Mesri et al. 1994). For example, hypothesis B predicts larger actual primary settlements in the field than would be predicted using the EOP e versus σ'_v from oedometer tests on 20 mm thick specimens. However, until 1972 "no such cases" had come to Leonards' attention. During the past two decades, Mesri and co-workers have carried out settlement and porewater pressure analyses, using the ILLICON procedure (Mesri and Choi 1985a, Mesri and Lo 1989), for large-scale embankment construction projects on soft clay deposits (Mesri et al. 1994, Table 3). ILLICON procedure assumes for each sublayer EOP e versus σ'_v relationship independent of the duration of primary consolidation. Therefore, EOP e versus log σ'_v curves including EOP σ'_p from oedometer tests on undisturbed specimens are used in settlement analysis. For these cases, predictions have been compared with observations as a function of time for 37 surface settlements (SS), 91 subsurface settlements (SUS), and 169 porewater pressures. Figures 6, 7, and 8 summarize settlement data on 10 embankments on soft clays with and without vertical drains. For these cases the predicted values of t_p range from several hundred days to 35,000 days, and the longest period of observation (EOO) during primary consolidation is 12,000 days (e.g., Mesri and Choi 1985a). Figure 6 includes 33 EOP or EOO surface settlements. Computed and measured surface settlements at three values of elapsed time, roughly corresponding to 20, 50, and 80 percent of EOP settlement are compared in Fig. 7. A total of 91 computed and measured subsurface settlements are compared in Fig. 8 at four elapsed times corresponding to 20, 50, 80, and 100 percent EOP settlement. The agreement between computed and measured settlements during primary consolidation in Figs. 6, 7, and 8 suggests that the EOP e versus σ'_v from laboratory tests also correctly models the e versus σ'_v path during primary consolidation of sublayers in the field.

These data for a wide range of soft clay types and loading and drainage conditions show computed and observed surface as well as subsurface settlements that are in general equal at the end as well as during primary consolidation. Therefore, three decades after Leonards (1972), one can conclude that primary settlements computed using the EOP e versus σ'_v from laboratory tests are similar to those observed in the field.

4.   Compressibility parameters during primary consolidation

Ladd et al. (1977) remarked that an "important question is whether creep acts as a separate phenomenon while ... excess pore pressures dissipate during primary consolidation. If it does, ... the strain corresponding to the end of primary consolidation ... depends upon the duration of primary consolidation and hence the thickness of the consolidating layer". Because, as is clearly shown by Eq. 4, creep (i.e., $\left(\partial e / \partial t \right)_{\sigma'_v}$) does contribute to compression during primary consolidation, then the question is why does not EOP strain depend upon the duration of primary consolidation? The answer is that, in fact, $\left(\partial e / \partial t \right)_{\sigma'_v}$ does not act as "a separate phenomenon". As was speculated by Mesri (1977), $\left(\partial e / \partial t \right)_{\sigma'_v}$ and $\left(\partial e / \partial \sigma'_v \right)_t$ are interrelated, and both depend on $d\sigma'_v / dt$ and t_p which are in turn influenced by permeability of soil and drainage boundary conditions (Feng 1991, Mesri et. al. 1995).

The isochrones of excess porewater pressure, such as those in Fig. 3a, together with the value of effective stress at the beginning of the pressure increment, provide information on effective stress at any depth and time. The isochrones of axial compression, such as those in Fig. 3b, together with void ratio at the beginning of the pressure increment and total volume change measurements, provide data on void ratio at any depth and time. Therefore, it is possible to develop relationships between void ratio and effective stress at fixed values of time, as well as relationships between void ratio and time at constant values of effective stress. These relationships were used to compute $(\partial e / \partial \sigma'_v)_t$ and $(\partial e / \partial t)_{\sigma'_v}$ throughout the consolidation process for 20 sublayers within the 500 mm thick specimen. The computed values of $(\partial e / \partial \sigma'_v)_t$ and $(\partial e / \partial t)_{\sigma'_v}$ together with de/dt and $d\sigma'_v / dt$ were used to confirm Eq. 2 (Mesri et al. 1995) which had been previously validated by Mesri and Choi (1979a). The values of $(\partial e / \partial \sigma'_v)_t$ and $(\partial e / \partial t)_{\sigma'_v}$ during primary consolidation were also computed for

the consolidation specimens with maximum drainage distance of 125 mm and 50 mm. For those specimens, excess porewater pressure at the bottom, axial compression, and total volume decrease were measured with time. Therefore, values of excess porewater pressure and compression were known only at the top and bottom of the specimen. The shapes of the isochrones of excess porewater pressure and compression between the top and bottom of these specimens were estimated using the shapes of the corresponding isochrones for the 500 mm thick specimens. For this purpose, distance of a sublayer from the drainage boundary was normalized with respect to maximum drainage distance, and elapsed time was expressed in terms of average degree of consolidation. The computed values of $(\partial e / \partial \sigma'_v)_t$ and $(\partial e / \partial t)_{\sigma'_v}$

for one pressure increment of Batiscan clay are plotted in Fig. 9 versus the decrease in void ratio during primary compression. These data show that during primary consolidation, $(\partial e / \partial \sigma'_v)_t$ and $(\partial e / \partial t)_{\sigma'_v}$ do not have a constant value. They change

with both the degree of compression and time, and have different values for sublayers that are located at different distances from the drainage boundary. More importantly, at the same degree of compression, $(\partial e / \partial \sigma'_v)_t$ and especially $(\partial e / \partial t)_{\sigma'_v}$ have

different values for specimens with different maximum drainage distance and, therefore, with different duration of primary consolidation. The highest values of $(\partial e / \partial \sigma'_v)_t$ and $(\partial e / \partial t)_{\sigma'_v}$ correspond to layers with the smallest values of maximum

drainage distance and therefore, the smallest t_p. Thus, in spite of the contribution of $(\partial e / \partial t)_{\sigma'_v}$ for a much longer period of primary consolidation time in the field as compared to that in the laboratory, together with smaller values of $(\partial e / \partial t)_{\sigma'_v}$ in the

field as compared to larger values in the laboratory, it is possible to have an EOP e versus σ'_v independent of the duration of primary consolidation.

Concluding Remarks

The empirical evidences 1-4 from laboratory and field observations support hypothesis A, and suggest that for all practical purposes the EOP e versus σ'_v is independent of the duration of primary consolidation. Only misleading interpretations have been used in the past to dismiss hypothesis A or support hypothesis B. Two examples are described in the following.

According to Ladd et al. (1977) hypothesis B is predicted by reheological models wherein "...creep acts as a separate phenomenon...". It was illustrated in the previous section that "creep" does act during primary consolidation stage, however, not as a separate phenomena because $(\partial e / \partial t)_{\sigma'_v}$ and $(\partial e / \partial \sigma'_v)_t$ are interrelated and both depend on $d\sigma'_v / dt$ and t_p. Therefore, hypothesis B need not be true.

Unfortunately, Jamiolkowski et al. (1985) reworded definitions of hypotheses A and B and stated that "hypothesis A assumes that creep occurs only after the end of primary consolidation." It was shown in the previous section that hypothesis A in fact does not require such an assumption about "creep" during primary consolidation stage. Referring to the assumption by Jamiolkowski et al. (1985) and noting that "creep" does in fact occur during primary consolidation stage, Leroueil (1994) concluded that "hypothesis A is thus wrong".

Mesri and Choi (1985b) cautioned that there is a serious potential for reaching a misleading conclusion when a single pressure increment on thin and thick samples is used to investigate the uniqueness of the EOP void ratio versus effective stress relationship. Mesri and Choi (1985b) pointed out that "in preparing to compare the behavior of a thin and thick sample under a pressure increment from σ'_{vj} to σ'_{vj+1} if the samples are allowed an arbitrary period of secondary compression under σ'_{vj}, then the compression versus time curves of the two samples for the pressure increment from σ'_{vj} to σ'_{vj+1} cannot be directly compared." Mesri and Choi (1985b) proceeded to illustrate that for an arbitrary period of consolidation under σ'_{vj}, the thin sample is likely to experience larger secondary compression under σ'_{vj} than the thick sample. Then the thin sample is expected to experience less primary compression under the pressure increment from σ'_{vj} to σ'_{vj+1}. "Without an e versus log σ'_v plot it may not be realized that the two samples were not identical at σ'_{vj}." Imai and Tang (1992) carried out two sets of experiments on clay specimens with different thicknesses. In the first series of tests on 10, 20 and 40 mm thick specimens, each specimen was first consolidated under a pressure of 157 kPa "until the end of the primary consolidation defined as the moment when 99% of excess porewater pressure had dissipated at the undrained boundary". Then, the consolidation pressure was increased to 314 kPa. The void ratio versus log t plots for this properly conducted series of tests completely supported hypothesis A in spite of the fact that EOP compression was reached in about 32, 128, and 512 minutes for 10, 20 and 40 mm specimens. Unfortunately, however, in a second series of tests on 5, 10, 20, and 35 mm thick specimens "each specimen was first consolidated for 24 hours under the load 157 kPa, then the load was increased to 314 kPa". The duration of primary consolidation under 157 kPa (estimated from the first series of tests) was 8, 32, 128 and 392 minutes for the 5, 10, 20 and 35 mm thick specimens, respectively. Therefore, during the remaining part of the 24 hours, different amounts of secondary compression took place under 157 kPa. As expected, under 314 kPa, EOP compression systematically increased with specimen thickness. Imai and Tang (1992) concluded that the results of the second series of tests "support the hypothesis curve type B". Both hypothesis A and B are expected to be examined in terms of initially identical thin and thick layers. In the second series of tests, the four specimens were systematically different. Therefore, interpretation of Imai and Tang (1992) is incorrect and misleading. In summary, no reliable data from laboratory or field measurements supporting hypothesis B are currently available.

Secondary Compression

In a Review Paper for a conference on Settlement of Structures held in April 1974 in University of Cambridge, Simons (1974) concluded that "After many years of research work into secondary consolidation no reliable method is yet available for calculating the magnitude and rate of such settlement, ...". The only empirical conclusion of general validity in relation to secondary compression that could be reached in early 1970's was a fairly good correlation between $C_\alpha /(1 + e)$ and natural water content (Mesri 1973). However, by the late 1970s, the unique interrelationship between $C_\alpha = \Delta e / \Delta \log t$ and $C_c = \Delta e / \Delta \log \sigma'_v$, throughout the secondary consolidation stage, was discovered (Mesri and Godlewski 1977) which began to explain many direct and indirect implications of secondary compression phenomenon and provided a simple and reliable method for calculating secondary settlement (Mesri and Godlewski 1979; Mesri and Choi 1984; Mesri 1986, 1987; Mesri and Castro 1987; Mesri and Feng 1991; Mesri and Shahien 1993; Mesri et al. 1994, 1997).

Assuming, in accordance with Buisman (1936), a straight line relationship between secondary settlement and log time, Walker and Raymond (1968), Walker (1969) and Ladd (1973) computed either C_α or $C_\alpha /(1 + e_0)$. Walker and Raymond (1968) and Walker (1969) plotted C_α versus C_c for different consolidation pressures. Ladd (1973) plotted $C_\alpha /(1 + e_0)$ versus "virgin compression" $C_c /(1 + e_0)$; however, he noted that "In the normally consolidated range $C_\alpha /(1 + e_0)$ remains almost constant or decreases slightly for soils with a constant $C_c /(1 + e_0)$. For soils exhibiting a marked decrease in $C_c /(1 + e_0)$ with increasing stress, $C_\alpha /(1 + e_0)$ also decreases." Thus, Walker and Raymond (1968), Walker (1969) and Ladd (1973), assuming a constant C_α or $C_\alpha /(1 + e_0)$ during secondary consolidation stage, noted a correlation between C_α and C_c or $C_\alpha /(1 + e_0)$ and $C_c /(1 + e_0)$.

In an attempt to explain an observed significant increase in C_α with time, Mesri and Godlewski (1977) discovered that a unique interrelationship exists between C_α and C_c throughout the secondary consolidation stage, and for all consolidation pressures from the recompression to the compression range. It is now well established that for any soil C_α / C_c is a constant applicable both to the recompression and compression ranges (Mesri and Choi 1984; Mesri 1987; Mesri and

Castro 1987; Mesri et al. 1997, 2001). Each corresponding pair of C_α and C_c at any instant (e, σ'_v, t) during secondary compression, represent, respectively, slope of the e versus log t and e versus log σ'_v curves passing through that point. This law of compressibility is illustrated in Fig. 10 in a plot of e versus log σ'_v and log t for a constant C_α / C_c. Examples of C_α versus C_c are shown in Fig. 11 where values of C_α were obtained from the first log cycle of secondary compression and corresponding values of C_c were determined from EOP e versus log σ'_v relationship. Table 4 shows that the values of C_α / C_c for all geotechnical materials are in the range of 0.01 to 0.07.

The magnitude and behavior of C_α with time t is directly related to the magnitude and behavior C_c with consolidation pressure σ'_v. In general, C_α remains constant, decreases or increases with time, in the consolidation pressure range at which C_c remains constant, decreases, or increases with σ'_v, respectively, Fig. 12 (Mesri and Castro 1987). Examples of secondary settlement versus log t are shown in Fig. 13-17. The pressure increment in Fig. 13 loads the Broadback clay sample from eastern Canada to a pressure near the preconsolidation pressure σ'_p, where C_c is dramatically increasing with σ'_v. The excess porewater pressure measurements identify the primary and secondary consolidation stages. The settlement data show that C_α is dramatically increasing with time. A similar behavior is shown in Fig. 14 by Middleton peat from Wisconsin. For this pressure increment in the recompression range, a primary consolidation period of 2 minutes is followed by significant secondary compression, with C_α increasing with time. The pressure increment in Fig. 15 is in the compression range beyond σ'_p. Primary compression is large and is reached in 45 minutes, and this is followed by secondary compression with a C_α that is practically constant with time (Mesri et al. 1997). The pressure increment on Mexico City clay in Fig. 16 is in the compression range where C_c is gradually decreasing with σ'_v. Therefore, C_α gradually decreases with time.

In Fig. 17, a Mexico City clay sample was loaded to the compression range, and 90 days of secondary compression was allowed, thus developing a preconsolidation pressure (Mesri 1987). Then, a pressure increment was applied to bring the clay near the preconsolidation pressure. On the recompression to compression EOP e versus log σ'_v curve, C_c first increases with σ'_v and then it decreases with σ'_v. Therefore, on the compression versus log t, C_α first increases with time and then it decreases with time.

The C_α / C_c law of compressibility has provided rational explanations and predictions for many implications of secondary compression phenomena (Mesri 1987, 1993; Mesri and Castro 1987; Mesri and Feng 1991; Mesri et al. 2001). Two implications are described in this paper.

Secondary Compression After Loading

The most direct implication of secondary compression is secondary settlement of soft ground after the completion of primary consolidation. The value of C_α / C_c together with the EOP e versus log σ'_v completely defines the secondary compression behavior of any soil. This is illustrated in Fig. 18 for Middleton peat from Wisconsin, and in Fig. 19 for James Bay peat from Québec, Canada. The EOP ε_v versus log σ'_v curve (i.e., ε_v versus log σ'_v at $t = t_p$) together with the value of C_α / C_c was used to construct, in Figs. 18 and 19, the ε_v versus log σ'_v curves corresponding to consolidation times greater than t_p. At any σ'_v the value of $C_c / (1 + e_o) = \Delta \varepsilon_v / \Delta \log \sigma'_v$ is determined and is used to compute a vertical strain during secondary compression as follows:

$$[\varepsilon_v]_t = [\varepsilon_v]_{t_p} + \frac{C_c}{1 + e_o} \frac{C_\alpha}{C_c} \log \frac{t}{t_p} \qquad (6)$$

For the first calculation, $t = 10 \, t_p$ was used. The process is repeated at other values of σ'_v and the resulting $(\varepsilon_v, \sigma'_v)$ points define the ε_v versus log σ'_v curve at $10 \, t_p$. Then, the ε_v versus log σ'_v curve at $100 \, t_p$ is constructed from ε_v versus log σ'_v at $10 \, t_p$ together with C_α / C_c, so on. Figures 18 and 19 show vertical strain resulting from primary compression plus secondary compression at any consolidation pressure σ'_v and at any time equal or greater than t_p. Predictions of secondary compression using the graphical construction are compared with the measurements in Figs. 14 and 15. A graphical construction such as those in Figs. 18 and 19, together with layer thickness is used to compute settlement at any final pressure σ'_{vf} and any time t equal or greater than t_p, without making any assumptions about the relationship of C_c with σ'_v and C_α with time.

Secondary Compression After Unloading

When the final effective vertical stress σ'_{vf} is at or beyond the preconsolidation pressure σ'_p, and t_p is small either because the load increment is in the recompression range or the load increment extends to the compression range, however, vertical drains have been installed to speed up primary consolidation, then post-construction secondary settlement can be significant and possibly unacceptable. Secondary settlement can be reduced by surcharging. A surcharge is applied, and an effective vertical stress σ'_{vs} greater than σ'_{vf} is reached before removing the surcharge. The surcharging effort is expressed in terms of effective surcharge ratio $R'_s = (\sigma'_{vs}/\sigma'_{vf})-1$. The surcharging concept is illustrated in Figs. 20 and 21.

The removal of surcharge leads to rebound including primary rebound up to t_{pr} and secondary rebound that levels off at t_ℓ and is followed by post-surcharge secondary compression. Figure 21 shows settlement observations for an embankment in Québec, Canada (Samson and LaRochelle 1972; Samson 1985; Mesri 1986). Because the initial undrained shear strength of the peat deposit was only 10 kPa, to minimize lateral deformation, the embankment was constructed in three stages. At each stage of construction, primary compression and some secondary compression were allowed.   The duration of primary consolidation for the third stage of construction for different locations at the site ranged from 77 to 280 days; however, the surcharge was removed after 365 days. The surcharging pressure and duration at location P-3 correspond to an effective surcharge ratio $R'_s = 0.60$.

Figure 20 shows that without surcharging secondary compression of a James Bay peat sample at $\sigma'_{vf} = 50\,\text{kPa}$ begins at point a and develops along aℓ according to a C_α that is directly proportional to C_c of the compression curve at point a, which is both large and remains practically constant with σ'_v. On the other hand, after surcharging, post-surcharge secondary compression begins at point c and develops along cm according to a C'_α which is much less than C_α at point a.

Ladd (1973) summarized post-surcharge secondary compression data showing that C'_α/C_c decreased from about 1 to 0.1 as effective surcharge ratio increased from 0.07 to 0.5. Field observations in Figure 21 as well as laboratory surcharging test results such as those in Figure 22 (Feng 1991, Ajlouni 2001) show that post-surcharge secondary compression starts with a very small C'_α at t_ℓ, which then gradually increases with time.

The C_α/C_c law of compressibility explains post-surcharge secondary compression behavior of inorganic and organic soils (Mesri, 1986, 1987; Mesri and Feng 1991; Mesri, et al. 1997).   Post-surcharge secondary compression develops along cm according to a C'_α that is directly proportional to C_c values related to the slopes of the recompression curve cn. The recompression curve cn starts with a very small slope C_c at point c, which continuously increases with σ'_v until the recompression curve merges with the EOP e versus log σ'_v curve of the James Bay peat. Therefore, C'_α is expected to start with a very small value and first increase with time and then either remain practically constant or gradually decrease with time according to the shape of EOP e versus log σ'_v compression curve. Laboratory surcharging test data in Fig. 22 show that when R'_s is small, t_ℓ is small and post-surcharge secondary compression appears soon after the removal of surcharge, and C'_α increases rapidly with time. On the other hand, when R'_s is large, t_ℓ is also large and post-surcharge secondary compression appears long after the removal of surcharge, and C'_α increases gradually with time. This behavior is predicted by the C_α / C_c law of compressibility. When R'_s is small, the recompression range from c to h is small, and as σ'_v increases, the slope C_c of the recompression curve rapidly increases to that of the compression curve beyond point h. Therefore, C'_α starts with a small value and rapidly increases with time. On the other hand, when R'_s is large, the recompression range from c to h is large, and the slope C_c of the recompression curve starts very small and gradually increases with the increase in σ'_v. Therefore, C'_α starts with a small value and only gradually increases with time.

Because C'_α is not a constant with time, for a practical method of settlement analysis a secant C''_α is defined, as in Fig. 21, from t_ℓ at which post-surcharge secondary compression begins to any t at which post-surcharge secondary compression is to be evaluated. Data on C''_α from laboratory surcharging tests on clays and peats are shown in Fig. 23. The values of C''_α have been normalized with respect to the C_α at σ'_{vf} without surcharging, and elapsed time has been normalized with t_ℓ.

For inorganic and organic clays, the empirical correlation between t_ℓ / t_{pr} and R'_s is $(t_\ell / t_{pr}) = 100 \; R'^{1.7}_s$, whereas the correlation for peats is $(t_\ell / t_{pr}) = 10 \; R'_s$ (Mesri and Feng 1991; Mesri et al. 1997, 2001). Therefore, the values of (t_ℓ / t_{pr}) for peats are in general significantly smaller than those for soft clay and silt deposits. The explanation appears to be that the fundamental tendency for rebound in comparison to the tendency for compression is less in peats than in clays. A significant part of the water in fibrous peat fabric is held as free water outside and within the peat particles. The free porewater that is squeezed out upon loading, does not have a physico-chemical tendency to return to peat fabric following unloading.

A theory of rebound similar to the theory of consolidation such as the ILLICON is used to determine t_{pr}, and the computed t_ℓ together with data on C''_α is used to compute post-surcharge secondary settlement (Mesri and Feng 1991, Mesri et al. 1994).

Conclusions

The two fundamental discrepancies between actual consolidation behavior of soils and that predicted by the Terzaghi consolidation theory, that Donald Taylor noted some six decades ago, have been explained. Many implications of secondary compression phenomenon, including secondary settlement, have been explained and predicted by the C_α / C_c law of compressibility. The proper role of compressibility with time, $\left(\partial e / \partial t\right)_{\sigma'_v}$, during primary consolidation, has been empirically established by the uniqueness of EOP void ratio versus effective vertical stress relationship. However, because the general behavior of $\left(\partial e / \partial \sigma'_v\right)_t$ and $\left(\partial e / \partial t\right)_{\sigma'_v}$ during primary consolidation is not yet known, a perfect theory of primary consolidation does not yet exist. Charles C. Ladd has played a leading part in this significant progress in geotechnical engineering.

References

Ajlouni, M.A. (2001). « Geotechnical properties of peat and related engineering problems. Ph.D. thesis, University of Illinois at Urbana-Champaign, Urbana, Illinois.

Barden, L. (1969). "Time-dependent deformation of normally consolidated clays and peats." *J. Soil Mech. Found. Engrg.*, ASCE, 95(1), 1-31.

Bourges, F., Carissan, M., and Mieussens, C. (1973). Etude des tassements: remblai de Palavas les Flots. *Bulletin de Liaison des Laboratoires Ponts et Chaussées*, Paris, numéro spécial T, 119-138

Brigando, M., and Simon, A. (1976). Etude du remblai de préchargement d'Arles. *Rapport de recherche 1-06-21-2*, Laboratoire de l'Equipement d' Aix en Provence, France.

Brinch-Hansen, J. (1961). "A model law for simultaneous primary and secondary consolidation." *Proc. 5th Int. Conf. Soil Mech. Found. Engrg.*, Paris, 1, 133-136.

Buisman, A.S.K. (1936). "Results of long duration settlement tests." *Proc. 1st Int. Conf. Soil Mech. Found. Engrg.*, 1, 100-106.

Chapeau, C. (1975). Contribution à l'étude des tassements instantanés des remblais sur argiles sensibles. M.Sc. thesis, Dept. of Civil Eng., Laval Univ., Québec.

Choi, Y.K. (1982). "Consolidation behavior of natural clays." Ph.D. thesis, University of Illinois at Urbana-Champaign, Urbana, IL, 422 p.

Clausen, C.J.F. (1970). Resultater av et belastnings forsøk på Mastemyr I Oslo, *NGI Publication No. 84*, 29⁻40, Oslo, Norway.

Croce, A., Viggiani, C., and Calabresi, G. (1973). In situ investigation on pore pressures in soft clays. *Proc. 8th Int. Conf. on Soil Mech. and Found. Eng*, Moscow, 2(2), 53-60.

Crooks, J.H.A., Becker, D.E., Jefferies, M.G., and McKenzie, K. (1984). "Yield behaviour and consolidation-1: pore pressure response." *Symp. Sedimentation Consolidation Models: Predictions and Validation*, ASCE, 356-381.

D'Appolonia, D.J., Lambe, T.W., and Poulos, H.G. (1971). "Evaluation of pore pressures beneath an embankment." *J. Soil Mech. Found. Engrg.*, ASCE, 97(6), 881-897.

Dascal, O., and Tournier, J.P. (1975). "Embankments on soft and sensitive clay foundation." *J. Geotech. Engrg.*, ASCE, 101(3), 297-314.

Engesgaar, H. (1970). Resultater av to belast-ningsforsøk på Sundland I Drammen. Oslo, Norway, *NGI Publication No. 84*, 41ˉ47.

Feng, T.W. (1991). "Compressibility and permeability of natural soft clays and surcharging to reduce settlements." Ph.D. thesis, University of Illinois at Urbana-Champaign, Urbana, IL.

Hoeg, K., Andersland, O.B., and Rolfsen, E.N. (1969). "Undrained behavior of quick clay under load tests at Asrum." *Géotechnique*, 19(1), 101-115.

Imai, G., and Tang, Y.X. (1992). "A constitutive equation of one-dimensional consolidation derived from inter-connected tests." *Soils and Foundations*, 32(2), 83-96.

Jamiolkowski, M., Ladd, C.C., Germaine, J.T., and Lancellotta, R. (1985). "New developments in field and laboratory testing of soils. *Proc. 11th Int. Conf. Soil Mech. Found. Engrg.*, San Francisco, 1, 57-153.

La Rochelle, P., Sarraith, J., Tavenas, F., Roy, M., and Leroueil, S. (1981). "Causes of sampling disturbance and design of a new sampler for sensitive soils." *Canadian Geotech. J.*, 18(1), 52-66.

Ladd, C.C. (1973). "Settlement analysis for cohesive soils." Res. Rpt. R71-2, No. 272, Dept. of Civil Engrg., Mass. Inst. of Tech., Cambridge, MA, 115 p.

Ladd, C.C., and Foott, R. (1974). "New design procedure for stability of soft clays." *J. Geotech. Engrg.*, ASCE, 100(7), 763-786.

Ladd, C.C., Foott, R., Ishihara, K., Schlosser, F., and Poulos, H.G. (1977). "Stress-deformation and strength characteristics." General Report, *Proc. 9th Int. Conf. Soil Mech. Found. Engrg.*, 2, 421-494.

Lefebvre, G., and Poulin, C. (1979). "A new method of sampling in sensitive clay." *Canadian Geotech. J.*, 16(1), 226-233.

Leonards, G.A. (1972). Discussion, "Shallow foundations." *Proc. Spec. Conf. on Performance of Earth-Supported Structures*, ASCE, Lafayette, IN, 3, 169-173.

Leonards, G.A., and Deschamps, R.J. (1995). "Origin and significance of the quasi-preconsolidation pressure in clay soils." *Proc. Compression and Consolidation of Clay Soils*, Hiroshima, Japan, 1, 341-347.

Leroueil, S. (1988). "Recent developments in consolidation of natural clays." *Canadian Geotech. J.*, 25(1), 85-107.

Leroueil, S. (1994). "Compression of clays: fundamental and practical aspects." *ASCE Conf. on Vertical and Horizontal Deformations of Foundations and Embankments*, College Station, TX, 1, 57-76.

Leroueil, S. (1995). "Could it be that clays have no unique way of behaving during consolidation?" *Proc. Conf. Compression and Consolidation of Clayey Soils*, 2, 1039-1048.

Leroueil, S., Tavenas, F., Mieussens, C., and Peignaud, M. (1978). "Construction pore pressures in clay foundations under embankments, Part II: Generalized behavior." *Canadian Geotech. J.*, 16(1), 66-82.

Leroueil, S., Tavenas, F. Samson, L., and Morin, P. (1983). "Preconsolidation pressure of Champlain clays. Part II; Laboratory determination." *Canadian Geotech. J.*, 20(4), 803-816.

Magnan, J.P., Mieussens, C., and Queyroi, D. (1977). Etude en place des tassements - remblai B, Cubzac-les-Ponts. *Report No. FAER-1-06-12-7*, Laboratoire Central des Ponts et Chaussées, Paris, France.

Mesri, G. (1973). "Coefficient of secondary compression." *J. Soil Mech. Found. Engrg.*, ASCE, 99(1), 123-137.

Mesri, G. (1977). Discussion, "Stress-deformation and strength characteristics." *Proc. 9th Int. Conf. Soil Mech. Found. Engrg.*, Tokyo, 3, 354-355.

Mesri, G. (1986). Discussion, "Post-construction settlement of an expressway built on peat by precompression" by L. Samson. *Canadian Geotech. J.*, 23(3), 403-407.

Mesri, G. (1987). "Fourth law of soil mechanics: A law of compressibility." *Proc. Int. Symp. Geotech. Engrg. of Soft Soils*, Mexico City, May, 2, 179-187.

Mesri, G. (2001). "The general principle of effective stress." Geotech – Year 2000, Developments in Geotechnical Engineering, Asian Institute of Technology, Bangkok, Thailand, 100-105.

Mesri, G., Ajlouni, M.A., Feng, T. W., and Lo, D.O.K. (2001). "Surcharging of soft ground to reduce secondary compression." *Proc. 3rd Int. Conf. on Soft Soil Engineering*, December 6-8, Hong Kong, 55-65.

Mesri, G., and Castro, A. (1987). "The C_α / C_c concept and K_o during secondary compression." *J. Geotech. Engrg.*, ASCE, 113(3), 230-247.

Mesri, G., and Choi, Y.K. (1979a). Discussion, "Strain rate behavior of Saint- Jean-Vianney clay" by Y.P. Vaid, P.K. Robertson, and R.G. Campanella. *Canadian Geotech. J.*, 16(4). 831-834.

Mesri, G., and Choi, Y.K. (1979b). "Excess porewater pressures during consolidation." *Proc. 6th Asian Regional Conf. Soil Mech. Found. Engrg.*, 1, 151-154.

Mesri, G., and Choi, Y.K. (1984). Discussion, "Time effects on the stress-strain behavior of natural soft clays" by J. Graham, J.H.A. Crooks, and A.L. Bell. *Géotechnique*, 34(3), 439-442.

Mesri, G., and Choi, Y.K. (1985a). "Settlement analysis of embankments on soft clays." *J. Geotech. Engrg.*, ASCE, 111(4), 441-464.

Mesri, G., and Choi, Y.K. (1985b). "The uniqueness of the end-of-primary (EOP) void ratio-effective stress relationship." *Proc. 11th Int. Conf. Soil Mech. Found. Engrg.*, San Francisco, 2, 587-590.

Mesri, G., and Feng, T.W. (1991). "Surcharging to reduce secondary settlement." *Proc. Int. Conf. Geotech. Engrg. for Coastal Development*, Yokohama, 1, 359-364.

Mesri, G., Feng, T.W., and Benak, J.M. (1990). "Postdensification penetration resistance of clean sands." *J. Geotech. Engrg.*, ASCE, July, 116(7), 1095-1115.

Mesri, G., Feng, T.W., and Shahien, M. (1995). "Compressibility parameters during primary consolidation." Invited Special Lecture. *Int. Symp. Compression and Consolidation of Clayey Soils*, Hiroshima, Japan, May, Lectures and Reports Volume, 201-217.

Mesri, G., and Godlewski, P.M. (1977). "Time- and stress-compressibility interrelationship." *J. Geotech. Engrg.*, ASCE, May, 103(5), 417-430.

Mesri, G., and Godlewski, P.M. (1979). Closure, "Time- and stress-compressibility interrelationship." *J. Geotech. Engrg.*, ASCE, Jan., 105(1), 106-113.

Mesri, G., and Lo, D.O.K. (1989). "Subsoil investigation: the weakest link in the analysis of test fills." *The Art and Science of Geotechnical Engineering: At the Dawn of the Twenty-First Century*, A Volume Honoring Ralph B. Peck. Prentice-Hall, Inc., Englewood Cliffs, NJ, 309-335.

Mesri, G., Lo, D.O.K., and Feng, T.W. (1994). "Settlement of embankments on soft clays." *ASCE Conf. on Vertical and Horizontal Deformations of Foundations and Embankments*, College Station, TX, 1, 8-56.

Mesri, G., and Rokhsar, A. (1974). "Theory of consolidation for clays." *J. Geotech. Engrg.*, ASCE, 100(8), 889-904.

Mesri, G., and Shahien, M. (1993). Discussion, "Long-term consolidation characteristics of diluvial clay in Osaka Bay" by K. Akai, M. Kamon, K. Sano, and K. Soga. *Soils and Foundations*, 33(1), 213-215.

Mesri, G., Stark, T.D., Ajlouni, M.A., and Chen, C.S. (1997). "Secondary compression of peat with or without surcharging." *J. Geotech. Engrg.*, ASCE, 124(5), 411-421.

Mieussens, C. (1973). Etude des tassements: remblai de la Plaine de l'Aude. *Bulletin de Liason des Laboratoires des Ponts et Chaussées*, Paris, France, numéro spécial T: 139-151.

Mieussens, C., and Ducasse, P. (1973). Etude des tassements: remblai de Narbonne. *Bulletin de Liason des Laboratoires des Ponts et Chaussées*, Paris, France, numéro spécial T, 152-167.

Morin, P., Leroueil, S., and Samson, L. (1983). "Preconsolidation pressure of Champlain clays, Part I: In situ determination." *Canadian Geotech. J.*, 20(4), 782-802.

Pilot, G., Moreau, M., and Paute, J.L. (1973). "Etude à la rupture: remblai de Lanester. Bulletin de Liaison des Laboratoires des Ponts et Chaussées, Paris, France, numéro spécial T., 194-206.

Ramalho-Ortigão, J.A., Werneck, M.L.G., and Lacerda, W.A. (1983). Embankment failure on clay near Rio de Janeiro. *ASCE, J. of Geotech. Eng.* 109(11): 1460-1479.

Sällfors, G. (1975). "Preconsolidation pressure of soft, high-plastic clays." Ph.D. thesis, Chalmers University of Technology, Goteborg, Sweden.

Samson, L. (1985). "Postconstruction settlement of an expressway constructed over peat by precompression." *Canadian Geotech. J.*, 22(2), 308-312.

Samson, L, and La Rochelle, P. (1972). "Design and performance of an expressway constructed over peat by preloading." *Canadian Geotech. J.*, 9(4), 447-466.

Simons, H.E. (1974). "Normally consolidated and lightly over-consolidated cohesive materials." *Proc. Conf. Settlement of Structures*, John Wiley & Sons, 500-530.

Tavenas, F., Blanchet, R., Garneau, R., and Leroueil, S. (1978). The stability of stage-constructed embankments on soft clays. *Canadian Geotech. J.*, 15(2): 283-305.

Taylor, D.W. (1942). "Research on consolidation of clays." *Publ. Serial 82,* Dept. of Civil and Sanitary Engrg., Mass. Inst. of Tech.

Taylor, D.W., and Merchant, W. (1940). "A theory of clay consolidation accounting for secondary compression." *J. Maths. and Physics*, 19(3), 167-185.

Terzaghi, K. (1923) "Die Berechnug der Durchlassigkeitsziffer des Tones aus dem Verlauf der hydronamischen Spannungserscheinungen." Akademie der Wissenschaften in Wien. Sitzungsberichte. Mathematischnaturwissenschaftliche Klasse. Part IIa, 132(3-4), 125-138. (Reprinted in *From Theory to Practice in Soil Mechanics*, Wiley, New York, 1960.)

Terzaghi, K., Peck, R.B., and Mesri, G. (1996). *Soil Mechanics in Engineering Practice*, Third Edition, John Wiley & Sons, New York, 549 p.

Trak, B. 1974. Contribution à l'étude de la stabilité à court terme des remblais sur fondations argileuses. Thèse de M.Sc., Département de Génie Civil, Université Laval, Québec (Qué).

Vogien, 1975. Etude du comportement avant la rupture d'un remblai expérimental construit sur sol mou à Cubzac-les-Ponts. Thèse de Docteur-Ingenieur, Université de Paris, France.

Walker, L.K., and Raymond, G.P. (1968). "The prediction of consolidation rates in a cemented clay." *Canadian Geotech. J.*, 5(4), 192-216.

Walker, L.K. (1969). "Secondary settlement in sensitive clays." *Canadian Geotech. J.*, 6, 219-222

Table 1. Natural soft clay deposits used in consolidation tests on thin and thick undisturbed specimens.

Clay	w_o (%)	w_ℓ (%)	w_p (%)	C_α / C_c	σ'_p / σ'_{vo}	σ'_{PI} / σ'_p
St. Alban	39-57	31	18	0.024	1.9-2.7	0.65
Bay Mud	96-98	89	37	0.05	1.2	0.71
Louiseville	64-71	65	28	0.03	2.6-2.9	0.74
Batiscan	71-88	49	22	0.03	1.65	0.59
St. Hilaire	62-84	55	23	0.03	1.45	0.60

Table 2. Preconsolidation pressure mobilized in full-scale field situation, deduced from porewater pressure measurements during embankment loading of soft clay deposits.

Site	Ref.	Depth (m)	σ'_{vo} (kPa)	σ'_p [1] (kPa)	σ'_p [2] (kPa)	σ'_v (kPa)	σ'_p / σ'_{vo}
Asrum I	Hoeg et al. 1969	3	10	24	26	24	2.64
Drammen II	Engesgaar 1970	4.4	49	63	69	60	1.41
Mastemyr	Clausen 1970	3.5	16	24	26	22	1.65
		6	32	48	53	41	1.65
Interstate 95	D'Appolonia et al. 1971	16.8	168	268	295	234	1.75
		24.5	240	240	264	320	1.10
		32.1	300	300	330	375	1.10
Lanester	Pilot et al. 1973	3	12	22	24.2	22	2.02
Narbonne	Mieussens & Ducasse 1973	5	48	58	64	78	1.32
		10	100	110	121	110	1.21
Palavas-les-Flots	Bourges et al. 1973	7	41	50	55	66	1.34
		14	73	93	102	86	1.40
Plaine de l'Aude	Mieussens 1973	10	90	95	105	110	1.16
		12	105	110	121	121	1.15
		14	120	120	132	139	1.10
		16	134	134	147	154	1.10
		18	148	156	172	168	1.16
		20	160	160	176	193	1.10
		11.5	103	108	119	119	1.15
		15.1	128	130	143	149	1.12
		18.6	152	165	182	182	1.19
Porto Tolle	Croce et al. 1973	15-18	140	140	154	140	1.10
		27-29	225	225	248	250	1.10
St. Alban A	Trak 1974	1.5	16	50	55	40	3.44
Backebol	Sälfors 1975	2.7	21		45	46	2.15
		3	24		44	46	1.84
		3.6	26		42	44	1.60
		4	28		53	49	1.87
		4.6	31		47	43	1.50
		5.5	37		49	48	1.32
		7.1	46		64	57	1.38
		10	62		65	62	1.06
Valen	Sällfors 1975	1.5	17		47	50	2.76
		2	16		26	33	1.64
		2.5	15		22	25	1.47
		3	14		19	22	1.32
		4	17		17	17	1.00
		5	27		32	27	1.19

Kristianstad	Sällfors 1975	3.2	29		86	85	2.97
		4.2	36		104	93	2.89
		5.2	44		112	116	2.54
		6.3	51		135	140	2.63
St. Alban B	Chapeau 1975	2.5	19	46	51	44	2.66
		5	32	73	80	66	2.50
St. Alban C	Chapeau 1975	2.5	19	46	51	44	2.66
St. Alban D	Chapeau 1975	2.5	19	46	51	43	2.66
Cubzac-les-Ponts A	Vogien 1975	2	18	53	58	50	3.23
		4	28	53	58	53	2.08
		6	38	55	61	54	1.59
		6	38	78	86	72	2.25
		8	48	62	68	72	1.42
Rupert A	Dascal & Tournier 1975	7.5	51	133	146	77	2.87
		10.7	78	148	163	141	2.09
Arles	Brigando & Simon 1976	4.5	35	100	110	68	3.14
		7.5	50	100	110	74	2.20
Cubzac-les-Ponts B	Magnan et al. 1977	2	21	34	37	33	1.78
		4	30	42	46	40	1.54
		6	38	48	53	49	1.38
		8	46	62	68	59	1.48
Rang de la Concession	Tavenas et al. 1978	9	86	103	113	116	1.32
Rang du Brulé	Tavenas et al. 1978	9	80	92	101	115	1.27
Rang Saint Georges	Tavenas et al. 1978	9	72	72	79	104	1.10
Rang du Fleuve	Tavenas et al. 1979	9.2	68	68	75	80	1.10
Fluminense Plains	Ramalho-Ortigão et al. 1983	2	6	20	22	19	3.40
		3	10	22	24	23	2.54
		4	13	26	29	26	2.24
		5	16	30	33	24	2.06
		8	26	42	46	34	1.80
Joliette	Morin et al. 1983	3.5	23	95	105	94	4.50
		5.9	37	123	135	108	3.60

[1] Preconsolidation pressure from 'conventional' 24-hour IL Oedometer test;
[2] End-Of-Primary preconsolidation pressure.

Table 3. Case Histories of Embankment Construction on Soft Clay Deposits

Case Record	w_o %	w_ℓ %	σ'_p / σ'_{vo}	C_α / C_c	SS	SUS
Olga B & C, Canada	30-92	30-68	2.0-3.0	0.033	2	-
Rupert 7, Canada	27-62	30-36	1.6-2.6	0.030	3	-
Vasby, Sweden	43-122	40-125	1.1-1.3	0.060	1	5
Gloucester, Canada	25-90	37-64	1.4-1.9	0.031	1	4
Ellingsrud, Norway	30-45	24-27	1.1-2.4	0.044	1	1
Skå-Edeby, Sweden	50-130	50-150	1.2-1.3	0.050	7	21
Changi, Singapore	30-75	35-95	1.5-3.1	0.035	11	26
ChekLapKok, Hong Kong	25-125	40-100	1.1-6.6	0.03-0.065	7	28
LaGrande, Canada	34-62	25-43	1.3-2.5	0.054	2	6
Bangkok, Thailand	50-120	72-100	1.1-2.0	0.050	2	-

Table 4. Values of C_α / C_c for Geotechnical Materials

Material	C_α / C_c
Granular soils including rockfill	0.02 ± 0.01
Shale and mudstone	0.03 ± 0.01
Inorganic clays and silts	0.04 ± 0.01
Organic clays and silts	0.05 ± 0.01
Peat and muskeg	0.06 ± 0.01

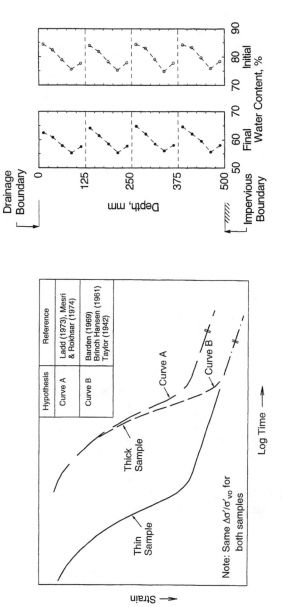

Figure 1. Hypotheses A and B proposed by Ladd et al. (1977).

Figure 2. Initial and final water content profiles in the 500 mm-thick St. Hilaire clay consolidation specimen (Mesri et al. 1995).

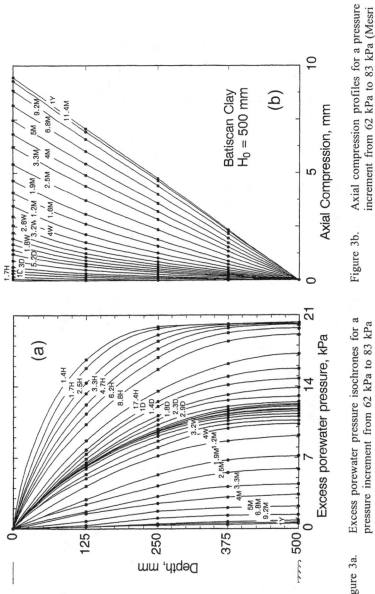

Figure 3a.   Excess porewater pressure isochrones for a pressure increment from 62 kPa to 83 kPa (Mesri et al.1995).

Figure 3b.   Axial compression profiles for a pressure increment from 62 kPa to 83 kPa (Mesri et al. 1995).

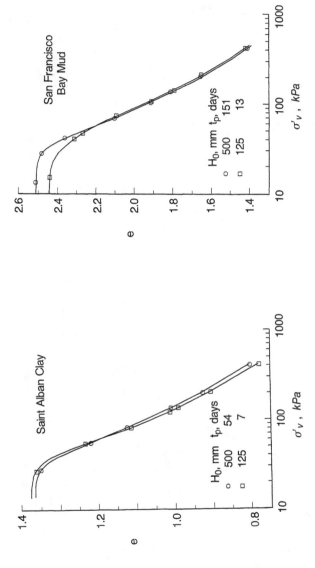

Figure 4a. EOP e versus log σ'_v curves of a thin and a thick layer of Saint Alben clay (Mesri and Choi 1985b).

Figure 4b. EOP e versus log σ'_v curves of a thin and a thick layer of San Francisco Bay mud (Mesri and Choi 1985b).

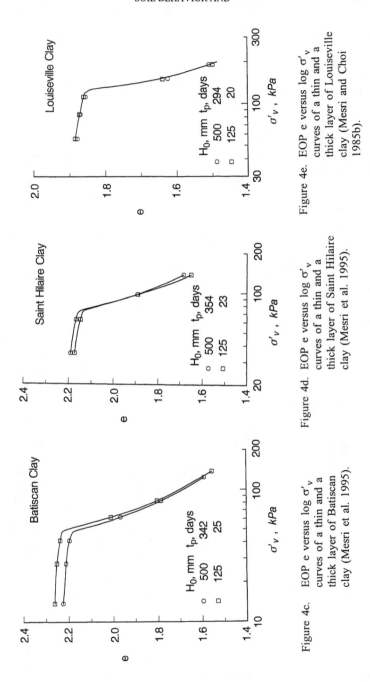

Figure 4c.   EOP e versus log σ'_v curves of a thin and a thick layer of Batiscan clay (Mesri et al. 1995).

Figure 4d.   EOP e versus log σ'_v curves of a thin and a thick layer of Saint Hilaire clay (Mesri et al. 1995).

Figure 4e.   EOP e versus log σ'_v curves of a thin and a thick layer of Louiseville clay (Mesri and Choi 1985b).

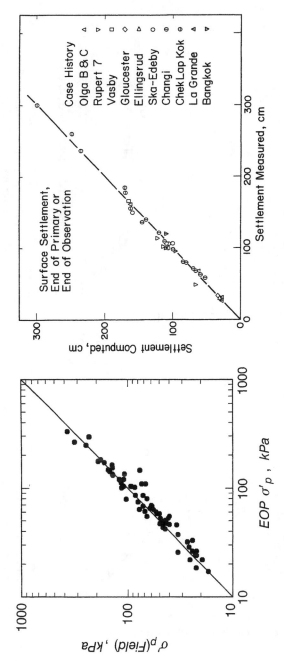

Figure 6.   Computed and measured EOP or EOO surface
settlement of embankments on soft clay and
silt deposits (Mesri et al. 1994).

Figure 5.   Preconsolidation pressure mobilized
in the field compared with EOP σ'_p
from 20 mm thick oedometer specimens
(Mesri et al. 1995).

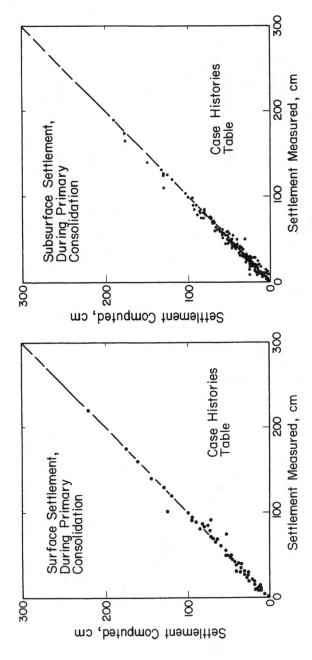

Figure 7. Computed and measured surface settlement during primary consolidation for embankments on soft clay and silt deposits (Mesri et al. 1994).

Figure 8. Computed and measured subsurface settlement during primary consolidation for embankments on soft clay and silt deposits (Mesri et al. 1994).

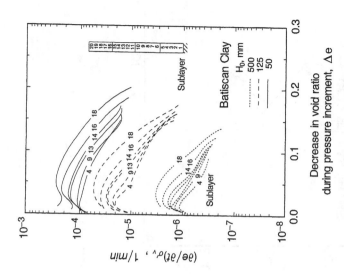

Figure 9b. Values of $(\partial e/\partial t)_{\sigma'_v}$ during primary consolidation for a pressure increment from 62 kPa to 83 kPa (Mesri et al. 1995).

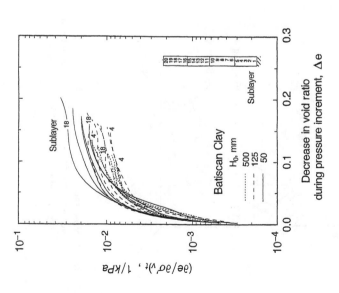

Figure 9a. Values of $(\partial e/\partial \sigma'_v)_t$ during primary consolidation for a pressure increment from 62 kPa to 83 kPa (Mesri et al. 1995).

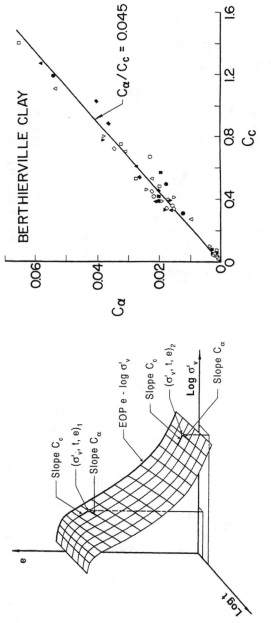

Figure 11a. C_α versus C_c for Berthierville clay (Mesri and Castro 1987).

Figure 10. Void ratio versus log σ'_v and void ratio versus log t during secondary consolidation (Mesri and Godlewski 1977).

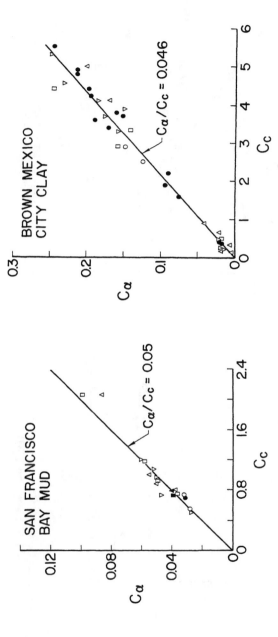

Figure 11c. C_α versus C_c for Brown Mexico City clay (Mesri 1987).

Figure 11b. C_α versus C_c for San Francisco Bay mud (Mesri 1987).

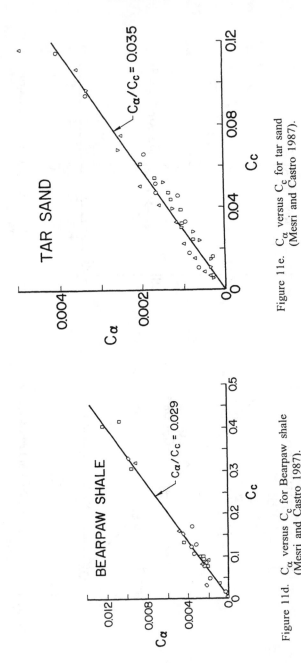

Figure 11e. C_α versus C_c for tar sand (Mesri and Castro 1987).

Figure 11d. C_α versus C_c for Bearpaw shale (Mesri and Castro 1987).

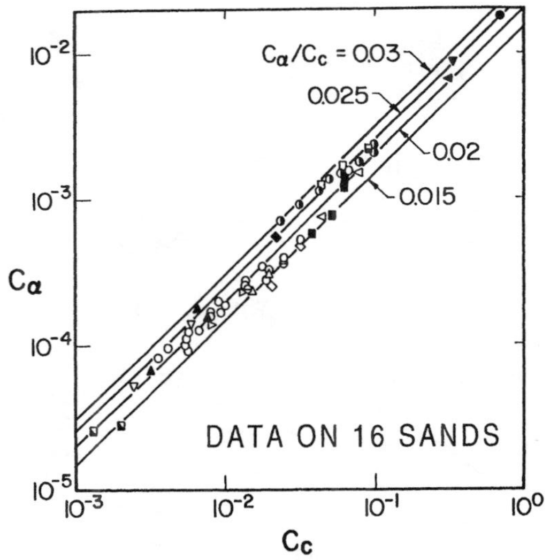

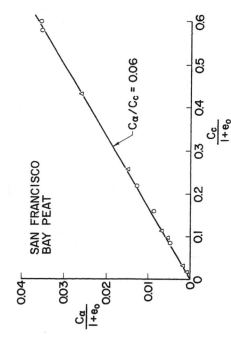

Figure 11g. $C_\alpha/(1+e_o)$ versus $C_c/(1+e_o)$ for San Francisco Bay peat.

Figure 11f. C_α versus C_c for clean sands (Mesri et al. 1990).

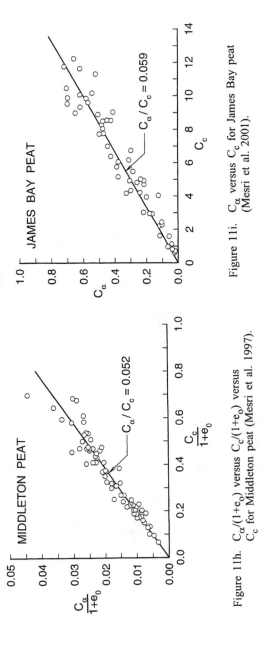

Figure 11i. C_α versus C_c for James Bay peat (Mesri et al. 2001).

Figure 11h. $C_\alpha/(1+e_o)$ versus $C_c/(1+e_o)$ versus C_c for Middleton peat (Mesri et al. 1997).

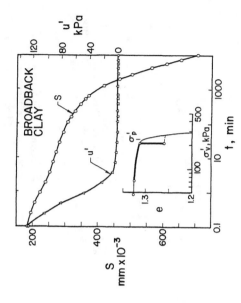

Figure 13.   An example of C_α of a soft clay increasing with time (Mesri 1987).

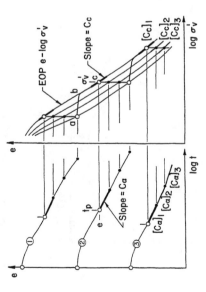

Figure 12.   Behavior of C_α with time reflecting the behavior of C_c with effective vertical stress (Mesri and Castro 1987).

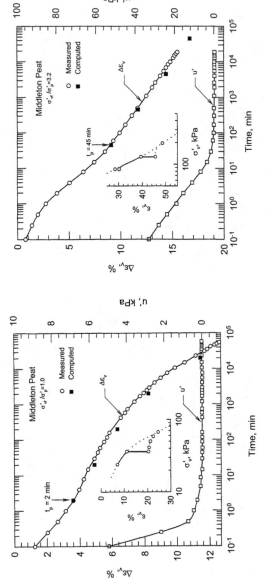

Figure 15. An example of C_α of a peat remaining practically constant with time (Mesri et al. 1997).

Figure 14. An example of C_α of a peat increasing with time (Mesri et al. 1997).

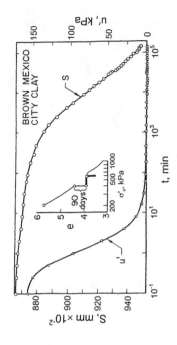

Figure 17. An example of C_α of a soft clay first increasing with time and then decreasing with time (Mesri 1987).

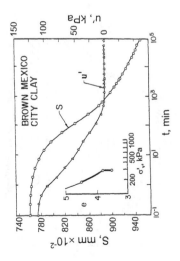

Figure 16. An example of C_α of a soft clay decreasing with time (Mesri 1987).

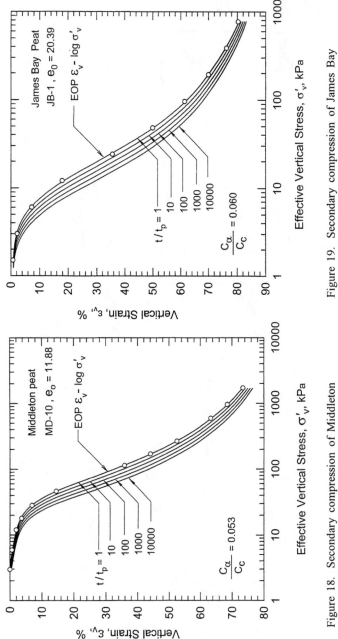

Figure 19. Secondary compression of James Bay peat predicted by the C_α/C_c law of compressibility (Mesri et al. 2001).

Figure 18. Secondary compression of Middleton peat predicted by the C_α/C_c law of compressibility (Mesri et al. 1997).

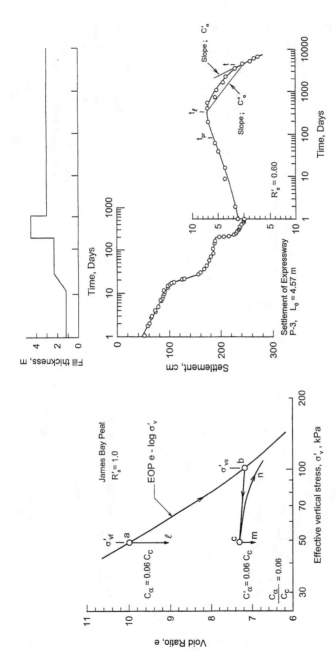

Figure 21. Post-surcharge secondary compression behavior, and definition of t_{pr}, t_ℓ, C'_α and C''_α (Mesri et al. 2001).

Figure 20. Interpretation of surcharging effect on secondary compression, using the C_α/C_c law of compressibility (Mesri et al. 2001).

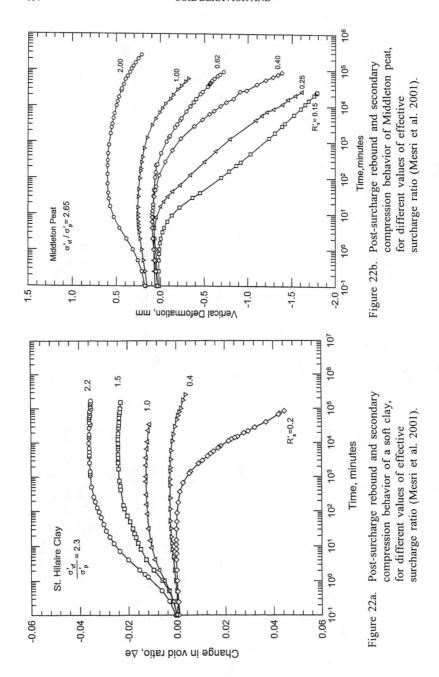

Figure 22a. Post-surcharge rebound and secondary compression behavior of a soft clay, for different values of effective surcharge ratio (Mesri et al. 2001).

Figure 22b. Post-surcharge rebound and secondary compression behavior of Middleton peat, for different values of effective surcharge ratio (Mesri et al. 2001).

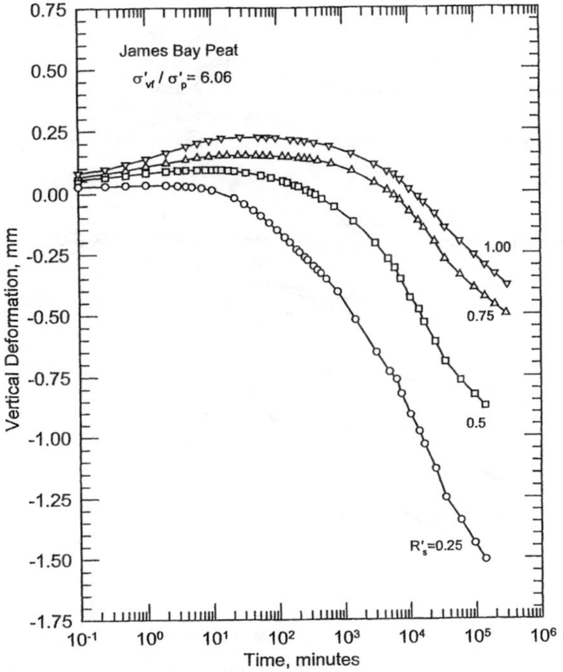

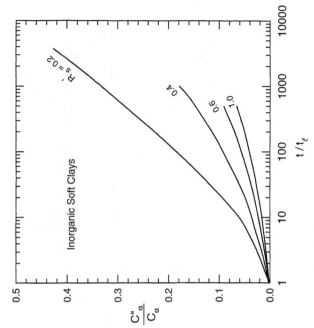

Figure 23a.   Post-surcharge secondary compression index of inorganic soft clays expressed in terms C''_α/C_α where C_α is the secondary compression index at σ'_{vf} without surcharging (Mesri et al. 2001).

Figure 22c.   Post-surcharge rebound and secondary compression behavior of James Bay peat, for different values of effective surcharge ratio (Mesri et al. 2001).

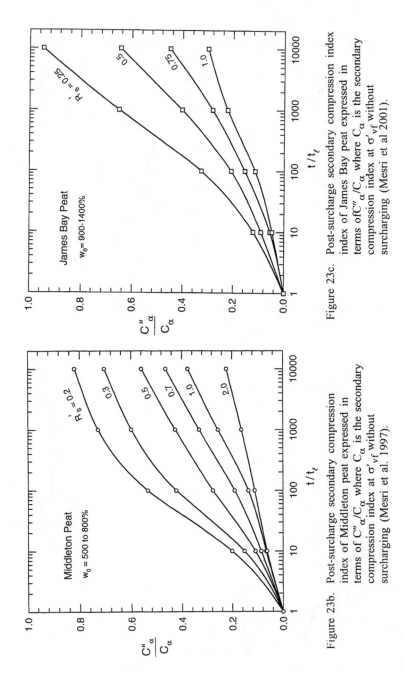

Figure 23c. Post-surcharge secondary compression index of James Bay peat expressed in terms of C''_α/C_α where C_α is the secondary compression index at σ'_{vf} without surcharging (Mesri et al 2001).

Figure 23b. Post-surcharge secondary compression index of Middleton peat expressed in terms of C''_α/C_α where C_α is the secondary compression index at σ'_{vf} without surcharging (Mesri et al. 1997).

LABORATORY MEASUREMENT AND INTERPRETATION OF SOFT CLAY MECHANICAL BEHAVIOR

Don J. DeGroot[1], M. ASCE

Abstract

This paper describes laboratory methods for measuring soft clay stress-strain-strength behavior and interpretation of test results for selection of design parameters. The paper traces the major practical contributions of experimental research to our understanding of soft clay mechanical behavior. Recommendations are given for implementation of research results in practice with the goal of improving the engineering state-of-the-practice, which often lags well behind the research state-of-the-art. Particular emphasis is given to the results of work produced by researchers working at the Massachusetts Institute of Technology under the leadership of Professor Charles C. Ladd. Significant advances in the understanding of clay behavior have been realized in the past 40 years including such issues as the influence of anisotropy, stress history, rate effects, and sample disturbance. Of these issues, sample disturbance remains by far the most difficult one to deal with in practice and in fact is often ignored. Field sampling and laboratory techniques that can reduce sample disturbance are described together with methods for quantifying sample disturbance. The merits of laboratory reconsolidation methods are discussed and are demonstrated through the use of three case studies ranging in sophistication in sampling and laboratory testing from the highest level down to the level more commonly encountered in practice.

Introduction

Soft clays challenge geotechnical engineers because of their high compressibility and low undrained shear strength. Soft clay deposits often have highly varied geological histories, thus making systematic quantification of their stress-strain-strength behavior complex. They exhibit significant stress history

---

[1] Assoc. Professor, University of Massachusetts, Amherst, MA 01003, degroot@ecs.umass.edu

effects, can have a high degree of anisotropy, and are strain rate sensitive. They are difficult to sample and test without causing excessive and irreversible sample disturbance. Given all these challenges, site characterization of clays is best done through a combination of in situ testing and high quality undisturbed sampling for laboratory testing. Each approach has advantages and limitations, but when properly used provide complementary information. The key disadvantage to in situ testing is the reliance on empirical correlations for estimating soil parameters whereas accurate laboratory measurements are largely dependent on the quality of samples available. The focus of this paper is on the laboratory side of site characterization programs. The paper reviews laboratory methods for measuring the static behavior of soft saturated clays and interpretation of measured data for design. Particular emphasis is given to one-dimensional compressibility and undrained shear strength properties.

This paper was prepared as part of the Massachusetts Institute of Technology (MIT) Symposium honoring Professor Ladd's contributions to geotechnical engineering practice. Particular emphasis is therefore given to the results of experimental soft clay behavior research led by Professor Ladd at MIT during the past 40 years. In a paper of such limited length it is clearly not possible to describe all the research contributions, the details of experimental methods, and the significance of their results. Therefore the paper will focus on highlighting those important research results that should be more widely implemented in geotechnical engineering practice. The goal is to make the case that significant improvement in the reliability and economics of geotechnical design are often possible if research results such as the ones presented herein are more routinely used. In many instances the state-of-the-practice lags well behind what has already been, for some time, the state-of-the-art. Some portions of the paper are abstracted from Ladd and DeGroot (1992), DeGroot and Sheahan (1995), and DeGroot (1999).

Historical Perspective

When Professor Ladd started his graduate work at MIT in the late 1950s, the state-of-the-art in geotechnical engineering laboratory testing included such devices as dial gauges, proving rings, and null indicators. Today, data acquisition systems coupled with electronic transducers can reliably take far more readings per second than required for static testing of soils, and with high resolution. All this capability is available at a modest price for the data acquisition system and accompanying PC. Computer controlled load frames and flow pumps for displacement, force, or stress control are also now readily available for commercial applications. Through the use of electronics, our ability to test soils has surpassed the level of implementation of many proper laboratory test procedures in geotechnical engineering practice.

In the early years of the profession, our basic understanding of soft clay behavior was predicated on the concept that any in situ or laboratory test can provide a measure of undrained shear strength as long as no change in water content occurs during shear. Over the years, research has showed us that soft clay behavior is not so simple and that there are many other important factors to consider in developing design strengths. Such factors include anisotropy, strain rate effects, stress history, and the disturbance to soil samples prior to laboratory testing. A major focus of Professor Ladd's research has been on these topics and the results have greatly extended our knowledge well beyond the aforementioned historical simplification of clay behavior. Equally important is that many of these findings have been clearly shown to be important to the sound and economic practice of geotechnical engineering. Some landmark publications by Professor Ladd and his colleagues that highlight these issues include the SHANSEP paper (Ladd and Foott 1974), the Tokyo Conference paper (Ladd et al. 1977), the San Francisco Conference paper (Jamiolkowski et al. 1985), the American Society for Testing and Materials (ASTM) paper on triaxial testing (Germaine and Ladd 1988), and the 22nd Terzaghi Lecture (Ladd 1991).

This historical perspective serves as a background for the major goals of this paper: (1) review the status of reliable and practical laboratory equipment for measurement of soft clay behavior; (2) discuss the results of research on fundamental clay behavior and their implications for design; and (3) give recommendations for implementation of key research results in practice.

Experimental Capabilities

Numerous devices have been developed to measure the consolidation and stress-strain-strength behavior of clays. They range from simple-to-use devices to sophisticated ones that are only suitable for research environments. Many papers have reviewed the capabilities of laboratory testing equipment including the comprehensive reviews by Ladd et al. (1977), Saada and Townsend (1981), Jamiolkowski et al. (1985), and Ladd (1991). This paper considers only devices and test methods that are acceptable for determining design parameters for soft clays and that are realistically available to engineers. This limits the list to incremental load (IL) and constant rate of deformation or strain (CRS) consolidation, triaxial compression (TC) and extension (TE), and direct simple shear (DSS) equipment.

The one-dimensional consolidation test is the most effective laboratory method for determining the preconsolidation stress and consolidation properties of clays. The test is typically performed using an oedometer cell with application of incremental loads. This equipment is widely available and the test is relatively easy to perform. However, the CRS test (Wissa et al. 1971) is a significant improvement over that of IL equipment as it allows for backpressure saturation and continuous measurement of deformation, vertical load, and pore pressure for direct calculation of the stress-strain curve and coefficients of permeability and consolidation. Furthermore, recently developed computer-controlled flow pumps and load frames allow for automation of most of the test. Capital investment in CRS equipment is higher than IL equipment, but in the broader picture, the improved data quality and test efficiency can result in significant cost benefits.

A wide range of devices is available for measuring the undrained shear strength of clays and this seems to often create confusion. The inexpensive and simple to use devices such as the torvane, pocket penetrometer, fall cone, and unconsolidated undrained triaxial compression test (UUC) are popular in practice. However, these tests are greatly affected by sample disturbance, use fast shear rates, and different modes of shear. As a result, the data from these devices at best represent relative strength rather than values suitable for design. Undrained shear strength profiles developed using data from these devices often show significant scatter. Therefore they are more suitably referred to as strength index tests and in general should only be relied upon to indicate the general consistency of soil layers. Reliable determination of undrained shear strength values for design must focus on use of equipment that can conduct consolidated-undrained (CU) tests. In this regard, laboratory equipment options include conducting CU tests in TC, TE, and DSS equipment. However, even data from these devices can be suspect if proper sampling and test procedures are not followed as discussed in the section on Test Methods.

Great advances have been made in the past 15 years in computer automation of triaxial equipment. Triaxial stress path cells have emerged through computer automation of flow pumps and load frames. Once sophisticated and time consuming tests such as K_0 consolidated undrained shear (CK_0U) can now be reliably conducted through computer control with a significant reduction in potential for operator error and with much shorter testing periods as compared to traditional manual procedures. Many of the basic features of top-level triaxial stress path cells that were developed at research institutions, including MIT (e.g., Sheahan and Germaine 1992, Sheahan et al. 1994), are being commercially manufactured, thus making this equipment available to geotechnical laboratories. Like the CRS device, the capital investment in automated stress path cells is not trivial, but the benefits from improved data quality and test efficiency cannot be overstated. Unfortunately, there are some features of high quality triaxial equipment (e.g., Baldi et al. 1988, Lacasse and Berre 1988, Germaine and Ladd 1988) such as internal load cells and tie rods, smooth or lubricated end platens, and internal small strain measurements (e.g., Da Re et al. 2001) that are modestly used or practically nonexistent in practice. Use of modern instrumentation systems, and especially computer control systems, require operation in well-controlled constant temperature chambers or rooms.

The DSS device has the unique ability to test soil specimens wherein the major principal stress is free to rotate during simple shear strain conditions. DSS tests are relatively easy to run and use less soil compared with triaxial specimens. Newer generation machines are benefiting from automation by making use of electrically controlled screw jacks for application of vertical stress and horizontal shear stress. In the more common Geonor DSS device, a circular specimen is trimmed into a wire-reinforced membrane resulting in a K_0 compression curve that is the same as those measured in IL oedometer tests. Unfortunately, the device has non uniform stress conditions and does not conveniently allow for backpressure saturation and direct measurement of pore pressure. In addition the complete state of stress at failure is unknown, although it is common to assume that the undrained shear strength is equal to the measured peak horizontal shear stress. These problems may lead one to question the utility of the DSS device. However, the device has been found to produce a reliable measure of the in situ mobilized undrained shear strength s_u(mob) for stability problems in non-varved sedimentary clays (Ladd 1991). At the present, no other practical laboratory device can produce such results for design as will be discussed in the section on Anisotropy.

Test Methods

Having all the right test equipment does not guarantee success in measuring the engineering properties of soft clays. Variations in test procedure can make significant differences in the measured behavior. In spite of advances in automation, there is still no substitute for skilled, knowledgeable testing personnel for field collection of samples, trimming of specimens, and monitoring test progress. This section highlights test procedures that are targeted towards improving the quality of laboratory consolidation and strength data. General recommendations for some of the tests are given in ASTM (2000) standards for IL consolidation (D2435), CRS (D4186), and DSS (D6528).

The first, and arguably the most crucial, step in laboratory testing of clays is sample selection and specimen trimming. Since the late 1970s, MIT and the Norwegian Geotechnical Institute (NGI) have radiographed tube samples to help in selecting the highest quality and most representative portions of each tube for consolidation and CU strength testing. Good quality radiographs can readily show many important features including variations in soil type, macrofabric, inclusions, and voids. It provides an important non-destructive visual detection of features and variations in degree of sample disturbance prior to testing that often cannot be determined from direct inspection and handling of samples. ASTM D4452 describes the necessary equipment and techniques for conducting radiography of soil samples. It is such a valuable tool that the cost for its use should be routinely incorporated as part of any exploration program in clays. Yet in spite of its value, and the availability of the ASTM standard, very few geotechnical engineering firms use radiography and fewer still have the equipment.

With time soil samples bond to the inside of tubes and in many cases subsequent extrusion can result in significant additional disturbance. Therefore, samples should not be extruded from tubes (except if done immediately after sampling) without first breaking any bonding at the soil-tube interface. Once a location within a tube is selected for testing, the tube should be cut adjacent to the desired sample location using a horizontal band saw or simply by hand (tube cutters should not be used since with only two or three contact points they tend to distort the tube during cutting). A hypodermic tube can be used to feed a thin wire into the soil/tube interface and rotated several times around the perimeter to break the soil/tube bond. With experience, the whole process only takes a few minutes. Another option that can be considered, as is sometimes done by NGI, is to immediately extrude the sample in the field, cut it into sections, and then set the sample sections inside an oversized tube made of coated rigid cardboard. Once in the tube, the annulus between the soil and tube is filled with wax. This procedure makes subsequent tube cutting in the laboratory even simpler and quicker. Soil within 1.5 times the tube diameter from the top and bottom of the tube should not be used for consolidation and strength testing because of greater disturbance near the sample ends (Lacasse and Berre 1988). Sample sides should also be trimmed during specimen preparation to also remove potentially disturbed material.

Some laboratories advocate the use of dry stones during specimen set-up for all consolidation and strength specimens regardless of the in situ overconsolidation ratio (OCR) of the soil (e.g., Sandbækken et al. 1986, Lacasse and Berre 1988). The reasoning behind the use of the dry stones, even in the case for soft clays is that the negative pore pressure in the specimen will draw in water from the stones resulting in swelling and a further reduction in the effective stress beyond that already caused by sample disturbance. Whatever method of filter stone preparation is used, the specimen should not be allowed to swell, otherwise destructuring can occur. Water bath and pore pressure system water salinity can be matched to that of the natural soil by mixing a NaCl solution that has approximately the same electrical conductivity as that measured for the natural pore water.

Consolidation Testing. IL oedometer and CRS tests should ideally be conducted by first loading the specimen beyond the preconsolidation stress (σ'_p) onto the virgin compression line, have an unload-reload cycle (to better define the behavior of OC clay), load to the maximum desired stress, and finally unload back to the seating load. If high quality samples are available then the unload-reload cycle is not necessary. In some cases the load increment ratio is reduced from the typical value of one (e.g., ½) for better definition of the σ'_p and virgin compressibility. However, for more structured soils even this may not be sufficient to properly define σ'_p in IL tests as compared to continuous CRS data. In addition, reduced load increment ratios result in deformation-time curves that are not well defined and cannot be interpreted using graphical construction methods (e.g., $\sqrt{t}$ or logt) for estimating the coefficient of consolidation (c_v). In the case of CRS tests, continuous k and c_v data can be computed providing an acceptable rate of deformation is selected as discussed in the following paragraph. Figure 1 plots data from IL and CRS tests conducted on Sherbrooke block samples of the Gloucester Clay in Canada and Boston Blue Clay from Newbury, MA. These data show that the IL test has poor definition of σ'_p and a too low compression ratio (CR = $\Delta\varepsilon/\Delta log\sigma'_v$) just beyond σ'_p as compared to the CRS data. For the CRS tests in Figure 1, the base excess pore pressure (Δu_b) was kept greater than 1%, but well below 10%, of the total vertical stress (σ_v) to avoid any significant rate effects on σ'_p (Leroueil 1994). A typical CRS test for most soft clays with back pressure saturation takes only 3 to 4 days (without an unload-reload cycle or pausing for measurement of the coefficient of secondary compression $c_\alpha = \Delta\varepsilon/\Delta\log t$). Because of the various advantages listed here, the CRS test should be used in place of IL consolidation testing whenever possible.

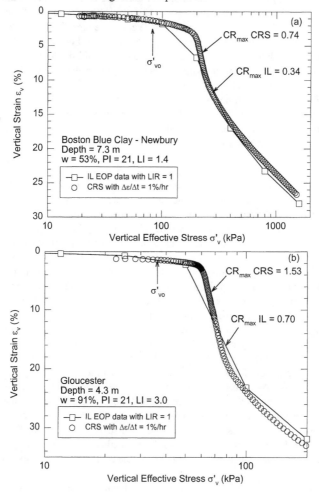

Figure 1. IL Consolidation and CRS Compression Curves for Sherbrooke Block Samples of (a) Boston Blue Clay – Newbury, (b) Gloucester Clay - Canada.

One problematic aspect of CRS testing is selecting an acceptable rate of loading. If tests are run too slowly, then appreciable secondary compression strains will occur. If tests are run too fast, high excess base pore pressures will lead to significant variations of void ratio and σ'_v in the specimen. Mesri and Feng (1992) and Ladd and DeGroot (1992) suggest that the strain rate be selected such that the normalized base excess pore pressure ($\Delta u_b/\sigma_v$) is not greater than 15 to 20%. Mesri and Feng (1992) give recommendations for selecting an appropriate strain rate that gives practically the same compression curve as the EOP curve from IL tests. However, if c_v data are required, some base excess pore pressure is necessary and the recommended rate should be increased by a factor of 10. For typical soft clays, this gives a rate of about 0.7 % strain per hour and it should produce $\Delta u_b/\sigma_v$ less than 15%, although the resulting σ'_p will be about 10% greater than the EOP σ'_p (Mesri et al. 1994). Sandbækken et al. (1986) report that for many of the clays tested at NGI, rates of 0.5 to 1% strain per hour are adequate and maintain $\Delta u_b/\sigma_v$ between 2 and 7% throughout the test. The ASTM D4186 recommendations produce strain rates that are often too high, especially during virgin compression, and can result in overpredicting σ'_p in some soils.

Triaxial Testing. Baldi et al. (1988), Germaine and Ladd (1988), and Lacasse and Berre (1988) give thorough recommendations on use of triaxial equipment and test procedures. The key steps in the process are specimen set-up, saturation, consolidation, and undrained shear. Ideally, specimens should be trimmed down from the original sample diameter to allow for removal of sample perimeter soil that is usually more disturbed. Very weak soft clays that cannot stand unsupported require the use of a special trimming jig to maintain continuous support during specimen set-up (e.g., Baldi et al. 1988). Backpressure saturation is essential for accurate measurement of volume change during consolidation and pore pressure changes during undrained shear. It should ideally take place at the measured or estimated sampling effective stress to minimize specimen volume changes and without allowing any swelling to occur. A final backpressure of approximately 200 to 300 kPa is typically sufficient to obtain saturation of most soft clays, although Skempton's B value should always be checked to ensure saturation has occurred. Consolidation should follow a K_0 effective stress path to the final required effective state of stress. As already noted, modern stress path cells make the task of conducting K_0 consolidation much simpler and more efficient than manual methods. If such capabilities are not available, which is still the norm, then at least an anisotropic effective stress path should be used to reach the target final effective stress (e.g., Lacasse and Berre 1988). This method does, however, require an estimate of the value of K_0 that corresponds to the estimated OCR of the specimen at the final vertical effective consolidation stress, σ'_{vc}. Although the consequences of using anisotropic consolidation, rather than following a true K_0 stress path are not completely resolved, it is a much more realistic preshear state of stress as compared to simple isotropic consolidation. The rate of strain for undrained shear should be selected taking into account the strain rate sensitivity of clays and typical field loading rates. Many leading laboratories use a strain rate of 0.5 to 1.0% per hour for CK_0U triaxial tests on soft clays (Germaine and Ladd 1988).

Direct Simple Shear. Bjerrum and Landva (1966), Ladd and Edgers (1972), and DeGroot et al. (1992) provide comprehensive reviews of DSS test equipment, data reduction/interpretation, and typical results for a large variety of clays. Undrained shear is usually performed by running constant volume tests that can either be performed manually or by computer automation. A typical test only takes a few days depending on the consolidation increments required to reach the preshear effective stress. Undrained shear is usually conducted at approximately 5%/hour, with the peak shear stress for most clays being reached within 1 to 3 hours. Recompression tests that reconsolidate the specimen to σ'_{vo} typically require the use of stones with embedded pins to prevent slippage. Unfortunately, these stones are more difficult to seat and do create an unknown degree of disturbance to the specimen. Another issue for Recompression tests is the lack of sufficient horizontal stress developed during recompression to σ'_{vo} such that the resulting laboratory K_0 is typically much lower than exists in situ (Dyvik et al. 1985). This can produce measured results that are markedly different from the correct behavior corresponding to the in situ OCR. Therefore specimens must first be loaded up to a stress level beyond σ'_{vo} and unloaded back to σ'_{vo} to develop additional horizontal stress, which can be a tricky procedure. As a guideline, NGI typically loads to approximately 80% of the best estimate of σ'_p and then unloads back to σ'_{vo} prior to undrained shear. Obviously, this cannot be done for very low OCR soils.

Soil Behavior Issues

The single most important engineering property to determine for clay deposits is stress history as expressed through σ'_p and OCR. All significant aspects of clay behavior from consolidation to undrained shear strength are influenced by stress history. The single most important issue in accurately measuring stress history, and undrained shear strength too, is sample disturbance. Other important soil behavior issues include anisotropy and rate effects. While we do not fully understand all these effects, we have made significant progress in dealing with some of them in a practical manner for design. For example, using a combination of shear modes in the laboratory (i.e., TC, DSS, and TE) allows us to measure the strength anisotropy of clays. In another example, using reasonable laboratory shear rates allows for measurement of strengths that are consistent with anticipated field loading events. Through research we have been able to develop reasonable, practical means for taking these factors into account, even if many of these developments are not yet fully appreciated by the profession. However, there remains a significant lack of practical knowledge in how to assess sample disturbance and how to deal with its adverse effects to obtain accurate soil parameters in the laboratory. This section of the paper discusses the four soil behavior issues noted above, although greater emphasis is given to sample disturbance.

Sample Disturbance. The most important effects of sample disturbance are a significant reduction in the sample effective stress (σ'_s) and a reduction in σ'_p during one-dimensional compression. Figure 2 shows a classic example of how the reality of sampling and testing can vary unpredictably from the ideal. This figure shows the anticipated stress paths experienced by a normally consolidated clay starting from its in situ state of stress (Point A) to its state of stress when ready for laboratory testing (Point F). While we design for in situ stress states, disturbance caused by sampling and subsequent storage and handling can significantly alter the state of stress of the samples. Clearly the state of stress at Point F is very different from the in situ state at Point A, and yet much of standard geotechnical engineering practice relies on strengths determined from samples starting from Point F (e.g., UUC test). Figure 2 further shows the significant difference in potential stress paths for soil elements during undrained compression shear starting at points A and F. The situation is further complicated when anisotropy, shear strain rate and soil structure are considered. Based on this simple depiction of what can happen to the soil's stress state during sampling, it should come as no surprise that there is often a gross mismatch between design performance based on laboratory derived strengths and field performance.

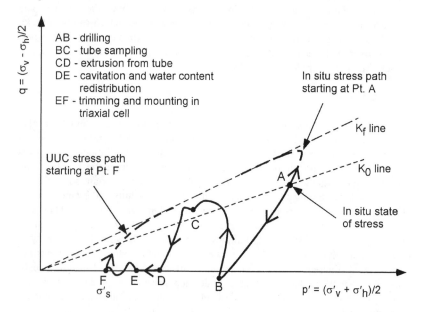

Figure 2. Hypothetical Stress Path for an OCR = 1 Clay Element During Tube Sampling/Specimen Preparation and Undrained Shear (after Ladd and Lambe 1963)

Many researchers have theoretically and experimentally studied the influence of sampling techniques and sampler design on the quality of soft clay samples (e.g., Hvorslev 1949, Baligh et al. 1987, Clayton et al. 1998). Some aspects of sample disturbance such as stress relief are unavoidable, but there are several measures in field sampling procedures and laboratory handling of samples that can help to minimize additional degrees of disturbance. Use of an appropriately weighted drilling mud is essential for maintaining bottom borehole stability prior to sample collection. For tube sampling, the best methods are to use large diameter, sharp edged, small area ratio tubes (e.g., the 75 mm diameter or larger thin-walled Shelby tube) in a fixed piston sampler. The writer prefers to use 89 mm diameter stainless steel tubes in a fixed piston sampler for trimming of conventional size 64 mm diameter oedometer/CRS and 67 mm diameter DSS specimens since it allows for more perimeter soil to be trimmed away from the specimen, but this size tube is not as commonly available. The 127 mm diameter tube is occasionally used and provides even more room for trimming specimens from the core of the tube, but these tubes are heavy and awkward to handle, causing potential for more disturbance during handling.

Of all possible sampling methods, research has shown that block sampling is considered the best method. Originally, block sampling consisted solely of hand-carved pieces of soil, usually taken from the floor of test pits or other large excavations, limiting the sampling to shallow depths. The Sherbrooke University (Lefebvre and Poulin 1979) and Laval University (LaRochelle et al. 1981) samplers were developed to collect block samples at greater depths and below the groundwater table.   Research results from samples collected using these specialized block samplers have contributed greatly to our understanding of the detrimental effects of sample disturbance, especially for more structured soils. Unfortunately, this equipment is not commonly available and considerable expertise is required to use these samplers. However, in recent years their use is slowly finding its way into practice and in some cases resulting in significant cost saving because of the higher quality samples that have been collected as compared to conventional sampling methods. For example, Lacasse et al. (2001) report a cost saving of 1.5 million USD, or 20% of total project costs, on a 1.5 km long rail crossing in Norway that directly resulted from the combined use of Sherbrooke block samples and piezocone penetration testing (CPTU). During the past year the writer has successfully conducted Sherbrooke block sampling at four sites in North America and in all cases used the sampler together with conventional drilling equipment. The daily productivity at these sites ran as high as being able to collect six block samples measuring approximately 250 mm diameter by 330 mm tall. There are difficulties when block sampling in deep boreholes because of bottom instability in the large diameter borehole, and more research needs to be conducted to determine if this problem can be overcome. While this sampling technique will likely not find its way into routine practice, it is prudent for engineers to appreciate its potential economic benefits.

Sample disturbance adversely influences all-important engineering design properties of soft clays from modulus to undrained shear strength. As shown in Figure 2, the disturbance from sampling and handling results in a decrease in the sample effective stress. Ladd and Lambe (1963) showed that this decrease can be well above 80% from the in situ effective stress and the reported laboratory stress-strain-strength properties at the sampling effective stress cannot possibly match the correct in situ properties. Even if samples are subsequently reconsolidated in the laboratory to in situ stresses, disturbed samples will undergo significant volumetric changes and further destructuring, again resulting in unrealistic stress-strain behavior. An example of this behavior is shown in Figure 3 for anisotropically consolidated (CAU) triaxial compression tests reported by Lunne et al. (1997) on samples reconsolidated to σ'_{vo} and σ'_{ho} (based on an estimate of the in situ K_o) prior to undrained shear. The samples were collected using three types of samplers: the NGI 54 mm diameter sampler, a standard 75 mm diameter fixed piston, thin-walled tube sampler, and the Sherbrooke block sampler. There are clear differences in measured stress-strain behavior and undrained shear strength that would have a significant impact on selection of design parameters.

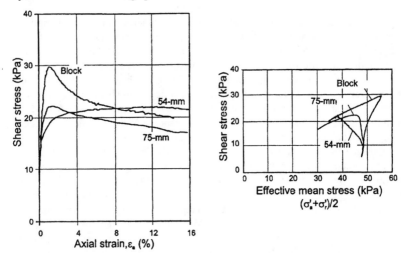

Figure 3. CAU Triaxial Recompression Tests Results for Specimens of Lierstranda, Norway Clay Collected with Three Different Samplers (from Lunne et al. 1997)

Whenever evaluating consolidation and strength data it is essential to evaluate sample quality and yet this is not commonly done in practice. As noted previously, x-rays are considered a critical step in any important test program and provide valuable information on sample quality. However, x-rays are limited to visual information and it is also important to be able to quantify the degree of sample disturbance. Many researchers have worked on this topic and some methods that have been developed include measurement of soil suction to get σ'_s (Ladd and Lambe 1963), volumetric strain (ε_{vol}) during laboratory reconsolidation to σ'_{vo} (Andresen and Kolstad 1979), and shear wave velocity (Shibuya et al. 2000). The simplest and most effective of these methods is the measure of ε_{vol} at σ'_{vo}. Andresen and Kolstad (1979) first developed this method with a ranking system that assigns a description of sample quality ranging from poor to excellent. Terzaghi et al. (1996) adapted this method and coined the term Specimen Quality Designation (SQD) with sample quality ranging from A (best) to E (worst) as listed in Table 1. Terzaghi et al. (1996) suggest that reliable estimates of engineering parameters such as σ'_p and s_u require samples with SQD equal to B or better. Recently, Lunne et al. (1997) updated Andresen and Kolstad (1979) sample quality method to use the more fundamental measure of $\Delta e/e_0$ for reconsolidation to σ'_{vo} as listed in Table 1. Although based on the same measurements, the $\Delta e/e_0$ criteria is considered a better parameter because it is reasonable to assume that a certain amount of change in pore volume during reconsolidation will be increasingly detrimental to the particle skeleton as the initial pore volume decreases (Lunne et al. 1997). It is also expressed as a function of OCR since an equal amount of volumetric strain is more detrimental in higher OCR samples. The ε_{vol} or $\Delta e/e_0$ measurements are objective and easy to perform on laboratory specimens and should be reported for every consolidation and CU strength test conducted on clays.

Table 1 – Quantification of Sample Disturbance Based on Specimen Volume Change During Laboratory Reconsolidation to σ'_{vo}

Specimen Quality Designation (SQD) (Terzaghi et al. 1996)		$\Delta e/e_0$ Criteria (Lunne et al. 1997)		
Volumetric Strain (%)	SQD	OCR = 1 – 2	OCR = 2 – 4	Rating*
		$\Delta e/e_0$	$\Delta e/e_0$	
< 1	A	< 0.04	< 0.03	Very good to excellent
1 – 2	B	0.04 – 0.07	0.03 – 0.05	Good to fair
2 – 4	C	0.07 – 0.14	0.05 – 0.10	Poor
4 – 8	D	> 0.14	> 0.10	Very poor
> 8	E			

* Refers to use of samples for measurement of mechanical properties

Stress History. The importance of stress history (σ'_p and OCR) on deformation and undrained shear behavior of soft clays is well documented and its assessment should be a mandatory part of any site characterization. This information is required for settlement calculations and is considered essential for evaluation of undrained shear strength data. Laboratory reconsolidation methods should as far as possible mimic field stress history. Numerous methods have been proposed for estimating σ'_p but they all depend on reliable data from good to high-quality samples. The data used for interpretation should be from CRS tests conducted at an acceptable rate of deformation (as noted previously) or end of primary data from IL tests with an appropriate LIR (less than unity for structured clays). As an added advantage of automation, computer controlled CK_0U stress path cell tests for specimens consolidated beyond σ'_p give reliable data for estimating σ'_p (Ladd et al. 1999). Casagrande's method is the oldest, simplest, and most widely used technique for estimating σ'_p. However, it is quite subjective and is difficult to perform for relatively stiff soils, often leading to a significant range in estimated values. The strain energy method of Becker et al. (1987) uses work per unit volume as the criterion for estimating σ'_p, using a plot of strain energy versus effective stress in linear scales. The method is easy to use and typically results in more reliable and consistent estimates of σ'_p as compared to Casagrande's procedure, especially for stiffer clays with more rounded compression curves (Ladd and DeGroot 1992, Ladd et al. 1999).

Undrained Shear Strength Anisotropy. Research has clearly shown the important effects of soil anisotropy on selection of design strengths for stability problems involving soft clays, especially for low plasticity clays as shown in Figure 4a for OCR = 1 (i.e., laboratory final vertical effective consolidation stress $\sigma'_{vc} > \sigma'_p$). The difference in undrained shear strength between TC and TE averages approximately a factor of 2 for low plasticity clays. Undrained shear strength data from test programs using a combination of CK_0U TC, DSS, and TE tests conducted at OCR = 1 and which are treated for strain compatility (Koutsoftas and Ladd 1985), show good agreement with strengths backcalculated from field failures, as shown in Figure 4b.

Isotropically consolidated (CIU) triaxial compression tests are routinely used to conduct because of the widespread availability of basic triaxial equipment and the relative simplicity of this test procedure. However, CIUC undrained shear strengths are most often much too high when used alone for stability problems. Germaine and Ladd (1988) report that based on data from 30 soils, $s_u(CIUC) = 0.33\sigma'_c$ on average, independent of plasticity index. Considering this value relative to the field data in Figure 4b highlights the danger of this approach to design.

It is appreciated that at present, test programs involving TC, DSS, and TE modes of shear are comprehensive and likely to only be possible for final design of major projects. An alternative and cost effective approach for modest projects is use of DSS data alone. As shown in Figure 4b, $s_u(DSS)$ for nonvared clays that plot above the A-line averages approximately

$$s_u(DSS) = 0.23\sigma'_{vc} \tag{1}$$

with very little trend with plasticity index (DeGroot et al. 1992, Ladd 1991). This compares well with the average trend for the CK_0U TC, DSS, and TE data in Figure

4b. It also compares well with Mesri's (1975) analysis of Bjerrum's field vane data in which Bjerrum's field vane correction factor was applied to field vane data normalized by σ'_p, resulting in the

$$s_u(mob) = 0.22\sigma'_p \qquad (2)$$

independent of plasticity index.

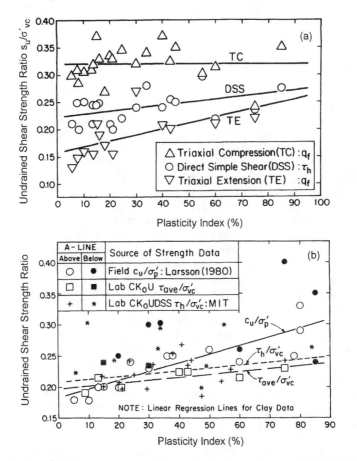

Figure 4. Undrained Shear Strength Ratios for OCR = 1 Clays and Silts: (a) Lab CK_0U Tests; (b) Comparison of Field and Lab CK_0U Data (from Ladd 1991).

Rate Effects. Research has documented that significant undrained shear strength changes can occur with changes in strain rate. Sheahan et al. (1996) show that this rate dependence varies with both OCR and strain rate level based on CK_0UC tests on resedimented BBC. Changes on the order of 10% increase in s_u per log cycle increase in strain rate are commonly reported, but this may be even higher at very high rates and with increasing OCR (Ladd 1991). Although this aspect of soil behavior is not fully understood, it is reasonably easy to control by using recommended laboratory shear rates as quoted in the section on Laboratory Methods (i.e., 0.5 to 1%/hr for TC/TE and 5%/hr for DSS). These rates represent a balance between expected field behavior and selection of practical shear rates for the laboratory. It is significant to note that in the popular UUC test, the typical strain rate is 60%/hr (ASTM D2850 or 2166) and for other index type tests such as Torvane and Fall Cone, it is even higher.

Laboratory Reconsolidation

The long popular practice of conducting UU type of strength tests for s_u design parameters does a very poor job of addressing the important soil behavior issues listed in the previous section. For example, use of UUC undrained shear strength data for stability problems violates all of these issues, i.e., (1) the effective stress prior to undrained shear is generally much less than in situ stresses because of sample disturbance; (2) with the incorrect preshear effective stresses, no consideration is given to stress history; (3) use of the compression mode of shear only ignores anisotropy; and (4) the fast rate of shear ignores strain rate effects. Thus one can only conclude that the successful (defined here as acceptable performance, but not necessarily economical) use of UUC data in practice relies on a combination of compensating errors, development of empirical correlations that are soil and test method specific, or overly conservative design. The potential economic impacts of investing in more accurate and reliable means of measuring soft clay behavior can be significant. For example, Terzaghi et al. (1996) report that UUC tests without cell pressure on SQD = D quality tube samples can result in s_u values less than 50% of that measured for A quality samples. However, the error is not always on the safe side. The use of improved sampling techniques may offset a significant portion of the decrease in s_u due to sample disturbance that compensates, to some unknown degree, errors in $s_u(UUC)$ from anisotropy and rate effects. Germaine and Ladd (1988) present data from UUC tests for four well studied cases with high quality sampling and well defined in situ reference strengths for which the $s_u(UUC)$ strengths were considered unsafe. These data further point out the problem with UUC testing in that it is not a rational framework within which engineers can control the important soil behavior issues that influence s_u.

The Recompression and SHANSEP laboratory testing methods for clays were developed to address all of the soil behavior issues mentioned above with particular emphasis on how to minimize the adverse effects of sample disturbance. Table 2 gives the basic procedure, advantages, and limitations of both methods. Additional details are given by Ladd (1991). The methods are identical except for an important difference in how to deal with sample disturbance. Both approaches advocate the use of CK_0U tests with shearing in different modes of failure (i.e., TC, DSS, and TE) at appropriate strain rates to account for anisotropy and strain rate effects.

Table 2 Recompression and SHANSEP Techniques (after Ladd 1991)

RECOMPRESSION	SHANSEP
Basic Procedures	
1. Perform CK_0U tests on specimens reconsolidated to the in situ state of stress, i.e., $\sigma'_{vc} = \sigma'_{vo}$ and $\sigma'_{hc} = K_0\sigma'_{vo}$ **2.** Select appropriate combination of TC, DSS and TE tests to account for anisotropy. **3.** Use strain rates of 0.5 to 1 %/hr for triaxial tests and 5 %/hr for DSS tests. **4.** Plot depth specific strength values versus depth to develop s_u profile.	**1.** Establish the stress history profile, i.e., σ'_{vo}, σ'_p, OCR from consolidation and automated CK_0U tests **2.** Perform CK_0U tests on specimens consolidated well beyond in situ σ'_p to measure NC behavior and also on specimens rebounded to varying OCR to measure OC behavior. **3.** Select appropriate combination of TC, DSS and TE tests to account for anisotropy. **4.** Use strain rates of 0.5 to 1 %/hr for triaxial tests and 5 %/hr for DSS tests. **5.** Plot results in terms of s_u/σ'_{vc} vs. OCR to obtain values of S and m for the equation $s_u/\sigma'_{vc} = S(OCR)^m$, where $S = s_u/\sigma'_{vc}$ for OCR = 1 and m is strength increase exponent. **6.** Use above equation with stress history to compute s_u profile.
Advantages/Limitations/Recommendations	
1. Preferred method for high quality samples. **2.** More accurate for highly structured clays. **3.** Preferred for strongly cemented clays and for highly weathered and heavily OC crusts. **4.** Should not be used for NC clays. **5.** Reloads soil in laboratory. **6.** Gives depth specific strength data. **7.** Should be accompanied by evaluation of stress history to check if s_u/σ'_{vo} values are reasonable.	**1.** Strictly applicable only to mechanically OC and truly NC clays exhibiting normalized behavior. **2.** Preferred for conventional tube samples of low OCR clays having low sensitivity. **3.** Should not be used for highly structured, brittle clays and strongly cemented clays. **4.** Difficult to apply to heavily OC clay crusts. **5.** Unloads soil in laboratory to relevant OCR. **6.** Forces user to explicitly evaluate in situ stress history and normalized soil parameters.

In the Recompression method, Bjerrum (1973) recognized the unreliable behavior of the standard UU test and attempted to improve upon this using a CU test. Thus the Recompression method anisotropically reconsolidates the specimen to the in situ state of stress (σ'_{vo}, σ'_{ho}) as shown in Figure 5. This does require an estimate of the in situ K_0 for application of σ'_{ho} since one-dimensional recompression to σ'_{vo} will result in too low a K_0. This procedure assumes that the reduction in water content during reconsolidation to σ'_{vo} is small enough so as to produce s_u data that are representative of in situ conditions. Bjerrum acknowledged that a reduction in water content would result in an increase in undrained shear strength but this would be compensated by the loss in strength due to sample disturbance. Berre and Bjerrum (1973) recommended that the volumetric strain during recompression should be less than 1.5 to 4 percent. The method works best when using high quality samples.

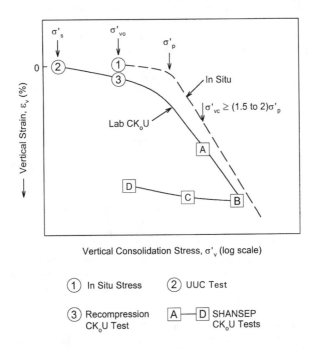

Figure 5. Consolidation Procedures for Lab CK_0U Testing (after Ladd et al. 1977).

The SHANSEP method (Ladd and Foott 1974, Ladd 1991) is based on the experimental observation that the undrained stress-strain-strength behavior of most "ordinary" clays, for a given mode of shear, is controlled by the stress history of the soil deposit. The method assumes that these clays exhibit normalized behavior and uses mechanical overconsolidation to represent all preconsolidation mechanisms. The procedure explicitly requires the stress history profile for the soil to be evaluated, which is usually done based on one-dimensional consolidation tests. This typically requires a separate consolidation testing program but with the advent of computer controlled triaxial equipment, stress history information is automatically obtained during the K_0 consolidation phase of strength tests. Tests are laboratory consolidated to stress levels greater than σ'_p to measure the normally consolidated behavior (Figure 5). Additional specimens are also loaded in this manner but then unloaded to varying OCRs to measure overconsolidated behavior. Undrained shear takes place after approximately one cycle of secondary compression has elapsed at both the maximum and final consolidation stress. Normalized soil parameters are used to relate the in situ s_u to the normally consolidated s_u and the increase in normalized s_u with an increase in OCR as

$$s_u/\sigma'_{vo} = S(OCR)^m \tag{3}$$

where $S = s_u/\sigma'_{vc}$ for OCR = 1. Due to its inherent assumptions, the method is not applicable to highly structured, naturally cemented clays and is questionable for highly weathered clay crusts in which mechanical overconsolidation does not represent the primary overconsolidation mechanism. Ladd and Foott (1974) noted that the SHANSEP technique will destroy some important aspects of soil structure that has developed during and after formation of the clay deposit. They feel, however, that the procedure provides a reasonable estimate of the in situ strength properties, especially when compared with other more traditional techniques. For modulus data (i.e., E_u) the SHANSEP method usually gives values that are too low (e.g., Ladd 1991)

The SHANSEP method has been criticized because of its consolidation procedure (e.g., Mesri 1975, Tavenas and Leroueil 1980). Mesri (1975) stated that "all natural clays possess a structure that is destroyed, and therefore, in situ undrained shear strength is altered whenever the clay is consolidated to stresses in excess of the critical pressure [preconsolidation pressure]". Mesri (1975) further suggests that the SHANSEP reconsolidation technique "does not overcome sample disturbance; on the contrary it further remolds and disturbs the natural structure of the clay." While this may be the case, there are several important issues to consider with respect to the philosophy behind the SHANSEP method: (1) there is probably universal agreement that all soils are structured (Leroueil and Vaughan 1990) but the more important point is that the degree of soil structure and hence its importance on soil behavior varies greatly among soils; (2) the SHANSEP method makes the important contribution of the use of normalized soil parameters for design; (3) the SHANSEP method clearly identifies the importance and need for evaluation of stress history; and (4) the SHANSEP method was developed as a *practical* tool for dealing with the adverse effects of sample disturbance which is pervasive in routine practice.

Items 2 and 3 are very powerful tools whether using SHANSEP or Recompression data, or in situ data for that matter. Those who criticize the SHANSEP method often overlook item 4. Whenever high quality samples are available most would agree that a Recompression test program is usually most appropriate. However, the reality is that practicing engineers are often faced with inferior quality samples but must make design decisions based on what information they can get out of the samples at hand. Clearly our goal is to improve sampling practice but in the mean time it is necessary to consider the day-to-day decisions engineers face. The important question of the SHANSEP method is the degree to which it destroys soil structure and in spite of this, whether or not this method still provides reasonable design estimates for certain soils. As such, it provides a systematic method for assessing undrained shear strength parameters and has shown, empirically, that it provides reasonable results on engineering projects involving a variety of sedimentary clays of low to moderate sensitivity (Ladd and Foott 1980).

Case Histories

There are few case histories in the literature that make a detailed evaluation of the application of Recompression and SHANSEP methods in practice. Most of the comparisons presented are for cases where tests were conducted at a few specific depths and not over a larger profile. Results from three full profile case histories are presented here with three levels of sophistication and scope of testing. The first case history is a comprehensive research project involving the highest level of sophistication in field sampling and laboratory testing. This research project was supervised by MIT and the results were applied to the design of a major transportation system in Boston, MA. Ladd et al. (1999) present details of the test program and its results, therefore only the major conclusions are presented herein. The second case history involves use of block samples and gives an example of how to apply the Recompression and SHANSEP methods in practice based on consolidation and DSS testing alone. The third case history involves conventional tube sampling only and highlights difficulties with applying either Recompression or SHANSEP methods for poor quality samples. The last project is arguably the more common situation engineers face. It also demonstrates the use of CPTU testing as a valuable complementary tool to laboratory testing for site characterization.

Boston Blue Clay – CA/T. This test site was developed in connection with design and construction of the Central Artery and Third Harbor Tunnel (CA/T) project in Boston, MA. This multi-billion USD project involves depressing a major interstate highway through downtown Boston and construction of a new harbor tunnel to Logan International Airport. Because of the unprecedented complexity of the CA/T project, a Special Testing Program (STP) was developed with the objective to conduct a detailed field and laboratory investigation of the stress-strain-strength behavior of the primary soil unit, Boston Blue Clay (BBC), involved in the CA/T project. The STP was developed as a joint effort between Haley & Aldrich, Inc, of Boston, MA and MIT, (the latter under the supervision of Professor Ladd and Dr. Germaine). The field program consisted of conducting a variety of in situ tests and collection of fixed piston tube and Sherbrooke block samples. The advanced laboratory testing consisted of an extensive series of IL, CRS, and lateral stress consolidation tests, and computer automated K_0 consolidated undrained (CK_0U) TC, DSS, and TE tests using the SHANSEP and Recompression techniques. Details of the STP and results are given in Ladd et al. (1999). The information to follow is abstracted from this reference.

The BBC deposit at the site is covered by about 10 meters of fill and a natural marine sand. BBC is a marine clay, approximately 33 meters thick, and is underlain by either glacial till or a glaciomarine deposit. Radiography of sample tubes show the BBC to be layered with occasional sand and silt partings. Atterberg limits plot parallel to and above the A-line in a Casagrande Plasticity chart with a range in liquid limit of 40 to 60% and a range in plasticity index of 20 to 35%. The liquidity index ranges from 0.3 to 0.8, increasing with depth. The clay fraction (%<2µm) averages about 50%. Stress history data determined from laboratory tests show a thick, dessicated clay crust overlying a near normally consolidated clay (Figure 6b). The crust is approximately 17 meters thick and the near normally consolidated clay shows significant structure relative to the crust material. Figure 6a plots the ε_{vol} at σ'_{vo} from the consolidation tests and the consolidation phase of the CK_0U tests. It was not possible to collect good quality block samples beyond 27 m below ground surface due to bottom instability in the large diameter borehole. The compression curves from the IL and CRS consolidation tests and the consolidation phase of the automated CK_0U SHANSEP tests were used to estimate σ'_p using a combination of Casagrande construction and the strain energy method. Compression curves for the deeper soil display a marked backwards "S" shape in semi-log space, and very high CR values just beyond σ'_p (similar to that shown in Figure 1a) which are clear indicators of a soil with significant structure.

Figure 7 plots the normalized s_u data as measured in CK_0U TC and TE modes of shear using both the SHANSEP and Recompression methods for samples from a combination of fixed piston and Sherbrooke block samples. The interpreted undrained shear strength parameters for these data are summarized in Table 3. The data reveal several important conclusions. The SHANSEP CK_0U TC data give a slightly lower S value, but identical m values as compared to the Recompression data. For the CK_0U TE data, the S values are identical, but the m parameter for the Recompression tests is much larger. One clear observation from the test program is that the modulus data for the SHANSEP tests are much lower; the Recompression stress-strain curves give a much stiffer and more realistic response. The S values are particularly interesting to study since for the SHANSEP tests these specimens were "destucturized" in the laboratory and yet the resulting parameters are similar to the Recompression TC value and identical to the Recompression TE value. Although not tested, it is reasonable to expect that if there were greater disturbance to the samples, the SHANSEP parameters would remain about the same whereas the Recompression parameters could very well change, but in an unknown manner. The reduction in volume from reconsolidation to σ'_{vo} for more disturbed samples may not necessarily result in a net increase in strength because an increase in destructuring would also occur which would reduce the strength.

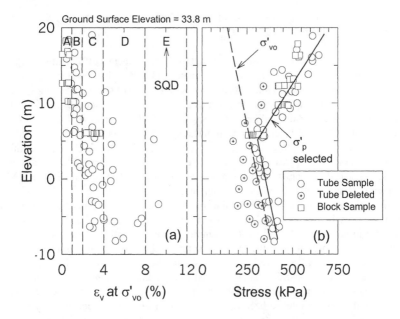

Figure 6. (a) Specimen Quality Designation and (b) Stress History for
Boston Blue Clay – CA/T Site (after Ladd et al. 1999).

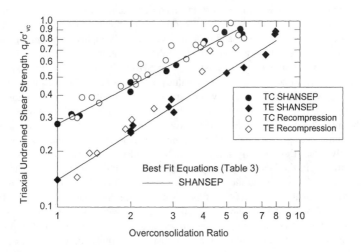

Figure 7. Undrained Shear Strength Data from SHANSEP and Recompression
CK_0U Tests for Boston Blue Clay – CA/T Site (after Ladd et al. 1999).

Table 3 Normalized Undrained Shear Strength Parameters from CK_0U
Triaxial Test Program – BBC CA/T (after Ladd et al. 1999).

Triaxial Test	Reconsolidation Method					
	SHANSEP			Recompression		
	n	S	m	n	S	m
CK_0UC	13	0.28	0.68	23	0.30	0.68
CK_0UE	17	0.14	0.83	9	0.14	0.98

1. n = number of tests; 2. $s_u/\sigma'_{vc} = S(OCR)^m$

Connecticut Valley Varved Clay – UMass Amherst. Soil samples for this case
history were taken from the University of Massachusetts Amherst (UMass Amherst)
US National Geotechnical Experimental Test Site (NGES). The site is located at the
UMass Amherst campus in the Connecticut River Valley of western Massachusetts.
Glacial Lake Hitchcock once covered the area during the late-Pleistocene Era of the
Wisconsin Ice Age approximately 15,000 years before present. The main lake
deposit consists of alternating silt and clay layers with occasional fine sand layers
and is locally known as Connecticut Valley Varved Clay (CVVC). The clay layers
were deposited during winter months while the silt and fine sand were deposited
during summer months.

Samples for the laboratory test program were obtained using the Laval
University block sampler. The subsurface stratigraphy of the site generally consists
of about 1 m of mixed fill overlying approximately 30 m of CVVC. The thickness of
individual silt or clay layers is typically on the order of 2 to 8 mm and the layers
generally lie in a horizontal direction. The upper 5 to 6 m of the deposit is
overconsolidated as a result of erosion, desiccation, and seasonal fluctuations in the
groundwater table. Below this weathered crust, the soils become soft and near
normally consolidated with increasing depth. Atterberg limits for the bulk soil are
approximately constant with depth with a liquid limit = 45%, plastic limit = 30%,
and plasticity index = 15%. The bulk water content increases with depth resulting in
a liquidity index greater than one for samples deeper than 4 m. Atterberg limits
typically straddle the A-line in a Casagrande plot (Lutenegger 2000).

IL oedometer tests with a reduced load increment ratio of 0.5 were conducted
for better definition of σ'_p. Estimates of σ'_p were made using the strain energy
method. Sample quality was assessed using the SQD as plotted in Figure 8a, which
shows that a majority of the specimens were of A and B quality with a few C quality
specimens. Figure 8b plots σ'_{vo} and σ'_p and Figure 9a shows that the soil is of
medium to high OCR at shallow depths and progresses towards an OCR just less
than 2 for the deeper samples. The variation in σ'_p is much greater for the shallow
more highly overconsolidated samples, which reflects the desiccated crust located in
the upper part of the deposit.

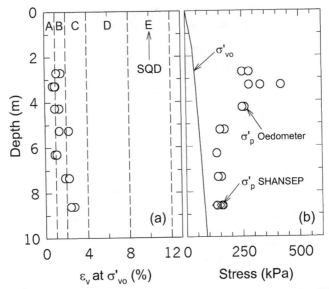

Figure 8. (a) Specimen Quality Designation and (b) Stress History for CVVC –
UMass Amherst Site (after DeGroot 1999).

Ten SHANSEP Geonor DSS tests were performed on samples taken from three of the blocks. Two tests were conducted on specimens consolidated to an OCR = 1 while the remaining tests were conducted on specimens that were first consolidated to an OCR = 1 and then mechanically unloaded to OCR values ranging from 2 to 8. Using the average normalized undrained shear strength of $s_u/\sigma'_{vc} = 0.18$ for the two OCR = 1 specimens, and a regression for the OC results, gives the following SHANSEP relationship

$$s_u/\sigma'_{vc} = 0.18(OCR)^{0.73} \tag{4}$$

Application of this equation together with the stress history profile gives estimates of s_u versus depth, as plotted in Figure 9b, by taking $\sigma'_{vc} = \sigma'_{vo}$ in Eq 4.

For Recompression tests, specimens were first loaded to approximately $0.8\sigma'_p$ and then unloaded to σ'_{vo} prior to undrained shear. Tests were performed on samples taken from the same block samples used for the oedometer tests. Typically two tests were conducted on each block sample to check for repeatability in the results (Figure 9b). The shallow more overconsolidated samples have greater differences between the companion tests as compared to the deeper lower OCR test pairs.

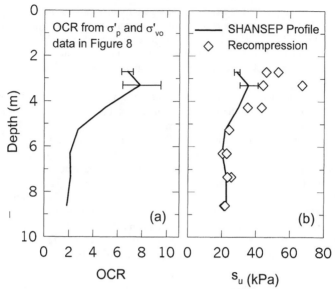

Figure 9. (a) OCR and (b) DSS Undrained Shear Strength Profiles for CVVC – UMass Amherst Site (after DeGroot 1999).

In comparing the results in Figure 9b, the SHANSEP method gives lower s_u values in the crust but near identical values below the crust in the softer clay below 5 m. In this case the Recompression results in the crust are considered more representative of the in situ s_u because of the high quality block samples available. Also as noted by Ladd and Foott (1974), the SHANSEP method is difficult to apply to desiccated crusts since it assumes mechanical unloading as the only preconsolidation mechanism. It is interesting to note that the liquidity index of the soil, based on bulk properties, is greater than 1 below 5 m and the sensitivity, based on field vane data, averages 13 below 5 m and yet the SHANSEP and Recompression s_u values are nearly identical.

Boston Blue Clay – Saugus. This test site is located in Saugus, just north of Boston, Massachusetts. The underlying soil stratum at the site consists primarily of Boston Blue Clay (BBC). Similar to the BBC in downtown Boston, this region has been subjected to significant weathering since deposition some 14,000 years before present (Kenney 1964). As a result, a stiff overconsolidated clay crust exists near the surface with a gradual transition to a lightly overconsolidated to normally consolidated clay. Geotechnical investigations at the site consisted of 75 mm fixed piston tube sampling, field vane, and CPTU with u_2 (behind the shoulder) pore pressure measurement. UMass Amherst conducted the laboratory and CPTU tests while the tube sampling and field vane testing were conducted by others for which apparently no effort was made to use an appropriate drilling mud during the field work. Atterberg limits are approximately constant with depth with a liquid limit = 45%, plastic limit = 25%, and plasticity index = 20% plotting above the A-line in a plasticity chart. The water content is less than the liquid limit in the crust and is approximately equal to the liquid limit throughout the rest of the deposit.

Laboratory DSS tests followed the same procedures described above for the CVCC. Figure 10 plots the SQD and σ'_p data from the IL oedometer tests and the σ'_{vo} profile assuming hydrostatic pore water stress conditions. The data show an overconsolidated soil near the top of the deposit and an apparent underconsolidated soil below an elevation of approximately –23 m. However, the geological history of the region (Kenney 1964) indicates the soil should not be underconsolidated and only a slight 1.5 m artesian condition exists in the underlying glacial till (Ladd et al. 1994). The apparent underconsolidation is therefore believed to be a result of sample disturbance causing a significant decrease in the measured σ'_p. This hypothesis is supported by the SQD data which shows that most of the specimens have an SQD of C or worse. Based on the recommendation of Terzaghi et al. (1996), most of these specimens would not produce reliable estimates of σ'_p or Recompression s_u.

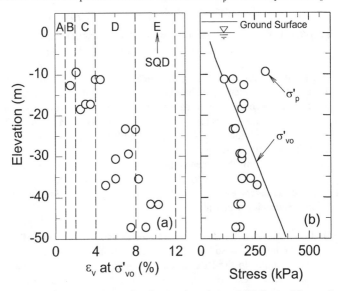

Figure 10. (a) Specimen Quality Designation and (b) Stress History for Boston Blue Clay – Saugus (after Mitchell et al. 1999).

Initial plans for the test program were to conduct Recompression tests but the poor sample quality, discovered during the oedometer test program, precluded extensive use of the Recompression procedure. The DSS SHANSEP test program consisted of tests conducted at OCRs = 1, 2, and 4 resulting in the relationship

$$s_u/\sigma'_{vc} = 0.19(OCR)^{0.75} \tag{5}$$

The next step in the SHANSEP procedure is to compute s_u using $\sigma'_{vc} = \sigma'_{vo}$ and the in situ OCR. This is where the SHANSEP procedure runs into similar problems as the Recompression procedure due to the poor quality samples. In this case, accurate estimates of OCR are not available based on the SQD data and the apparent OCR < 1 below elevation – 23 m (Figure 10). One approach to resolving this is to consider the site geology that infers the soil deposit at depth is likely to be normally to lightly overconsolidated. A *practical* assumption is therefore to assume the soil is normally consolidated starting at a depth where the oedometer σ'_p values are less than σ'_{vo}, i.e. approximately elevation -23 m. An estimate of the OCR profile for the crust can be made using the measured σ'_p data from the better quality shallow samples. Figure 11a plots this interpretation of the laboratory data. This analysis will provide s_u values that are likely to be less than the in situ values.

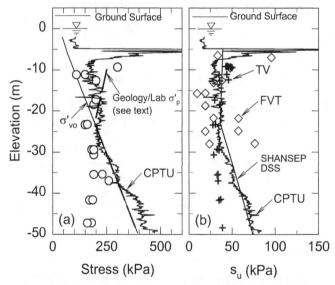

Figure 11. (a) Stress History and **(b)** Undrained Shear Strength for Boston Blue Clay – Saugus Site (after Mitchell et al. 1999).

An additional estimate of the σ'_p profile for the SHANSEP method is available in this case from the CPTU tests conducted at the site. Estimates of σ'_p were computed using the empirical correlation of Leroueil et al. (1995)

$$\sigma'_p = 0.28(q_t - \sigma_{vo}) = 0.28q_{net} \tag{6}$$

where q_t = corrected tip resistance and σ_{vo} = total in situ vertical stress. Figure 11a plots the interpreted CPTU data using Eq. 6. Figure 11b plots the s_u profile using Eq. 5 and the assumed laboratory stress history profile of Figure 11a. The comparison with the field vane (FVT) data is poor, although the field vane data are suspect given the large scatter and unusually low values for some locations – a possible consequence of a lack of proper drilling mud. Also plotted for reference are undrained shear strength values from Torvane tests (TV) conducted on the tube samples at locations adjacent to all the oedometer test specimens. In the deeper section of the deposit the s_u(TV) values correspond to a normalized undrained shear strength $s_u = 0.11\sigma'_{vo}$ which is much less than s_u(DSS) $= 0.19\sigma'_{vc}$. Interpretation of the CPTU data for s_u(DSS) estimates was done using (Lunne et al. 1997)

$$s_u = (q_t - \sigma_{vo})/N_{kt} \tag{7}$$

where N_{kt} = the cone factor. In this case it is the N_{kt} for DSS mode of shear that is of interest for comparison with the laboratory DSS data. Mitchell et al. (1999) analyzed these data and recommended N_{kt}(DSS) = 20 based on N_{kt}(TC) = 12 and $K_s = s_u$(DSS)/s_u(TC) = 0.6 for BBC. The resulting s_u(CPTU) profile for this value of N_{kt} is plotted in Figure 11b for comparison with the s_u(DSS) data. Both the σ'_p and s_u CPTU derived profiles were estimated prior to inspection of the laboratory data.

This case history highlights the difficulty in applying the Recompression or SHANSEP methods with poor quality samples. In the case of the Recompression method direct measurements of s_u were not considered reliable for the deeper sections of the deposit because of excessive sample disturbance. For the SHANSEP method, measurement of the relationship between s_u/σ'_{vc} and OCR was possible. However, accurate laboratory estimates of σ'_p were difficult to obtain and in this case relied on other information including site geology and corroborating evidence from CPTU data. It also emphasizes the fact that s_u at OCR = 1 is an important reference strength for soft clays regardless of the in situ OCR profile. Such a test can be conducted, as per the SHANSEP method, even if only poor quality samples are available.

Recommendations

The case studies serve to complement research findings for soft clay behavior and how such results should be implemented in practice. For laboratory consolidation and strength testing of soft clays to determine engineering design parameters, the following practices are recommended:

- tube sampling should involve sharp edged, large diameter ($\geq$ 75 mm) tubes in a fixed piston sampler and use of appropriately weighted drilling mud;
- for large/important projects, consideration should be given to block sampling;

- all tube samples being considered for selection of test specimens should be x-rayed;
- test specimens should be cut from tubes and debonded from the tube prior to extrusion and trimming;
- it is essential to evaluate sample quality; ε_{vol} or $\Delta e/e_0$ at σ'_{vo} are simple yet effective measures of sample disturbance;
- CRS consolidation should largely replace IL consolidation testing;
- significant advantages in test reliability and productivity can be obtained by using automated stress path equipment for conducting CK_0U triaxial tests;
- highly variable and unreliable s_u data can result from UU testing and therefore test programs should focus on CU testing with K_0 or at least anisotropic consolidation;
- any assessment of s_u for soft clays for design must consider the important effects of sample disturbance, stress history, anisotropy, and rate effects. The Recompression and SHANSEP methods explicitly address rational methods for dealing with these issues;
- when high quality samples are available preference should be given to the Recompression method except for truly normally consolidated clays for which the SHANSEP method should be used;
- the DSS mode of shear gives reasonable estimates of the in situ mobilized undrained shear strength $s_u(mob)$ for stability problems in nonvarved sedimentary clays over a large range of plasticity index;
- the CPTU with behind the tip measurement of pore pressure (u_2) is proving to be a valuable complementary tool to laboratory testing for site characterization.

Conclusions

Experimental research during the past 40 years has led to significant advances in our understanding of soft clay mechanical behavior. Important soil behavior issues that have been identified include sample disturbance, stress history, anisotropy, and rate effects. Many of the advances in our knowledge of soil behavior are important to design and can realistically be implemented in practice. Measurement of consolidation behavior should rely more heavily on constant rate of strain tests and/or the K_0 consolidation phase of consolidated undrained strength tests. Undrained shear strength behavior should rely on use of consolidated undrained triaxial and direct simple shear tests. Consolidation for shear testing should ideally follow a K_0 effective stress path and can become routine with the advent of modern computer controlled stress path systems. The use of triaxial compression/extension and direct simple shear stress systems and recommended rates of shear provide effective means for dealing with soil anisotropy and rate effects for design. The most difficult issue which remains to be dealt with is sample disturbance. It is therefore essential to evaluate sample quality through a combination of radiography and physical measurements such as volume change of specimens during laboratory reconsolidation to the in situ vertical effective stress. Better quality samples can be achieved through the proper use of drilling mud, large diameter – sharp, thin-walled sampling tubes in fixed piston sampling, and either field extrusion of samples or laboratory cutting of sample tubes prior to specimen trimming. The best quality samples can be obtained through block sampling and although this method is more complicated and the equipment is not as readily available, the potential cost benefits for some projects can be significant. The Recompression and SHANSEP methods were developed as a means to handle all of the important soil behavior issues noted above. When properly used, these methods provide a much more rational framework for determining design parameters than conventional unconsolidated undrained laboratory test programs.

Acknowledgements

The writer thanks Professor Charles C. Ladd and Dr. John T. Germaine for teaching him much about the subject of this paper. The writer thanks Professor Serge Leroueil for arranging to take the Laval University Block samples; Tom Lunne of NGI for arranging for collection of the Sherbrooke Block samples, Professor Alan J. Lutenegger, manager of the UMass Amherst NGES site; and the Massachusetts Highway Department for access to the BBC-Saugus site. Rune Dyvik, Charles C. Ladd, and Alan J. Lutenegger provided valuable review comments on the manuscript. Matthew M. Bonus conducted the CVVC tests; Travis J. Mitchell conducted the BBC-Saugus tests; and Robert M. Ryan conducted the CRS and IL consolidation tests. Some of the results presented herein are from research projects supported in part by US National Science Foundation.

References

Andresen, A. and Kolstad, P. (1979). "The NGI 54-mm Samplers for Undisturbed Sampling of Clays and Representative Sampling of Coarser Materials." *Proc. of the Int. Conference on Soil Sampling*, Singapore, 1-9.

ASTM (2000). *Annual Book of Standards, Vol. 4.08, Soil and Rock (I): D420 - D5779*. West Conshohocken, PA, USA.

Baldi, G., Hight, D.W., Thomas, G.E. (1988). "State-of-the-Art: A Reevaluation of Conventional Triaxial Test Methods." *Advanced Triaxial Testing of Soil and Rock*, ASTM STP 977, 219-263.

Baligh, M.M., Azzouz, A.S., Chin, C.T. (1987). "Disturbances Due to Ideal Tube Sampling." *J. of Geotech. Engrg.*, 113(7), 739-757.

Becker, D.E., Crooks, J.H., Been, K. and Jefferies, M.G. (1987). "Work as a Criterion for Determining In Situ and Yield Stresses in Clays." *Can. Geotech. J.*, 24(4), 549-564.

Berre, T. and Bjerrum, L. (1973). "Shear Strength of Normally Consolidated Clays." *Proc., 8th Int. Conf. on Soil Mech. and Found. Engrg.*, Moscow, 1, 39-49.

Bjerrum, L. (1973). "Problems of Soil Mechanics and Construction on Soft Clays." *Proc., 8th Int. Conf. Soil Mech. and Found. Engrg.*, Moscow, 3, 111-159.

Bjerrum, L. and Landva, A. (1966). "Direct Simple Shear Tests on Norwegian Quick Clay." *Geotechnique*, 16(1), 1-20.

Clayton, C.R.I., Siddique, A., and Hopper, R.J. (1998). "Effects of Sampler Design on Tube Sampling Disturbance – Numerical and Analytical Investigations." *Geotechnique*, 48(6), 847-867.

Da Re, G., Santagata, M.C., and Germaine, J.T., (2001). "LVDT Based System for Measurement of the Prefailure Behavior of Geomaterials." *Geotech. Testing J.*, ASTM, 24(3), 288-298.

DeGroot, D.J. (1999). "Laboratory Measurement of Undrained Shear Behaviour of Clays." *Int. Conference on Offshore and Nearshore Geotech. Engrg.*, Panvel, India, 133-140.

DeGroot, D.J., Ladd, C.C. and Germaine, J.T. (1992). *Direct Simple Shear Testing of Cohesive Soils*, Research Report No. R92-18, Center for Scientific Excellence in Offshore Engineering, MIT, Cambridge, MA.

DeGroot, D.J. and Sheahan, T.C. (1995) "Laboratory Methods for Determining Engineering Properties of Overconsolidated Clays." *Transportation Research Record*, No. 1479, 17-25.

Dyvik, R., Lacasse, S., and Martin, R. (1985). "Coefficient of Lateral Stress from Oedometer Cell." *Proc. 11th Int. Conf. on Soil Mech. and Foundation Engrg.*, San Francisco, 2, 1003-1006.

Germaine, J.T. and Ladd C.C. (1988). "State-of-the-Art: Triaxial Testing of Saturated Cohesive Soils." *Advanced Triaxial Testing of Soil and Rock*, ASTM STP 977, ASTM, 421-459.

Hvorslev, M.J. (1949). *Subsurface Exploration and Sampling of Soils for Civil Engineering Purposes*. Waterways Experiment Station, U.S. Army Corps of Engineers, Vicksburg.

Jamiolkowski, M., Ladd, C.C., Germaine, J.T. and Lancellotta, R. (1985). "New Developments in Field and Laboratory Testing of Soils." *Proc., 11th Int. Conference on Soil Mech. and Found. Engrg*, San Francisco, 1, 57-154.

Kenney, T. C. (1964). "Sea-Level Movements and the Geologic Histories of the Post-Glacial Marine Soils at Boston, Nicolet, Ottawa and Oslo." *Geotechnique*, 14(3), 203-230.

Koutsoftas, D.C., and Ladd C.C. (1985). "Design Strengths for an Offshore Clay." *J. of Geotech. Engrg.*, 111(3), 337-355.

Lacasse, S., Andersen, K.H., Hermann, S., Kvalstad, T., Kveldsvik, V., Madshus, C. (2001). "Design, Construction and Maintenance of Infrastructure." Proc. 15th Int. C. Soil Mech. and Found. Engrg., Istanbul, Post Conference Volume, in press.

Lacasse, S. and Berre. T. (1988). "State-of-the-Art: Triaxial Testing Methods for Soils." *Advanced Triaxial Testing of Soil and Rock*, ASTM STP 977, 264-289.

Ladd, C.C. (1991). "Stability Evaluation During Stage Construction." *J. of Geotech. Engrg.*, 117(4), 540-615.

Ladd, C.C. and DeGroot, D.J. (1992). *Guidelines for Geotechnical Experimental Program for Foundation Design of Offshore Arctic Gravity Structures*, Research Report No. R92-33, MIT, Cambridge, MA.

Ladd, C.C. and Edgers, L. (1972). *Consolidated-Undrained Direct Simple Shear Tests on Saturated Clays*. Research Report R72-82, MIT, Cambridge, MA.

Ladd, C.C. and Foott, R. (1974). "New Design Procedure for Stability of Soft Clays." *J. of the Geotech. Engrg. Div.,* 100(GT7), 763-786.

Ladd, C.C. and Foott, R. (1980). "Discussion of 'The Behavior of Embankments of Clay Foundations'", *Can. Geotech. J.,* 17, 454-460.

Ladd, C.C., Foott R., Ishihara K., Schlosser, F. and Poulos, H.G. (1977). "Stress-Deformation and Strength Characteristics: SOA Report." *Proc., 9th Int. Conf. on Soil Mech. and Found. Engrg.,* Tokyo, 2, 421-494.

Ladd, C.C. and Lambe, T.W. (1963). "The Strength of Undisturbed Clay Determined from Undrained Tests." *Symp. on Laboratory Shear Testing of Soils*, ASTM, 342-371.

Ladd, C.C., Young, G.A., Kraemer, S.R., and Burke, D.M. (1999). "Engineering Properties of Boston Blue Clay from Special Testing Program." *Special Geotechnical Testing: Central Artery/Tunnel Project in Boston, Massachusetts*. GSP No. 91., ASCE, 1-24.

Ladd, C.C., Whittle, A.J., and Legaspi, D.E., "Stress-Deformation Behavior of an Embankment on Boston Blue Clay." *Proc., Vertical and Horizontal Deformations of Foundations and Embankments.* ASCE, Vol. 2, 1730-1759.

LaRochelle, P. Sarraih, J., Tavenas, F., Roy, M., and Leroueil, S. (1981). "Causes of Sampling Disturbance and Design of a New Sampler for Sensitive Soils." *Can. Geotech. J.*, 18(1), 52-66.

Lefebvre, G. and Poulin, C. (1979). "A New Method of Sampling in Sensitive Clay." *Can. Geotech. J.*, 16(1), 226-233.

Leroueil, S. (1994). "Compressibility of Clays: Fundamental and Practical Aspects." *Proc., Vertical and Horizontal Deformations of Foundations and Embankments.* ASCE, Vol. 1, 57-76.

Leroueil, S. and Vaughan, P.R. (1990). "The General and Congruent Effects of Structure in Natural Soils and Weak Rocks." *Geotechnique*, 40(3), 467-488.

Leroueil, S., Demers, D., La Rochelle, P., Martel, G., and Virely, D. (1995). "Practical Use of the Piezocone in Eastern Canada Clays." *Int. Symp. on Cone Penetration Testing*, Linkoping, Sweden, 2, 515-522.

Lunne, T., Berre, T. and Strandvik, S. (1997). "Sample Disturbance Effects in Soft Low Plasticity Norwegian Clay." *Proc. of Conference on Recent Developments in Soil and Pavement Mechanics*, Rio de Janeiro, 81-102.

Lunne, T., Robertson, P. K., and Powell, J. J., (1997). *Cone Penetration Testing in Geotechnical Practice*, Blackie Academic & Professional, London.

Lutenegger, A.J. (2000). "National Geotechnical Experimental Site: University of Massachusetts." *National Geotechnical Experimental Sites*, GSP No. 93, ASCE, 102-129.

Mesri, G. (1975). "Discussion of 'New Design Procedure for Stability of Soft Clays." *J. of Geotech. Engrg.*, 101(GT4), 409-412.

Mesri, G., and Feng, T.W. (1992). "Constant Rate of Strain Consolidation Testing of Soft Clays." *Marsal Volume*, Mexico City, 49-59.

Mesri, G. Kwan Lo, D.O., and Feng, T.W. (1994) "Settlement of Embankments on Soft Clays." *Proc., Vertical and Horizontal Deformations of Foundations and Embankments.* ASCE, Vol. 1, 8-56.

Mitchell, T.J., DeGroot, D.J., Lutenegger, A.J., Ernst, H., and McGrath, V. (1999). "Comparison of CPTU and Laboratory Soil Parameters for Bridge Foundation Design on Fine Grained Soils: A Case Study in Massachusetts." *Transportation Research Record*, 1675, 24-31.

Saada, A.S. and Townsend, F.C. (1981). "State-of-the-art: Laboratory Strength Testing of Soils." *Symp. on Laboratory Shear Strength of Soil*, ASTM STP 740, 7-77.

Sandbeakken, G., Berre, T., and Lacasse, S. (1986). "Oedometer Testing at the Norwegian Geotechnical Institute." *Consolidation of Soils: Testing and Evaluation*, ASTM STP 892, 329-353.

Sheahan, TC., and Germaine, J.T. (1992). "Computer Automation of Conventional Triaxial Equipment." *Geotech. Test. J.*, ASTM, 15(4), 311-322.

Sheahan, T.C., DeGroot, D.J., and Germaine, J.T. (1994). "Using Device-Specific Data Acquisition for Automated Laboratory Testing." *Innovations in Instrumentation and Data Acquisition Systems*. Transportation Research Record 1432, 9-17.

Sheahan, T.C., Ladd, C.C., and Germaine, J.T. (1996). "Rate-Dependent Undrained Shear Behavior of Saturated Clay." *J. of Geotech. Engrg.*, 122(2), 99-108.

Shibuya, S., Mitachi, T. and Hwang, S.C. (2000). "Case Studies of In-Situ Structure of Natural Sedimentary Clays." *Soils and Foundations*, 40(3), 87-100.

Tavenas, F. and Leroueil, S. (1980). "The Behavior of Embankments on Clay Foundations." *Can. Geotech. J.*, 17, 236-260.

Terzaghi, K., Peck, R.B., and Mesri, G. (1996). *Soil Mechanics in Engineering Practice*. John Wiley and Sons, New York.

Wissa, A.E.Z., Christian, J.T., Davis, E.H. and Heiberg, S. (1971). "Consolidation at Constant Rate of Strain." *J. of the Soil Mech. and Found. Div.*, 97(SM10), 1393-1413.

Evaluation of Relative Density and Shear Strength of Sands from CPT and DMT

M. Jamiolkowski[1], D. C. F. Lo Presti[2] and M. Manassero[3]

Abstract

The paper summarises the experience gained by the writers in the interpretation of the cone penetration test (CPT) and flat dilatometer test (DMT) for the assessment of the geotechnical properties of sands. In the first part of the paper, the problem of determining the relative density (D_R) as function of the penetration test results and ambient stress (σ'), for silica sands, is dealt with. In the second part of the paper, the assessment of the peak angle of shearing resistance (φ'_p) is dealt with. The attention is given to the use of the Bolton's (1986) strength-dilatancy theory in order to estimate φ'_p. Engineering correlations, based on Bolton's (1986) work, are proposed allowing estimation of φ'_p as function of penetration resistance and σ', taking into account the compressibility and the curvilinear shear strength envelope.

Introduction

The concept of relative density (D_R) suggested by Burmister (1948), despite its intrinsic uncertainties and limitations [Tavenas and La Rochelle (1972), Tavenas (1972), Achintya and Tang (1979)], is still extensively used in geotechnical engineering as an index of the mechanical properties of coarse grained soils. Because of the well-known difficulties and the high costs in retrieving good quality undisturbed samples from sand and gravel deposits [Yoshimi et al. (1978), Hatanaka et al. (1988), Goto et al. (1992), Yoshimi (2000)], geotechnical engineers need to

[1] Professor, Department of Structural and Geotechnical Engineering, Politecnico di Torino, Corso Duca degli Abruzzi 24, 10129 Torino Italy; phone +39 011 5644840, sgi_jamiolkowski@studio-geotecnico.it
[2] Associate Professor, Department of Structural and Geotechnical Engineering, Politecnico di Torino, Corso Duca degli Abruzzi 24, 10129 Torino Italy; phone +39 011 5644842, diego.lopresti@polito.it
[3] Associate Professor, Department of Geoenvironmental Engineering, Politecnico di Torino, Corso Duca degli Abruzzi 24, 10129 Torino Italy; phone +39 011 5647705, manassero@polito.it

estimate the in situ D_R using empirical correlations between this parameter and penetration test results. This indirect way of evaluating D_R adds further uncertainties to those already faced when determining the relative density in the laboratory. The first attempt to correlate the blow-count of the Standard Penetration Tests (N_{SPT}) to the density of sands is linked to the works by Terzaghi and Peck (1948) and Gibbs and Holtz (1957). The continuous interest in this kind of correlation is testified by the more recent works by Skempton (1986) and Cubrinovski and Ishihara (1999). As far as the CPT is concerned, a pioneering work can be dated back to Schmertmann (1976). He presented the first comprehensive correlation between the cone resistance (q_c) and D_R on the basis of static Cone Penetration Tests (CPT) performed in Calibration Chambers (CC). Such correlation relates D_R to the effective overburden stress (σ'_{vo}) and is applicable to normally consolidated (NC) fine to medium unaged sands. Twenty five years later, based on the results of 484 CC-CPT's performed in three silica sands, the writers have attempted to present similar correlations considering the effect of CC size on the measured q_c and giving appropriate consideration to mechanically overconsolidated (OC) sands. The assessment of D_R represents the most common intermediate step in estimating the stress-strain-strength characteristics of sands and gravels.

    In the second part of the paper, we will deal with the assessment of the peak friction angle φ'_p of sands, making reference to the simplified Bolton's (1986) strength-dilatancy theory. Based on a theoretically sound framework of Rowe (1962), the input parameters required to estimate φ'_p are: D_R, the friction angle φ'_{cv} at critical state and a parameter Q related to sand compressibility. The presentation also includes a comprehensive discussion about the intrinsic parameters φ'_{cv}, Q.

Evaluation of relative density

The first attempt to correlate the penetration resistance of the cone penetration tests (q_c) to the density of sands dates back to the work by Schmertmann (1976). He presented the first comprehensive correlation between q_c and relative density (D_R) on the basis of CPTs performed in the Calibration Chamber (CC). Such a correlation is applicable to normally consolidated (NC) fine to medium, unaged, clean sands. Schmertmann (1976) suggested a correlation between the cone resistance (q_c) the relative density and the vertical effective stress (σ'_{vo}), using the results of CPT's performed on sands in the CC of the University of Florida. The analytical expression takes the form:

$$q_c = C_o \cdot \left(\sigma'_{vo}\right)^{C_1} \cdot \exp(C_2 \cdot D_R) \tag{1}$$

$$D_R = \frac{1}{C_2} \ln\left[\frac{q_c \cdot \left(\sigma'_{vo}\right)^{C_1}}{C_o}\right] \tag{2}$$

where: C_o, C_1, C_2 = empirical correlation factors

Since the pioneering work by Schmertmann, many CC's have been put into operation in North America, Europe Australia and Japan generating a large data-base of CPTs performed in different sands and providing a deeper insight into the merits and limitations of this kind of large-scale laboratory test and of the empirical correlations that can be obtained. The key points that have emerged from these experiments can be summarised as follows:

- The analysis of the variance (Tumay 1976), performed to investigate the relative importance of the different factors influencing the magnitude of the q_c of silica sands measured in CC tests, led to the conclusion that the relative density (D_R) and the consolidation stress tensor (i.e. the level of effective stress existing in the specimen, prior penetration) are the most important variables that influence q_c. (Harman 1976, Schmertmann 1976, Garizio 1997).

- The correlation of q_c vs. D_R and σ'_{vo} holds only for NC sands. A correlation for NC and OC deposits should refer to the effective mean in situ stress σ'_{mo} instead of σ'_{vo}.

- Stress and strain history that can be reproduced in the laboratory play a secondary role with the exception of increase of the horizontal effective consolidation stress as result of the mechanical overconsolidation which concurs to the value of the relevant stress tensor (Jamiolkowski et al. 1988).

- In the case of siliceous sands, their grain shape and crushability play a secondary role (Robertson & Campanella 1983, Lunne et al. 1997). The influence of grading on the penetration resistance has not been systematically investigated. However, the use of the correlations obtained from CC experiments leads to underestimation of D_R in the case of sand deposits containing more than 5 to 10 % of fines (Jamiolkowski et al. 1988)

- Thanks to the works by Dussealt and Morgenstern (1979) and Barton and Palmer (1989) which have investigated the effect of geological time on porosity, fabric and mechanical properties of coarse grained soil deposits, it is obvious that the empirical correlations based on the results of tests performed on laboratory reconstituted specimens, are applicable only in the case of young, unaged NC soils. Skempton (1986) has shown a certain influence of aging on the correlations between D_R and the blow-count of the Standard Penetration Test (N_{SPT}). Analogously, it is reasonable to suppose that aging influences the D_R vs. q_c correlations (see also Wride et al. 2000).

- Due to the finite dimensions of the CC, the measured cone resistance is affected by an error in comparison to that obtainable in the case of an infinite sand deposit with the same relative density. This phenomenon, named chamber size effect (Schmertmann 1976, Parkin and Lunne 1982, Baldi et al. 1986, Foray 1986, Mayne and Kulhawy 1991, Tanizawa 1992, Salgado 1993) leads, within some boundary conditions, to an underestimate or overestimate of the field q_c depending on the boundary conditions imposed on the CC specimen during the cone penetration. The magnitude of such an underestimation or overestimation depends on the crushability and compressibility of the test sand, the ratio of the CC specimen diameter (D_c)

to that of the cone (d_c), D_R and confining stresses applied to the CC specimen.
- The degree of saturation and boundary conditions imposed on the CC specimen during the cone penetration are much less influential on the q_c.

The previously illustrated considerations also apply to the penetration resistance (q_D) as obtained from blade thrust readings in the dilatometer tests (DMT). It is worthwhile to point out that this kind of measurement is not routinely performed in dilatometer tests (Marchetti 1980, 1997).

In light of what has been stated above, the writers propose empirical correlations, similar to that used by Schmertmann (1976) and based on the results of 484 CC-CPT's and 136 CC-DMT's that have been performed in three silica sands. As far as DMT's are concerned, correlations between the lateral stress index K_D and D_R are also shown. The practical use of the proposed correlations is the prediction of the relative density of granular deposits. This task can be accomplished by using the proposed correlations and keeping in mind the intrinsic limitations of such correlations as already discussed.

Experimental Data. The CPTs and DMTs have been performed in Calibration Chambers of ENEL of Milan and the research institute ISMES of Bergamo. The apparatus houses 1.2 m in diameter and 1.5 m in height specimen reconstituted by means of pluvial deposition in air (Bellotti et al. 1982, 1988, Garizio 1997, Felice 1997). After deposition, samples were subject to the one-dimensional compression in order to apply the desired consolidation stress level and stress-history. After the consolidation stage, the penetration test (CPT or DMT) was performed, applying to the CC specimen one of the four available boundary conditions (BC).
Most of the tests were performed under two BC's:
- BC-1: constant axial (σ_a') and radial (σ_r') effective stresses;
- BC-3: constant axial effective stresses (σ_a') and zero radial strain (ε_r);

In addition, a limited number of CC tests were carried out using either BC-2 (axial strain $\varepsilon_a =0$, $\sigma_r' =$ constant), or BC-4 $(\varepsilon_a = \varepsilon_r = 0)$ during the penetration stage. Table 1 indicates the percentage of tests performed under each BC's.

All CPT's were performed using the cylindrical Fugro-type electrical cone tips. In most tests, the standard cone tip 35.6 mm in diameter (Lunne et al. 1997) has been employed. A limited number of tests were also performed using cone tips

Table 1. Percentage of CPT's and DMT's performed under different boundary conditions

Test	BC-1 (%)	BC-2 (%)	BC-3 (%)	BC-4 (%)
CPT's (total 484 tests)	66	11	20	3
DMT's (total 136 tests)	86	0	11	1

having diameters (d_c) equal to 25.4, 20, 11 and 10 mm. These tests were aimed at investigating the influence of the CC diameter (D_c) to d_c ratio (R_d) on the q_c measured under different BC's.

Most of the DMTs were performed using a standard dilatometer (Marchetti 1980). The probe is 14 mm thick, 95 mm wide and 220 mm high. An expandable steel membrane, 60 mm in diameter, is located on one side of the probe; a load cell for the measurement of the penetration resistance (q_D) is located just above the probe. Few tests were performed using a research dilatometer (RDMT) (Fretti et al. 1992, 1996, Bellotti et al. 1997). The main differences between standard dilatometer and RDMT are: i) the expandable steel membrane of RDMT is equipped with strain gauges, so that it is possible to monitor the complete expansion curve, ii) the structure of the RDMT probe is much stiffer in comparison of the standard DMT, even though the dimensions of the two probes are identical.

The CC tests were carried out in three well-known silica sands: fine to medium Ticino (TS), Toyoura (TOS) and Hokksund (HS) sands. The index properties of these sands are reported in Table 2 and Figure 1.

Data Interpretation

Size effect. The tests run with different cone sizes confirmed the well-known fact that the penetration resistance measured in the CC is influenced by the imposed

Table 2. Index Properties of Test Sands

PARAMETER		TICINO	HOKKSUND	TOYOURA
γ_{max}	[kN/m³]	16.67	17.24	16.13
γ_{min}	[kN/m³]	13.64	14.10	13.09
G_S	[-]	2.68	2.72	2.65
U_c	[-]	1.30	1.91	1.31
D_{50}	[mm]	0.60	0.45	0.22
Quartz	[%]	30	35	90
Feldspar	[%]	65	55	8
Mica	[%]	~5	~10	~3
ϕ'_{cv}	[°]	33	34	32

γ_{max} and γ_{min}	= maximum and minimum dry density respectively
G_S	= specific density
U_c	= uniformity coefficient
D_{50}	= mean grain size
ϕ'_{cv}	= friction angle at critical state

SOIL BEHAVIOR AND

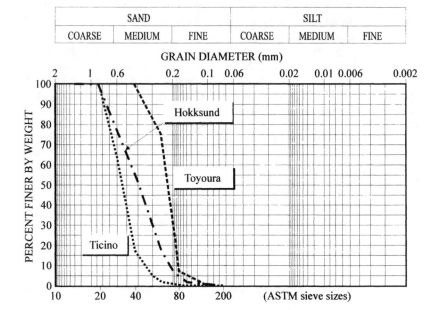

Figure 1. Grain size distribution of three test sands

BC's. Such effect is inversely proportional to the R_d and decreases with increasing the sand compressibility. Further details on such effect can be found in the already mentioned works by Schmertmann (1976), Parkin and Lunne (1982), Baldi et al. (1986), Foray (1986), Mayne and Kulhawy (1991), Tanizawa (1992), Salgado (1993), Salgado et al. (1998). The measured penetration resistance appears to be independent on the BC's when the R_d is sufficiently large, i.e. 70 to 100 for silica sands of moderate to low compressibility respectively. Under these conditions the penetration resistance measured in the CC matches the field value. The measured penetration resistance were therefore corrected for chamber size effect by means of the following empirical equation (Tanizawa 1992, Garizio 1997, Felice 1997):

$$CF = a[(D_R)^b]^m \quad [-] \tag{3}$$

where:
CF = correction factor by which the measured penetration resistance has to be multiplied
a, b = empirical coefficients function of R_d inferred from the CC performed in TS and TOS using CPT tips having different size
m = + 1 and –1 for BC-1 and BC-3 respectively.

The values of coefficients a and b for different R_d ratios are given in Table 3. The lower bound of $D_R = (D_R)_{min}$ below which CF=1 is also reported in Table 3. The trend of the empirical CF, yielded by Eq. 3, vs. R_d is similar, although generally lower, to what achieved by Salgado (1993) from numerical modelling.

The measured q_D have not been corrected to account for CC size effect. Based on limited CC evidence, it appears that there is no need to correct the penetration resistance of a plate blade because the experimentally determined q_D is not influenced by the finite dimensions of the chamber (Felice 1997), even though the reasons are not well understood.

Table 3. Coefficient a and b of Eq. 3

$R_d = \dfrac{D_c}{d_c}$	A	B	$(D_R)_{min}$ %
100	0	0	100
60	0.412	0.221	55.8
47.2	0.166	0.457	50.8
33.6	0.090	0.624	47.4
22.1([4])	0.054	0.827	34.1
$(D_R)_{min}$= D_R value in percent at which CF should be taken equal to one			

Proposed correlations. The writers adopted the following equations to fit the experimental data:

1) the same equation used by Schmertmann (1976)

$$q_c = C_o p_a \left(\frac{\sigma'}{p_a}\right)^{C_1} \exp(C_2 D_R) \qquad (4)$$

from which is it possible to obtain:

$$D_R = \frac{1}{C_2} \ln\left[\frac{q_c / p_a}{C_o (\sigma' / p_a)^{C_1}}\right] \qquad (5)$$

where:

q_c = measured cone resistance multiplied by CF of Eq. 3

σ' = an initial effective geostatic stress component or stress invariant $\left[FL^{-2}\right]$

D_R = relative density (as decimal)

([4])   tests performed in a smaller CC in Japan by Tanizawa (1992) in TOS

C_o, C_1, C_2 = non dimensional empirical correlation factors, see Table 4 for CPT's

p_a = atmospheric pressure expressed in the same unit system of stress and penetration resistance (i.e. 98.1 kPa or 1 bar etc.)

2) the equation proposed by Lancellotta (1983)

$$D_R = A_o + B_o \cdot X \tag{6}$$

where:

$$X = \ln\left[\frac{q_c}{(\sigma'_{vo})^\alpha}\right]$$

A_o, B_o and α = empirical correlation factors (see Table 5). The parameter α is obtained following an optimisation process which minimises the differences between computed and measured values of the penetration resistance in terms of standard deviation. In this case q_c and σ'_{vo} are in kPa.

The same equations have been used in the case of DMT's. The empirical correlation factors obtained from DMT results are reported in Tables 5 & 6.

As to the definition of the effective stress σ' to be introduced into Eqs. 4 and 5, the following should be taken into account:

- Zolkov and Weisman (1965) postulated that N_{SPT} is controlled by the horizontal in situ effective stress (σ'_{ho}).
- Similar experimental evidence emerged from CPT's performed in CC's (Schmertmann 1971, 1972, Baldi et al. 1986, Houlsby and Hitchman 1988, Mayne and Kulhawy 1991, Salgado 1993, Garizio 1997, Felice 1997). It shows that the magnitude of the penetration resistance is much more influenced by σ'_r than by σ'_a.
- The above statement suggests that any rational correlation between penetration resistance and relative density should be related to the mean (σ'_{mo}) or horizontal (σ'_{ho}) effective geostatic stresses rather than to the (σ'_{vo}). The lesson learned is that the correlation between the penetration resistance and relative density involving σ'_{vo} is applicable only to NC deposits of coarse grained soils in which K_o ranges from 0.4 to 0.5 remaining more or less constant with depth.

Given the above considerations, Figures 2 and 3 report the $D_R = f(q_c, \sigma')$ correlations for NC (Fig. 2) and (NC+OC) (Fig. 3) dry silica sands respectively. As to the latter, the writers fully appreciate the extreme difficulties linked with the estimation of σ'_{ho} in sand and gravel deposits. Overall, but particularly for coarse-grained soils, the determination of σ'_{ho} or of the coefficient of the earth pressure at rest (K_o) in situ is still an unsolved problem in geotechnical engineering.

Table 4. Coefficients C_0, C_1 and C_2 of Eqs. 4 and 5 (CPT's)

$$q_c = C_0 p_a \left(\frac{\sigma'}{p_a} \right)^{C_1} \exp(C_2 D_R) \qquad D_R = \frac{1}{C_2} \ln \left[\frac{q_c / p_a}{C_0 \left(\sigma' / p_a \right)^{C_1}} \right]$$

$\sigma' = \sigma'_{vo}$	TS	TS + TOS + HS
C_0	17.74	17.68
C_1	0.55	0.50
C_2	2.90	3.10
R	0.90	0.89
$\overline{\sigma}$	0.12	0.10
N	305	180

$\sigma' = \sigma'_{mo}$	TS	TS + TOS + HS
C_0	23.19	24.94
C_1	0.56	0.46
C_2	2.97	2.96
R	0.87	0.87
$\overline{\sigma}$	0.10	0.10
N	299	484
R = correlation coefficient		
$\overline{\sigma}$ = standard error		
N = number of CC tests considered		

Table 5. Coefficients A,B and α of eq. 6 (CPT's & DMT's)

$$D_R = A_0 + B_0 \cdot X \qquad X = \ln \left[\frac{q_c}{(\sigma'_{vo})^{\alpha}} \right]$$

CC test	A_0	B_0	α	R	$\overline{\sigma}$	N
CPT's	-1.292	0.268	0.52	0.94	7.9	456
DMT's	-1.082	0.204	0.36	0.92	6.6	100

R= Correlation coefficient

$\overline{\sigma}$ = standard error

N = number of CC test considered

Table 6. Coefficients C_0, C_1 and C_2 of Eqs. 4 and 5 (DMT's)

$$q_c = C_0 P_a \left(\frac{\sigma'}{P_a}\right)^{C_1} \exp(C_2 D_R) \qquad D_R = \frac{1}{C_2} \ln\left[\frac{q_c / P_a}{C_0 \left(\sigma' / P_a\right)^{C_1}}\right]$$

$\sigma' = \sigma'_{vo}$	TS	TS + TOS + HS
C_0	19.14	20.64
C_1	0.62	0.52
C_2	3.61	3.71
R	0.88	0.88
$\overline{\sigma}$	0.11	0.10
N	57	69

$\sigma' = \sigma'_{mo}$	TS	TS + TOS + HS
C_0	26.99	26.62
C_1	0.60	0.49
C_2	3.75	3.80
R	0.91	0.89
$\overline{\sigma}$	0.12	0.11
N	110	136
R = correlation coefficient		
$\overline{\sigma}$ = standard error		
N = number of CC tests considered		

Nevertheless, the estimate of σ'_{mo} in coarse-grained soils is facilitated by the following considerations:

- In NC deposits the upper limit of K_o can be taken as $1 - \sin\varphi'_{cv}$ (i.e. the angle of shearing resistance at critical state).
- In heavily OC sands (i.e. OCR = 15) K_o is not greater than 1.0 as suggested by the CC results (Jamiolkowski et al. 1988).

Figures 4 & 5 show Eq. 6 in the case of CPT's and DMT's respectively. Figures 4 & 5 enable one to appreciate the accuracy of Eq. 6 to fit the experimental data that are also plotted in the Figures together with the limits corresponding to $\pm 2\overline{\sigma}$.

Empirical correlations were also established between the lateral stress index (K_D) and D_R. K_D is computed from DMT results in the following way:

$$K_D = \frac{p_o - u_o}{\sigma'_{vo}} \qquad (7)$$

where:

p_o = lift-off pressure

u_o = pore pressure, prior to penetration and expansion

σ'_{vo} = vertical effective stress, prior to penetration and expansion

The dependence of K_D on D_R is clearly shown in Figure 6 for the three considered sands. Two different equations were used to fit the experimental data:

$$K_D = C_o(\sigma')^{C_1} p_a^{1-C_1} \exp(C_2 D_R) \qquad (8)$$

$$K_D = A \exp(B D_R) \qquad (9)$$

Eq. 8 is similar to Eq. 4 with the only difference that, instead of considering the penetration resistance, the lateral stress index appears in the formula. Eq. 9 is much more crude and does not take into account the stress level prior to penetration and expansion (σ'). Moreover, D_R is expressed as a fraction of one.

The accuracy of Eq. 8 is not influenced by the choice of the effective stress.

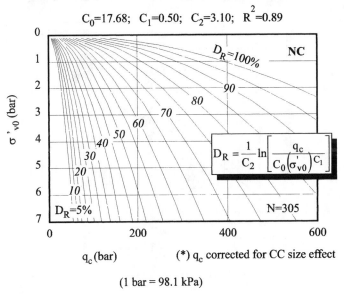

$C_0 = 17.68$; $C_1 = 0.50$; $C_2 = 3.10$; $R^2 = 0.89$

$$D_R = \frac{1}{C_2} \ln\left[\frac{q_c}{C_0\left(\sigma'_{v0}\right)^{C_1}} \right]$$

(*) q_c corrected for CC size effect

(1 bar = 98.1 kPa)

Figure 2. Relative density of NC siliceous sands

$C_0=24.94$; $C_1=0.46$; $C_2=2.96$; $R^2=0.87$

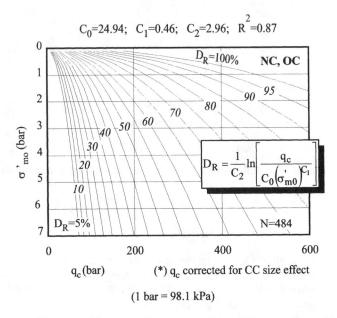

q_c (bar)   (*) q_c corrected for CC size effect

(1 bar = 98.1 kPa)

Figure 3. Relative density of NC and OC siliceous sands

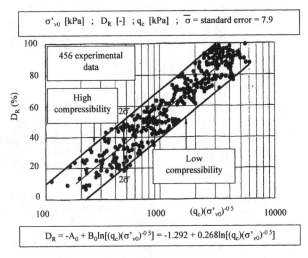

(*)q_c : corrected values for taking into account the CC size

Figure 4. Experimental correlation D_R-q_c-σ'_{v0} for mainly NC sands of different
compressibility (Lancellotta, 1983; Garizio, 1997)

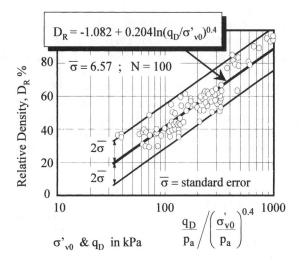

Figure 5. Relation between wedge resistance and relative density –
Calibration chamber tests in Ticino sands (Felice, 1997)

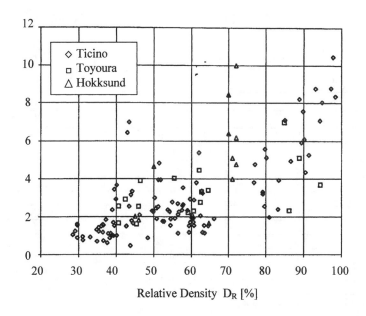

Figure 6. K_D vs D_R from CC-DMT's in three siliceous sands

In Table 7 the correlation coefficients that have been obtained for the case $\sigma' = \sigma'_{vo}$ (NC tests) are reported. It is worthwhile to point out that the stress exponent assumes negative values, irrespective of the selected effective stress. It is also important to notice that the use of a simpler formula. like Eq. 9 does not involve relevant reduction in accuracy. In any case, the correlations between K_D and D_R are less accurate than those between q_c (or q_D) and D_R. Figure 7 shows Eq. 9 $\pm 2\overline{\sigma}$ and the experimental data. The empirical correlation factors used in Figure 7 are slightly different from those reported in Table 7 because they have been obtained disregarding those test results with a deviation from the computed value higher than $\pm 2\overline{\sigma}$.

Degree of saturation. All CPT's and DMT's used to derive the previously shown correlations were carried out on dry specimens. Only a limited number of tests were performed in saturated TS (Bellotti et al. 1988) showing little influence of the saturation on the measured penetration resistance. Strictly speaking, these correlations are applicable to dry fine to medium clean, unaged, uncemented silica sands of low to moderate compressibility in which the static cone penetration process corresponds essentially to a drained process.

In order to overcome, at least partially, the limitations derived from the above specified condition the following indications can be helpful:

Table 7. Coefficients C_0, C_1, C_2 A,B eqs.8 and 9 (DMT's)

$$K_D = C_0 (\sigma')^{C_1} p_a^{1-C_1} \exp(C_2 D_R) \qquad K_D = A \exp(B D_R)$$

Eq. 8 $\sigma' = \sigma'_{vo}$	TS	TS+TOS+HS
C_0	$5.3 \cdot 10^{-3}$	$6.6 \cdot 10^{-3}$
C_1	-0.18	-0.25
C_2	2.60	2.29
R	0.78	0.76
$\overline{\sigma}$	12	12
N	58	73
Eq. 9	NC	NC+OC
A	0.53	0.57
B	2.42	2.56
R	0.71	0.71
$\overline{\sigma}$	13	13
N	73	136
R= Correlation coefficient $\overline{\sigma}$ = standard error N = number of CC test considered		

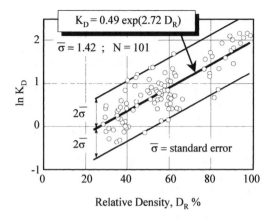

Figure 7. Relation between lateral stress index and relative density – Calibration chamber tests in Ticino sands (Felice, 1997)

-   The comparison between q_c(dry) and q_c(saturated) resulting from CC-CPT's carried out on almost identical CC specimens of TS shows very small differences. Similar conclusions have been reached by Schmertmann (1976) comparing CC tests performed on dry and nearly saturated quartz Ottawa sand. It is worthwhile to point out that the use of the previously shown correlations in saturated sands leads to an underestimation of D_R, whose magnitude can be inferred from the following empirical relationship:

$$\frac{D_R(\text{saturated}) - D_R(\text{dry})}{D_R(\text{dry})}100 = -1.87 + 2.32\ln\frac{q_c}{\left(\sigma'_{vo}p_a\right)^{0.5}} \qquad (10)$$

The above exposed formula becomes meaningless for $q_c/(\sigma'_{vo}p_a)^{0.5} \leq 2.24$; underestimation of D_R, for the sands considered in this paper, ranges between 7 and 10 %.

Aging and Cementation. As far as the effects of cementation and ageing on the penetration resistance are concerned, currently there is a lack of information able to estimate and quantify their influence. Schmertmann (1991) has shown that the accumulation of secondary compression in sands tends to moderately increase the cone resistance, see for example Kulhawy and Mayne (1990).

More relevant might be the impact of even light cementation on q_c (Puppala 1993, Puppala et al. 1995, Eslaamizzad 1997). It may be useful to mention that the use of correlations similar to those here presented and established on freshly deposited sands, in aged and/or cemented deposits leads to an overestimation of the relative density.

Evaluation of shear strength

When dealing with the shear strength of non-cemented granular materials, the friction angle, resulting from the secant slope of the failure envelope, is, in general, the reference parameter
for both the simplified design approaches (e.g. limit equilibrium and limit analysis methods) and the most complex multi-surface non-linear elasto-plastic work-hardening models.

The appropriate definition of the peak friction angle φ_p and the operational friction angle φ_{op}[5] referring to the simplified design approaches, appears to be even more difficult than when more complex models of soil behaviour are used. As a matter of fact, the operational value of φ_{op}, for a given boundary value problem, is a function, among the other state parameters, of all the components of stress and strain tensors which can be reliably assessed only through sophisticated theoretical approaches. On the other hand, using simplified design methods, the most appropriate operational value of φ_{op}, should be theoretically evaluated with reference to the average values of the significant state parameters within the yielding volume of soil. This kind of evaluation, using simplified design methods, is only possible in an approximate manner and for a limited number of the recurrent boundary values problems in Soil Mechanics.

In light of the previous considerations, in the following part of the paper, the basic principles governing the shear strength of sandy soils, with particular reference to the framework and the relationships proposed by Bolton (1986), are illustrated and partially worked out to widen the possible practical applications when simplified models of soil behaviour are adopted. In particular, an evaluation procedure to estimate the operational friction angle from CPT is illustrated. It allows the evaluation of operational friction angles of sands having different mineralogical composition and/or grain size distribution, once the point resistance q_c and the mean geostatic effective stress σ'_{mo} are known.

Basic principles governing shear strength of sandy soils. The basic principles that govern the shear strength of granular materials (i.e. the critical state concept, the energy dissipation by particle rearrangement, the dilatancy and the dependency of the latter on the current state parameters) and the influence of secondary factors such as strain conditions and anisotropy, have already been clearly pointed out by Casagrande (1936), Taylor (1948), Rowe (1962), Schofield & Wroth (1968), Bolton (1986), Mitchell (1976), Lade & Lee (1976), Ladd et al. (1977) and Tatsuoka et al. (1986). Looking at the framework and the related equations proposed by the aforementioned researchers, the main components of shear strength of sands can be split as follows:
-   The pure frictional resistance between smooth surfaces of the sand mineral, quantified by the interparticle friction angle φ_μ;
-   The particle rearrangement component determining the strength increase from φ_μ to the constant volume friction angle φ_{cv};

---

(5) Average mobilized friction angle in correspondence of the general failure for current geotechnical boundary value problems.

The dilation component determining the difference between the peak strength (represented by the peak friction angle φ_p) and the steady state strength corresponding to φ_{cv}.

Based on the results of triaxial (TX) and plane strain (PS) compression tests obtained for different sands, Bolton (1986) attempted an empirical correlation to assess the peak friction angle that takes into account the relative density (D_R), the mean effective stress at failure σ'_{mf}, and the sand type in terms of grain size distribution, mineralogy and grain shape.

The equation originally proposed by Bolton (1986) can be written as follows:

$$\varphi_p - \varphi_{cv} = m\{D_R[Q - \ell n(\sigma'_{mf})] - R\} \qquad \varphi_p \geq \varphi_{cv} \qquad (11)$$

where: φ_p = peak friction angle in [°]; φ_{cv} = constant volume friction angle [°]; m = coefficient equal to 3 or 5 for axisymmetric (TX) and plane strain (PS) conditions respectively; D_R = relative density; Q = particle strength parameter (reported in Table 8); σ'_{mf} = mean effective stress at failure in [kPa]; R = coefficient that in a first approximation is a function of (φ_{cv}-φ_μ) and normally it is assumed equal to 1 for sands; φ_μ = pure friction angle between smooth surfaces of the mineral forming the considered sand.

Table 8. Q values suggested by Bolton (1986)

GRAIN MINERAL	Q
Quartz and feldspar	10
Limestone	8
Anthracite	7
Chalk	5.5

Looking in detail at Eq. 11 it can be remarked (see also Fig. 8) that it represents a bunch of straight-lines in the plane φ_p; $ln(\sigma'_{mf})$ converging in the point with coordinates $ln(\sigma'_{mf})$=Q and φ_p= (φ_{cv}-m·R), moreover, the angular coefficient of the straight lines is equal to (m·D_R).

The common point of the aforementioned straight-lines is quite peculiar since its coordinates express some intrinsic features of the considered sand not depending on the state parameters (e.g. the typical combinations of D_R and σ'_{mf}).

In first approximation, the term (m·R) can represent the shear strength component due to the grains rearrangement at constant volume strains (Rowe, 1962), so it can be expressed by the difference between φ_{cv} and φ_μ.

The term Q can be expressed by the logarithmic function of an appropriate equivalent grain yield stress (σ'_c) that can be related to the grain crushing strength for coarse materials (p'_{fm}) as defined by Biliam (1967) and Marsal (1967) or can be

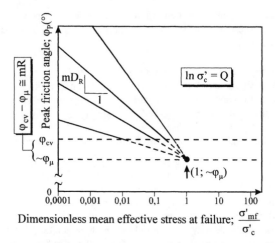

Figure 8. Re-plotting of Bolton's (1986) relationships referring to pure friction angle (φ_μ) and dimensionless confining stress

referred to the threshold confining stress level at which, for a given sand, a single value of void ratio is obtained independently from its initial relative density, fabric and arrangement of the solid skeleton (see Figures 9 and 10).

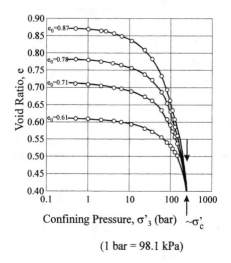

(1 bar = 98.1 kPa)

Figure 9. Isotropic compression tests on Sacramento river Sand (from Lee and Seed, 1967)

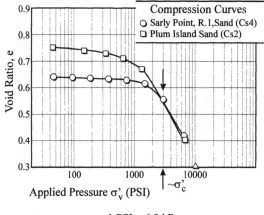

1 PSI = 6.9 kPa

Figure 10. K_0-compression tests for two similar sands (from Roberts, 1964)

In order to justify this simple model from a mechanical point of view it is possible to imagine that, when the isotropic stress level is able to destroy the internal strength of the sand grains, the shear stress increments can be only sustained by the pure friction mobilized between smooth surfaces of the mineral forming the grains, then without any contribution from the particle rearrangement and, of course, from the dilatancy.

Based on the above considerations, Eq. 11 can be rewritten, referring explicitly to the intrinsic and state parameters influencing the peak friction angle φ_p (see also Fig. 8):

$$\varphi_p - \varphi_\mu = m \cdot D_R \cdot \ell n(\sigma'_c / \sigma'_{mf}) \qquad \varphi_p \geq \varphi_\mu + m \qquad (12)$$

With reference to the silica sands considered by Bolton (1986), in order to obtain the same results of Eq. 11, it is necessary to introduce the following parameters into Eq. 12:

$\sigma'_c = 22026.5$ (kPa);

$\varphi_{cv} - \varphi_\mu = 3°$ for axisymmetric conditions (TX) and

$\varphi_{cv} - \varphi_\mu = 5°$ for plane strain conditions (PS).

As to the value of σ'_c, it is in good agreement with the expected values observed and/or extrapolated from the isotropic compression tests of Lee & Seed (1967) and Robertson (1964) reported in Figs. 9 and 10.

As far as the constant volume friction angle φ_{cv} is concerned, the reinterpretation of the fitting parameters by Bolton (1986), under the light of the proposed approach (Eq. 12), points out possible different values from TX and PS conditions.

Although many authors, e.g.: Hanna et al. (1987); Schanz (1998), have produced the experimental evidence that φ_{cv}(TX) and φ_{cv} (PS) are the same, the uncertainties linked to the large strain and strain non-uniformities still leave some open questions with respect to this problem..

Apart the practical problems of test equipment, other sources of uncertainties, that can influence the φ_{cv} values, might also arise from the fitting procedures of the experimental data carried out by Bolton (1986). Therefore, considering all the above aspects, the difference of 2° between φ_{cv} values from TX and PS conditions could be easily justified and accepted, also considering the empirical nature of Eq. 12.

In order to validate the proposed modification to the original Bolton formula, it can be interesting to note that Eq. 12 is very similar to the one proposed by Barton (1973) to describe the curved shear strength envelope of rock joints. Moreover, the equation by Baligh (1975, 1976), describing the curvature of the sands failure envelope at a given relative density, can also be re-written in the form of the above equations.

Of course the proposed simplification of the Bolton's (1986) formula, based on the use of the intrinsic parameters (φ_μ, σ'_c) characterizing the coarse granular media behaviour, must be validated by further experimental data. Nevertheless, it can be used as a reference framework for analysing the basic contributions to the shear strength of sandy soils.

Beside the intrinsic (φ_μ, or φ_{cv} and σ'_c) and state (D_R and σ'_{mf}) parameters, which mainly influence the shear strength of granular materials, other aspects can also play a significant role under some specific conditions.

Among them, the first to be mentioned is the intermediate principal stress σ'_2 (see Fig. 11) that has been already introduced in an indirect way referring to PS and TX strain conditions.

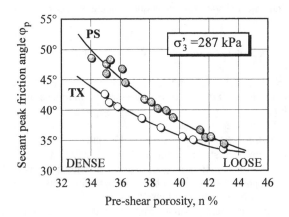

Figure 11. Comparison of φ_p (PS) with φ_p (TX) for a siliceous sand (Cornforth, 1964)

Most of the laboratory experimental data are the output of tests in the triaxial apparatus, whereas many geotechnical structures work in plane strain conditions. It can be, therefore, of practical interest to transform $\varphi_p(TX)$ in $\varphi_p(PS)$ or vice-versa, by means of a number of empirical formulae reported in the geotechnical literature. Lade and Lee (1976), for siliceous sands have suggested:

$$\varphi_p(PS)= 1.5 \cdot \varphi_p(TX)-17° \qquad \varphi_p(TX) \geq 34° \qquad (13)$$

Schanz & Vermeer (1996), considering the results on Hostun sand as well as those obtained by Cornforth (1964) and Leussink (1996), for siliceous sands have proposed:

$$\varphi_p(TX)= (1/5)(3\varphi_p(PS)+2\varphi_{cv}) \qquad (14)$$

This latter equation is exactly the same as that which can be obtained by equation (11) of Bolton (1986) evaluated for m=5 and m=3 in (PS) and (TX) conditions respectively.

Both Eqs. 13 and 14 hold when comparing TX and PS compression tests. A more general handling of the problem regarding the effects of σ'_2 on shear strength of granular soils must include the full range of variation of the following parameter: $b = (\sigma'_2 - \sigma'_3)/(\sigma'_1 - \sigma'_3)$. However, the analysis of the changes of φ_p with variation of b is beyond the scope of the present paper.

As documented in the last fifteen years by Tatsuoka and his co-workers (Tatsuoka et al., 1986, 1990; Pradhan et al., 1988; Park & Tatsuoka, 1994), the peak secant angle of shearing resistance shows a pronounced anisotropic response. As a matter of fact, in addition to the value of parameter, b, the peak friction angle is also affected by the angle, δ, existing between the direction of the major principal stress at failure (σ'_{1f}) and the direction of the bedding planes. The present φ_p anisotropy is not usually taken into account in the interpretation of in situ tests results for strength.

The previous approaches to the assessment of the shear strength envelope of coarse grained materials assume that there is no cohesion intercept (c') in terms of effective stress.

Such statement holds even in very dense and interlocked materials as recently argued by Schofield (1998).

However, lightly cemented coarse-grained soils are noticed within natural formations for which a c'>0 intercept is a consequence of the weak bond between the soil grains (Nader, 1983; Bachus, 1983).

The lightly cemented soil deposits generally have unconfined compressive strength less than 100 kPa and the c' resulting from drained TX compression tests falls in a range between 5 and 30 kPa.

At present there is a lack of well consolidated methods allowing to infer both φ'_p and c' from in situ tests, while some possibilities can be envisaged for SBPT (Bachus, 1983; Carter et al., 1986).

Currently the assumption of c'=0 when interpreting in situ CPT and DMT leads to an overestimate of φ'_p in case of lightly cemented sands.

Evaluation of the peak friction angle from CPT and DMT. Nowadays the CPT and DMT are among the tools most commonly used in design to evaluate φ_p. With this respect two basic different approaches can be envisaged (Jamiolkowski & Lo Presti, 2000):

A. The first approach can apply to both CPT and DMT results and refers to the use of the existing bearing capacity theories (e.g. Durgunoglu & Mitchell, 1975; Janbu & Senneset, 1974; Vesic, 1975, 1977; Salgado, 1993, Salgado et al., 1997). In this case, the ultimate bearing capacity is measured (i.e. the cone resistance q_c in case of CPT and the wedge resistance q_D in case of DMT), therefore, the bearing capacity formula is used to estimate φ_p. The summary of the input data required when using these approaches is shown in Table 9. Further details can be found in the works by Mitchell & Keaveny (1986) and Yu & Mitchell (1998).

B. The second approach consists of the in situ evaluation of D_R from the results of the considered penetration tests. Once the D_R has been assessed, the estimation of φ_p can be carried out by using correlations $\varphi_p = f(D_R,$ grading) like the one proposed by Schmertmann (1978), see Fig. 12 or, in a more refined manner, by means of an iterative use of Bolton's (1986) stress dilatancy Eqs. (11) and/or (12).

The main features and the use of the methods belonging to the groups A and B are summarized in the paper of Jamiolkowski & Lo Presti (2000).

As a matter of fact, the methods for estimating φ_p by the bearing capacity theories, in spite of the more elegant initial approach, require rather complex input data and /or are affected by important limitations and approximations of the original theoretical models so that, most of the procedures, practically applicable, must turn to "calibrating" coefficients and/or "operational" parameters that reintroduce empiricism into the initial equations (see for example Mitchell & Keaveny, 1986).

For such reasons the methods of group B, passing from the empirical evaluation of D_R by CPT and DMT before the final assessment of φ_p at different confining stress levels, can still be considered more robust, reliable, and useful for practical applications within the current geotechnical design practice.

Referring to the φ_p assessment by the methods of the group B, once D_R has been evaluated by means of one of the approaches outlined in the first section of this paper, an estimate of the $\varphi_p(TX)$ can be attempted with reference to Fig. 12 by selecting the line appropriate for the gradation curve for the soil layer in question. The main limitation which arises from the use of the functions of Fig. 12 is that φ_p only refers to triaxial and direct shear (DS) tests carried out at a confining stress range of $50 < \sigma_{oct} < 350$ kPa and at a normal stress range of $80 < \sigma'_{vc} < 400$ kPa respectively. Therefore, it is not possible to take into account the influence of parameters such as: different confining stress level at failure, strain conditions and last but not least sands characterized by different compressibility and mineralogy from the tested ones.

In order to overcome these limitations, Eqs. 11 or 12 can be used to assess $\varphi_p(TX)$ or $\varphi_p(PS)$ following the procedure outlined in Fig. 13. This approach has

Table 9. Friction angle from cone resistance – possible approaches

REFERENCE	STRESS-STRAIN RELATIONSHIP	COMPRESSIBILITY	STRENGTH ENVELOPE	STRESS TENSOR	OTHER INPUT REQUIRED
SCHMERTMANN (1978)	NO ASSUMPTION EMPIRICAL	NO	LINEAR	σ'_{v0} or σ'_{m0}	D_R GRADING CURVE
BEEN ET AL. (1987)	NO ASSUMPTION	YES (IMPLICITLY)	LINEAR	σ'_{m0}	$\xi = f(NCR)_0$
BOLTON (1986)	STRENGTH-DILATANCY THEORY SEMI-EMPIRICAL	YES (IMPLICITLY)	CURVILINEAR	σ'_{mf}	Q, φ'_{cv}
JANBU AND SENNESET (1974)	RIGID PLASTIC	EMPIRICALLY	LINEAR	σ'_{v0}	s_q
DURGUNOGLU AND MITCHELL (1975)	RIGID PLASTIC	NO	LINEAR	σ'_{v0}	s_q, K_0, δ
VESIC (1972, 1977)	ELASTIC PERFECTLY PLASTIC	YES	LINEAR	σ'_{m0} or σ'_{h0}	s_q, K_0, G
SALGADO (1993)	ELASTIC NON LINEAR PLASTIC	YES	CURVILINEAR	σ'_{h0}	s_q, φ'_{cv}, K_0, D_R, Q, $G(\gamma, \sigma'_m)$

The table is not exhaustive, for similar approaches see also Robertson & Campanella (1983), Jamiolkowski et al. (1988), Trofimenkov (1974)

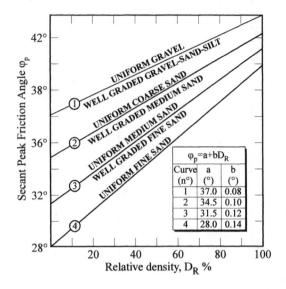

Figure 12. Correlation $\varphi_p = f(D_R$ Grading) (Schmertmann, 1978)

been validated for a number of siliceous sands by Bellotti et al. (1989), Jamiolkowski (1990), Yoon (1991) and others. This method is equally conditioned by the reliability of the D_R best estimate as that involving the use of Fig. 12, but has the advantage to be able to take into account the stress level relevant to the considered boundary value problem via the introduction of an appropriate value of σ'_{mf}. The higher rationality of this approach has, however, the limitation of a more elaborated input involving the knowledge and/or the assumption, in addition to D_R, of two intrinsic (φ_{cv} or φ_{μ} and Q or σ'_c) and one state (σ'_{mf}) parameter.

As far as $Q=\ln\sigma'_c$ is concerned, Table 8 and 10 report respectively the values of Q as suggested by Bolton (1986) and those resulting for a number of sands inferred from triaxial compression tests. Table 10 associates with each value of Q also the corresponding φ_{cv}. In the light of the preliminary considerations related to the possible difference of 2° existing between φ_{cv}(TX) and φ_{cv}(PS), in the authors opinion, the value reported in Table 10 should be considered as φ_{cv}(TX). Moreover it may be worthy to point out that the values of Q displayed apply to grains having dimensions corresponding to those of fine to medium sands. As documented by Lee (1992) for coarse sands and gravel particles constituted by the same mineral, Q tends to decrease with increasing the equivalent grains diameter.

Regarding the estimate of σ'_{mf} corresponding to the boundary value problem of practical interest, the issue is far from being solved in a rigorous manner. At present, only the following rules of thumb can be suggested to the readers:

For shallow foundations (De Beer, 1967):

$$\sigma'_{ff} \cong \frac{q_{lim} + 3\sigma'_{v0}}{4}(1 - \sin\varphi'_{op}) \qquad (15)$$

q_{lim} : limit bearing capacity of the considered shallow foundation;
φ'_{op} : operational friction angle for q_{lim} evaluation.

For deep foundations and penetration tests (Fleming et al., 1992):

$$\sigma'_{ff} \cong \sigma'_{v0} \cdot \sqrt{N_q} \qquad (16)$$

N_q : dimensionless bearing capacity factor for deep foundation, function of φ'_{op}.

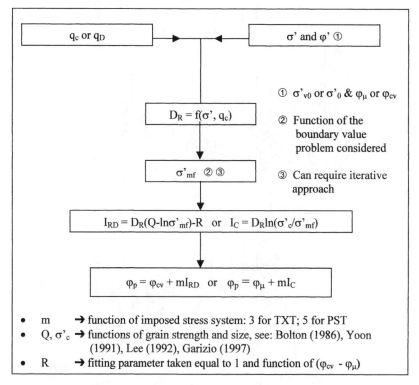

Figure 13. Friction angle of sand from penetration tests Bolton's (1986) stress-dilatancy theory

Table 10. Q-values of different uniform sands (*)

SAND	MINERALOGY		Q	φ_{cv}	REFERENCE
TICINO	SILICEOUS (**)		10.8	34.6	Jamiolkowski et al., 1988
TOYOURA	QUARTZ		9.8	32	
HOKKSUND	SILICEOUS		9.2	34	
MOL	QUARTZ		10	31.6	Yoon, 1991
OTTAWA	QUARTZ	FINES 0%	9.8	30	Salgado et al. 1997
		FINES 5%	10.9	32.3	Salgado et al. 2000
		FINES 10%	10.8	32.9	
		FINES 15%	10	33.1	
		FINES 20%	9.9	33.5	
ANTWERPIAN	QUARTZ & GLAUCONITE		7.8 to 8.3	31.5	Yoon, 1991
KENYA	CALCAREOUS		9.5	40.2	Jamiolkowski et al., 1988
QUIOU	CALCAREOUS		7.5	41.7	

(*) inferred from TX compression tests

(**) i.e.: containing a comparable amount of quartz and feldspar grains

The link between σ'_{ff} and σ'_{mf} for compression loading stress path is given by:

$$\sigma'_{mf} \cong \sigma'_{ff} \cdot \left(\frac{3 - \sin\varphi_{0_p}}{3\cos^2 \varphi_{0_p}} \right) \qquad (17)$$

As already pointed out, the method, based on the Bolton (1986) theory, sketched in Fig. 13 has been mainly validated referring to silica sands. An attempt has been carried out, by the writers, to extend the aforementioned procedure, for the assessment of D_R and then φ_p from q_c values, to sands characterized by different mineralogical compositions that results mainly in a different deformability under isotropic stress increments and in different values of φ_{cv} and φ_μ.

As a first step, the relationships q_c - D_R, reported in Fig. 4, has been considered. These correlations have been worked out by Lancellotta (1983) and subsequently have been revised by Garizio (1997) for taking into account the influence of CC dimensions, geometry and boundary conditions and referring to a much more extended data base.

To account for the influence of different sand compressibilities, the fitting parameters A_0 and B_0 of the equation in Fig. 4 have been evaluated for the average, upper and lower bond respectively of the data set displayed in the same figure. This data set is mostly related to different normally-consolidated (NC) and lightly over-consolidated (OC) sands for which, as previously mentioned, D_R may be related to q_c through σ'_{vo}.

In order to be able to include, within the proposed procedure, the OC sands, as a first attempt, the equation of Figure 4 has been modified as reported in Table 11, i.e. instead of the geostatic vertical stress (σ'_{vo}) the relative density has been expressed as a function of the mean geostatic effective stress $[\sigma'_{mo}=\sigma'_{vo}(1+2K_0)/3)]$ and the A_0 value has then been transformed by using the following equation:

$$A'_0 = A_0 - B_0 \ln \sqrt{\frac{1 - 2K_0(NC)}{3}} \qquad (18)$$

where $K_0(NC)$ is the coefficient of horizontal pressure at rest for NC sands assumed, in a first approximation, equal to 0.4.

The proposed modification of the equation of Figure 4 into the one of Table 11 does not change the results of the original correlations in the case of NC sands but via the K_0 values allows, in principle, to extend the proposed procedure to the OC sands.

The obtained values of A'_0 and B_0 for the different mineralogical compositions are reported in Table 11 together with the values of Q characterizing the same kind of sands as suggested by Bolton (see also Table 8).

Substituting the equation reported in Table 11 into Eqs. 11 or 12 and using the parameters given in Table 11 and the values of φ_μ or φ_{cv} within the ranges of

Table 11. Parameters A'_0, B_0 and Q used in Figs. 14, 15 and 16 for different sands

	MINERALOGICAL COMPOSITION	COMPRESSIBILITY	EXAMPLES IN THE LITERATURE	A'_0	B_0	Q
QUARTZ SANDS	QUARTZ	LOW	MONTEREY OTTAWA TOYOURA SYDNEY	-1.506	0.268	10.0
SILICA SANDS	FELDSPAR QUARTZ MICA	MEDIUM	TICINO HOKKSUND	-1.360	0.268	9.5
CALCAREOUS SANDS	SANDSTONE MICA	HIGH	QUIOU KENYA BASS STRAIT ANTWERPIAN CHATTAHOOCHEE	-1.214	0.268	8.5

$$D_R = A'_0 + B_0 \ln\left(\frac{q_c}{\sqrt{\sigma'_{m0}}}\right) \quad ; \quad q_c, \; \sigma'_{m0} \quad \text{in kPa}$$

Table 12. Range of φ_{cv} e φ_μ values for sands of different mineralogical compositions

	φ_μ	φ_{cv}
Quartz sands	$25° \div 30°$	$30° \div 34°$
Silica sands	$27° \div 32°$	$32° \div 36°$
Calcareous sands	$32° \div 38°$	$36° \div 42°$

Table 12, the peak friction angle $\varphi_p(TX)$ or $\varphi_p(PS)$ can be evaluated for the considered sand.

Figures 14, 15 and 16 show $\varphi_p(TX)$ trends, for silica, calcareous and quartz sands respectively, evaluated by the proposed procedure and referred to a mean confining stress at failure σ'_{mf} equal to the geostatic mean effective stress σ'_{mo}. In the same figures the adopted values of φ_{cv} and Q are also reported.

The values of $\varphi_p(TX)$ obtained from the aforementioned figures have been compared with the estimation of the same parameter at the same confining stress levels carried out in different ways by Durgunoglu & Mitchell (1975), Robertson & Campanella (1983) and Chen & Juang (1996), The comparison results have pointed out a very good agreement for all the mineralogical compositions of the considered sands.

Once the set of intrinsic and state parameters, characterising the mechanical behaviour of the considered sand, has been assessed, it is possible to evaluate the $\varphi_{op}(TX)$ and/or the $\varphi_{op}(PS)$ values for different confining stress levels at failure by using the Bolton (1986) Eqs. 11 or 12 and referring to Eqs. 15, 16 and 17 for some typical boundary value problems. However, since σ'_{mf} is a function of q_{lim} that in turn depends on φ_{op}, a trial and error procedure must be adopted.

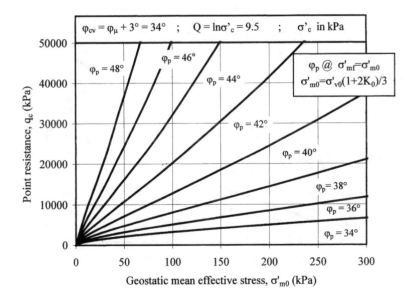

Figure 14. Peak friction angle from CPT for silica sands using Bolton (1986) theory

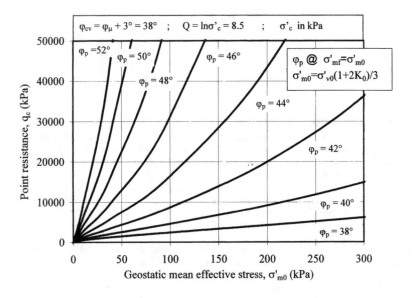

Figure 15. Peak friction angle from CPT for calcareous sands using Bolton (1986) theory

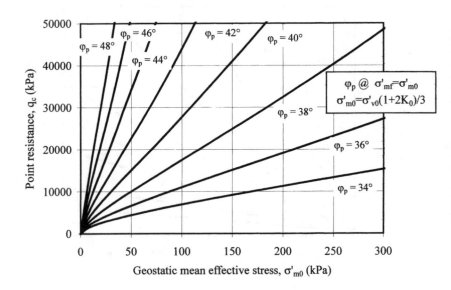

Figure 16. Peak friction angle from CPT for quartz sands using Bolton (1986) theory

For the estimation of q_{lim} and N_q of Eqs. 15 and 16 it is recommended to adopt $\varphi_{op}=(\varphi_p+\varphi_{cv})/2$ so as to take also into account, beyond the effect of the curvature of the strength envelope, the influence, on the mobilized average friction angle, of the strain level in correspondence of the limit pressure of the considered boundary value problem (i.e. progressive failure effect).

Finally, it is worthy to recall that the proposed procedure for the estimation of φ_p and φ_{op} allows also to consider the over-consolidation ratio (OCR) of the considered sand via the K_o value. Unfortunately, at the present state of the art, the quantification of the OCR degree and then of K_o, within coarse grained materials, is still very unreliable.

Closing remarks

1. Eq. 5, referring to the empirical coefficient C_0, C_1 and C_2, shown in Table 4, allows the estimation of D_R in deposits of unaged, uncemented silica sands of low to moderate compressibility.
2. The use of the above equations involves the effective overburden stress σ'_{vo} in case of NC sands but requires the estimation of the mean geostatic stress σ'_{mo} for OC deposits.
3. An alternative approach for the evaluation of D_R by Lancellotta (1983) and later reworked by Garizio (1997) is displayed in Fig. 4. This correlation, using the available data-base for both NC and OC sands makes reference to σ'_{vo}.
4. Making reference to the above mentioned $D_R = f(q_c, \sigma')$ correlation, an iterative approach based on the Bolton's strength-dilatancy theory is proposed to estimate φ'_p on the basis of the CPT results.
5. The proposed approach, at least in principle, allows evaluation, for a given sand, of φ'_p taking into account: sand compressibility, curvilinear nature of the shear strength envelope and imposed strain conditions.
6. Figures from 14 to 16 display the correlations $\varphi'_p = (q_c, \sigma'_{mo})$ for three qualitatively defined classes of sand compressibilities assuming $\sigma'_{mf} = \sigma'_{mo}$. For σ'_{mf} different than the σ'_{mo} iterative procedure outlined in Fig. 13 should be followed.
7. A comparison between $\varphi'_p = f(q_c, \sigma')$ yielded by Fig. 14 and φ'_p, obtained from drained compression loading CK_oD triaxial tests, suggests that the proposed procedure tends to underestimate φ'_p by 1° to 1°½, see Jamiolkowski (1990).
8. On the basis of the aforementioned approaches, an iterative procedure has been suggested for the estimation of the operational friction angle φ'_{op} referred to some current limit equilibrium boundary value problem. This approach allows also to take into account, in a first approximation, the progressive failure aspect.
9. This paper gives also engineering correlations between Marchetti's (1980) dilatometer lateral stress index K_D, D_R and φ'_p respectively as inferred from CC tests on siliceous sands.

References

Achintya, H. & Tang, W.H. (1979) "Uncertainty Analysis of Relative Density" *Journal of Geotechnical Engineering Division, ASCE, 105(GT7), 899-904.*

Bachus R.C. (1983) "An investigation of the strength deformation response of naturally occuring lightly cemented sands" *Ph.D. Thesis, Civil Engineering Stanford University, CA.*

Baldi, G., Bellotti, R., Ghionna, V., Jamiolkowski, M. & Pasqualini, E. (1986) "Interpretation of CPT's and CPTU's. 2nd Part: Drained Penetration" *Proceeding 4th International Geotechnical Seminar, Singapore, pp.143-156.*

Baligh M.M. (1975). "Theory of deep static cone penetration resistance" *Research Report R75-56, MIT, Cambridge, Mass.*

Baligh M.M. (1976) "Cavity Expansion in Sands with Curved Envelopes" *Journal of Geotechnical Engineering Division, ASCE, 102(GT11), 1131-1146.*

Barton N.R. (1973). "Review of a new shear strength criteria for rock joints" *Engineering Geology, 7, 287-332.*

Barton M.E. & S.N. Palmer (1989) The Relative Density of Geologically Aged British Fine and Fine-medium Samds. *Quaterly Journal of Engineering Geology.* 1: 49-58.

Been K., Jefferies M.G., Crooks, J. H. & Rothenburg L. (1987) "The cone penetration test in sands II: general inference of state" *Geotechnique, 37(3), 285-299.*

Bellotti, R., Bizzi, G. & Ghionna, V.N. (1982) "Design, Construction and Use of Calibration Chamber" *Proc. 2nd ESOPT, Amsterdam, Vol.2, pp.439-446.*

Bellotti, R., Crippa, V., Pedroni, S. & Ghionna, V.N. (1988) "Saturation of Sand Specimen for Calibration Chamber Tests" *Proc. ISOPT-1, Orlando, Vol.2, pp .661-672.*

Bellotti R., Ghionna V.N., Jamiolkowski M., & Lancellotta R. (1989) "Shear strength from CPT" *Proc. XII ICSMFE, Vol 1, pp. 165-170, Rio de Janeiro.*

Bellotti R., Benoit J., Fretti C. & Jamiolkowski M. (1997) "Stiffness of Toyoura sand from dilatometer Tests" *Journal of Geotechnical and Geoenvironmental Engineering, ASCE, 123(9), 836-846.*

Biliam J. (1967). "Some aspects of the behaviour of granular materials at high pressure" *Proc. Roscoe Memorial, Stress Strain Behaviour of Soils, Cambridge University.*

Bolton M.D. (1986) "The strength and dilatancy of sands" *Geotechnique, 36(1),* *65-78*

Brinch Hansen J. (1970) "A revised and extended formula for bearing capacity". *Danish Geotechnical Institute Bulletin, 11.*

Burmister, D.M. (1948) "The Importance and Practical Use of the Relative Density in Soil Mechanics" *ASTM Proc. 48, 1249-1268.*

Carter J.P., Booker J.R. & Yeung S.K. (1986) "Cavity expansion in cohesive frictional soils" *Geotechnique, 36(3), 349-353.*

Casagrande A. (1936) "Characteristics of cohesionless soils affecting the stability of slopes and earth fills" *Journal of Boston Society of Civil Engineers., Jan.: 257-276.*

Chen J.W. & Juang C.H. (1996) "Determination of drained friction angle of sands from CPT" *Journal of Geotechnical Engineering Division, ASCE, 122(5), 374-381.*

Cornforth D.H. (1964) "Some experiments on the influence of strain conditions on hte strength of sands" *Geotechnique, 14(2), 143-167.*

Cubrinovski, M. & Ishihara, K. (1999) "Empirical Correlation between SPT N-value and Relative Density of Sandy Soils" *Soils and Foundations, 39(5), 61-71.*

De Beer E.E. (1967) "Bearing capacity and settlement of shallow foundations in sand". *Proc., Bearing capacity and settlement of foundations, Duke University, Durham, N.C.*

Durgunoglu H.T. & Mitchell J.K. (1975). "Static penetration resistance of soils, I-Analysis, II-Evaluation of the Theory and implications for practice". ASCE Spec. Conference In situ measurements of soil properties, Vol. 1, Raleigh, NC.

Dussealt M.B. & Morgenstern N.R. (1979) "Locked Sands" *Quarterly Journal of Engineering Geology.* 2: 117-131

Eslaamizaad, S. & Robertson, P.K. (1996) "Estimation of In-Situ Lateral Stress and Stress History in Sands" *Proceeding 49th Canadian Geotechnical Conference, St. John's, Newfoundland, pp.439-448.*

Fleming W.G.K., Weltman A.J., Randolph M.F. & Elson W.K. (1992) "Piling engineering" *Blakie Academic and Professional, Glasgow.*

Foray, P. (1986) "First Results in the I.M.G. Calibration Chamber: the Boundary Condition Problem" *1st International Seminar on Research Involving Validation of In-Situ Devices in Large Calibration Chambers, Milano, unpublished.*

Felice A. (1997). Determinazione dei Parametri Geotecnici e in Particolare di Ko da Prove Dilatometriche. *M.Sc. Thesis, Department of Structural Engineering, Politecnico di Torino.*

Fretti C., Lo Presti D.C.F. & Salgado R. (1992) "The Research Dilatometer: in Situ and Calibration Chamber Test Results" *Rivista Italiana di Geotecnica, Vol. XXVI, n. 4, pp 237-243.*

Fretti C., Froio F., Jamiolkowski M., Lo Presti D.C.F., Olteanu A. & Pedroni S. (1996) "Dilatometer Tests in Calibration Chamber: Stiffness and Ko Assessment" *VIII Conferenza Nazionale Rumena di Geotecnica, Iasi, Romania, 25-27 September 1996. Vol. I, pp. 85-94.*

Garizio G.M. (1997) "Determinazione dei parametri geotecnici e in particolare di K0 da prove penetrometriche" *M.Sc. Department of Structural Engineering, Politecnico di Torino.*

Gibbs, H.J. & Holtz, W.G. (1957 "Research on Determining the Density of Sands by Spoon Penetration Testing" *Proceedings 4th ICSMFE, Vol.1, pp.35-39, London.*

Goto, S., Suzuki, Y., Nishio, S. & Oh Oka, H. (1992) "Mechanical Properties of Undisturbed Tone-River Gravel obtained by In-Situ Freezing Method" *Soils and Foundations 32(3), 15-25.*

Hanna A.M., Massoud N. & Youssef H. (1987) "Prediction of plane-strain angles of shear resistance from triaxial test results" *Proc. Prediction and Performance in Geotechnical Engineering, Univ. of Calgary, pp. 369-376.*

Harman, D.H. (1976) "A Statistical Study of Static Cone Bearing Capacity, Vertical Effective Stress and Relative Density of Dry and Saturated Fine Sands in Large Triaxial Test Chamber" *M.Sc. Dissertation, University of Florida, Gainesville, Fla.*

Hatanaka, M., Suzuki, Y., Kawasaki, T. & Endo, M. (1988) "Cyclic Undrained Shear Propeties of High Quality Undisturbed Tokyo Gravel" *Soils and Foundations, 28(4), 57-68.*

Houlsby, G.T. & Hitchman, R. (1988) "Calibration Chamber Tests of Cone Penetrometer in Sand" *Geotechnique, 38(1), 39-44.*

Jamiolkowski M. (1990) "Shear strength of cohesionless soils from CPT" *De Mello Volume, pp. 191-204, Editoria Edgar Blucker, San Paulo.*

Jamiolkowski M., Ghionna V.N., Lancellotta R. & Pasqualini E. (1988). "New correlations of penetration tests for design practice". *Proc., Penetration Testing 1988, ISOPT 1, Orlando, Florida, J. De Ruiter ed., Vol. 1 pp: 263-296*

Jamiolkowski M. & Lo Presti D. (2000). "Shear strength of coarse grained soils from in situ tests. A compendium" *Proc., 4th International Geotechnical Engineering Conference, Cairo, Egypt.*

Janbu N. & Senneset K. (1974) "Effective stress interpretation of in situ static penetration tests" *Proc., ESOPT I, Stockholm. Vol. 2.2, pp. 181-194.*

Kulhawy, F.H. & Mayne, P.H. (1990) "Manual on Estimating Soil Properties for Foundation Design" *Electric Power Research Institute, EPRI.*

Ladd C.C., Foott R., Ishihara K., Schlosser F. & Poulos H.G. (1977) "Stress-deformation and strength characteristics" *S.O.A. Report, 9th ICSMFE, Tokyo, 2: 421-494.*

Lade P.V. & Lee K.L. (1976). "Engineering properties of soils" Report UCLA-ENG-7652, University of Califirnia at Los Angeles.

Lancellotta R. (1983) "Analisi di Affidabilità in Ingegneria Geotecnica" *Atti dell'Istituto di Scienza delle Costruzioni, No 625, Politecnico di Torino.*

Lee D.M. (1992) "The angles of friction of granular fills" *Ph.D. Thesis, Cambridge University.*

Lee K.L. & Seed H.B. (1967) "Drained strength characteristics of sands". *Journal of Soil Mechanics and Foundation Division, ASCE, Vol. 93(SM6), Proc. Paper 5561, 117-141.*

Leussink H., Wittke W. & Weseloh K. (1966) "Unterschiede im scherverhalten rollinger erdstoffe und kugelschuttungen im dreiaxial und biaxialversuch" *Veroffentlichungen des Institut fur bodenmechanik und felsmechanik, Universitat Karlsruhe.*

Lunne, T., Robertson, P.K. & Powell, J.J.M. (1997) "Cone Penetration Testing in Geotechnical Practice" *Blackie Academic and Professional, London.*

Marchetti S. (1980) "In situ tests by Flat Dilatometer" *Journal of the Geotechnical Engineering Division, ASCE, Vol. 106(GT3), 299-321.*

Marchetti, S. (1997) "The Flat Dilatometer: Design, Applications" *Proceeding 3rd International Geo-technical Engineering Conference, Cairo, pp.421-428.*

Marsal R.J. (1967) "Large scale testing of rockfill materials" *Journal of Soil Mechanics Foundation Division, ASCE, 93(SM2), 27-43.*

Mayne, P.W. & Kulhawy, F.H. (1991) "Calibration Chamber Data Base and Boundary Effects Correction on CPT Data" *Proceedings International Symposium on Calibration Chamber Testing, Potsdam, NY, pp.257-264.*

Mitchell J.K. (1976). "Fundamentals of Soil Behaviour". *J. Wiley & Sons, New York.*

Mitchell J.K. & Keaveny J.M. (1986) "Determining sand strength by penetrometers" *Proc. ASCE GSP No. 6 Use of In Situ Tests in Geotechnical Engineering. Blacksburg, Va, pp. 823-929.*

Nader S.R. (1983) "Static and dynamic behaviour of cemented sands" *Ph.D. Thesis, Civil Engineering Stanford University, CA*

Park C.S. & Tatsuoka F. (1994) "Anisotropic strength and deformation of sands in plane strain compression" *Proc. XIII ICSMFE New Delhi, India, Balkema, Rotterdam, pp. 1-4.*

Parkin, A.K. & Lunne, T. (1982) "Boundary Effects in the Laboratory Calibration of a Cone Penetrometer in Sand" *Proc. 2nd ESOPT. Amsterdam, Vol.2, pp.761-768.*

Pradhan T.B.S. & Tatsuoka F., Horii N. (1988) "Simple shear testing on sands in a torsional shear apparatus" *Soil and Foundations 28(2), 95-112.*

Puppala, A.J. (1993) "Effect of Cementation on Cone Resistance in Sands: A Calibration Chamber Study". *Ph. D. Thesis, Louisiana State University, Baton Rouge.*

Puppala, A.J., Acar, Y.B. & Tumay, M.T. (1995) "CPT in Very Weakly Cemented Sand" *Proceedings International Symposium on Cone Penetration Testing, CPT '95, Vol.2, pp.269-276.*

Roberts J.E. (1964) "Sand compression as a factor in oil field subsidence". *Ph.D. Thesis, Massachusetts Institute of Technology.*

Robertson P.K. & Campanella R.G. (1983) "Interpretation of cone penetration tests. Part I: Sand" *Canadian Geotechnical Journal, 20(4), 718-733.*

Rowe P.W. (1962). "The stress-dilatancy realtion for static equilibrium of an assembly of particles in contact" *Proc. Royal Society London.*

Salgado R. (1993). "Analysis of penetration resistance in sands". *Ph.D. Thesis, University of California, Berkeley.*

Salgado R., Mitchell J.K. & Jamiolkowski M. (1997). "Cavity expansion and penetration resistance in sand" *Journal of Geotechnical Engineering Division, ASCE, 123(4), 344-354*

Salgado R., Mitchell J.K. & Jamiolkowski M. (1998). "Calibration Chamber Size effects on Penetration resistance in Sand" *Journal of Geotechnical Engineering Division, ASCE, 124(9), 878-888*

Salgado R., Bandini P. & Karim A. (2000) "Shear Strength and Stiffness of Silty Sand" *Journal of Geotechnical Engineering Division, ASCE, 126(5), 451-462.*

Schanz T. & Vermeer P.A. (1996) "Angles of friction and dilatancy of sands" Geotechnique, 46(1), 145-151.

Schanz T: (1998). "Zur modellierun des mechanischen verhaltens von reibungsmaterilaen" *Institute fur Geotechnik, Universitat Stuttgart, Mitteilung 45, Edit. P.A. Vermeer.*

Schmertmann J.H. (1971) "Discussions and Conference Record – Discussion Session 1" *Proceedings of the Fourth Panamerican Conference on Soil Mechanics and Foundation Engineering, San Juan, Puerto Rico, June 1971, ASCE, Vol. III*

Schmertmann J.H. (1972) "Effects of in situ lateral stress on friction – cone penetrometer data in sands" *Verhandelingen Fugro, Sonder Symposium 1972, pp: 37-39.*

Schmertmann, J.H. (1976) "An Updated Correlation between Relative Density D_R and Fugro-Type Electric Cone Bearing, q_c." *Contract Report DACW 39-76 M 6646 WES, Vicksburg, Miss., 1976*

Schmertmann J.H. (1978) „Guidelines for cone penetration test performance and design" *US Dept. of Transportation, FHWA, R.78-209. Washington D.C.*

Schmertmann, J.H. (1991) "The Mechanical Ageing of Soils" *Journal of Geotechnical Engineering Division, ASCE, 117(9), 1288-1330.*

Schofield A. (1998). "Mohr Coulomb error correction". *Ground Engineering, 31(8), 30-32.*

Schofield A.N. & Wroth C.P. (1968) "Critical state soil mechanics" *McGraw-Hill, London.*

Skempton, A.W. (1986) "Standard Penetration Test Procedures and the Effects in Sands of Overburden Pressure, Relative Density, Particle Size and Overconsolidation" *Geotechnique, 36(3), 425-447.*

Tanizawa, F. (1992) "Correlations between Cone Resistance and Mechanical Properties of Uniform Clean Sand" *Internal Report, ENEL-CRIS, Milano.*

Tatsuoka F., Sonoda F., Hara K., Fukishima S. & Pradhan T.B.S. (1986) "Failure and deformation of sands in torsional shear" Soil and Foundations 26(4), 79-97.

Tavenas, F.A. (1972) "Difficulties in the Use of Relative Density as a Soil Parameter" *ASTM, STP 523, Selig and Ladd Editors, pp.478-483.*

Tavenas, F. & La Rochelle, P. (1972) "Accuracy of Relative Density Measurements" *Geotechnique, 22(4), 549-562.*

Taylor D.W. (1948) "Fundamentals of soil mechanics". *J. Wiley & Sons, New York.*

Terzaghi, K. & Peck, R.B. (1948) "Soil Mechanics in Engineering Practice" *J. Wiley & Sons New York.*

Trofimenkov J.B. (1974) "Penetration Testing in USSR" State of the Art Report, *Proc., ESOPT I, Stockholm. Vol. 1, pp. 147-154.*

Tumay M. (1976) "Cone Bearing vs. Relative Density Correlation in Cohesionless Soils" Dozen Dissertation. Bogazici Univ. Istanbul.

Vesic A.S. (1972) "Expansion of cavities in infinite soil mass" Journal of Soil Mechanics Foundation Division, ASCE, 98($SM3$), 265-290.

Vesic A.S. (1975) "Bearing capacity of shallow foundations" *Foundation Engineering Handbook, Winterkorn and Fang eds., Van Nostrand Reinhold.*

Vesic A.S. (1977) "Design of pile foundations" *National Cooperative Highway Research programs. Synthesis of Highway Practice 42, Transportation Research Board, Washington D.C.*

Wride C.E., Robertson P.K., Biggar K. W., Campanella R.G., Hofmann B.A., Hughes J.M.O., Küpper A. & Woeller D.J. (2000) "Interpretation of in situ test results from the CANLEX sites" *Canadian Geotechnical Journal, 37(3), 505-529.*

Yoon Y. (1991). "Static and dynamic behaviour of crushable and non-crushable sands" *Ph.D. Thesis, Ghent, University.*

Yoshimi, Y. (2000) "A Frozen Sample that did not Melt" *Proceedings of Conference on Development in Geotechnical Engineering.* GEOTECH-YEAR 2000, pp.293-296, Bangkok.

Yoshimi, Y., Hatanaka, M. & Oh-Oka, H. (1978) "Undisturbed Sampling of Saturated Sands by Freezing" *Soils and Foundations, 18(3), 59-73.*

Yu H.S. & Mitchell J.K., (1998) "Analysis of cone resistance: review of methods" Journal of Geotechnical Engineering Division ASCE, 124(2), 140-149.

Zolkov, I. & Weisman, G. (1965) "Engineering Properties of Dune and Beach Sands and the Influence Stress History" *Proc. 6th ICSMFE, Montreal, Vol.I, pp.134-138.*

Improved engineering solutions
because of improved site characterization

Suzanne Lacasse, MASCE[1], Steinar Hermann[2], Tor Georg Jensen[3] and Vidar Kveldsvik[4]

Abstract: Increasing demand for sustainable transportation systems and for modernized infrastructure poses new challenges to the geotechnical engineer. More than before, the profession should aim at engineering more economical solutions, developing improved and safer transportation solutions and encouraging innovation while preserving the environment. To meet these challenges, an alliance of good practice and research is required. Such is illustrated with examples where improved *in situ* characterization of the foundation soil considerably reduced costs, and improved performance. The extent of site investigations should be a decision taken considering hazard and consequence, including cost, and therefore it is a risk-based decision.

1 INTRODUCTION

The contribution of geotechnical engineering to transportation infrastructure is indispensable, but it is often underestimated by both public and government. Geotechnical engineers influence alignment, planning, design and maintenance of traffic and transportation arteries. Geotechnical engineers decide location and orientation of runways, and are the key to the safe foundation of coastal and offshore structures. Geotechnical engineers have the required expertise for the evaluation of the risks

[1] Managing Director, Norwegian Geotechnical Institute, P.O.Box 3930 Ullevaal Stadion, N-0806 Oslo, Norway, Tel.: +47 22023000, Fax: +47 22230448, e-mail: suzanne.lacasse@ngi.no.
[2] Division Manager, Norwegian Geotechnical Institute, P.O.Box 3930 Ullevaal Stadion, N-0806 Oslo, Norway, Tel.: +47 22023000, Fax: +47 22230448, e-mail: steinar.hermann@ngi.no.
[3] Project Engineer, Norwegian Geotechnical Institute, P.O.Box 3930 Ullevaal Stadion, N-0806 Oslo, Norway, Tel.: +47 22023000, Fax: +47 22230448, e-mail: tor.georg.jensen@ngi.no.
[4] Discipline Manager, Norwegian Geotechnical Institute, P.O.Box 3930 Ullevaal Stadion, N-0806 Oslo, Norway, Tel.: +47 22023000, Fax: +47 22230448, e-mail: vidar.kveldsvik@ngi.no

involved in their designs because they have the knowledge, judgment and experience to evaluate the uncertainties entering the assumption and calculations.

Four all-encompassing objectives of our profession are (1) engineering more economical solutions, (2) developing improved and safer transportation solutions, (3) encouraging innovative solutions and (4) protecting and preserving the environment. To meet these challenges, an alliance of good practice with the results of recent research is required.

The paper first presents two examples of cost-savings achieved because of improved geotechnics. The paper then discusses contracting in general, suggesting possible improvement and recalls that the planning of a site investigation is a risk-based decision.

2   MORE ECONOMICAL SOLUTIONS THROUGH IMPROVED SITE CHARACTERIZATION

Our profession needs to focus on cost-effective solutions. Optimum designs, taking responsibility for sustainable solutions in accord with society's needs, should also promote innovation. Two examples are briefly given here of cost-savings contributions because of improved site characterization: the Nykirke rail crossing in southern Norway, and the Bergslia-Nydalen subway tunnel section in Oslo.

2.1 Rail crossing at Nykirke

The 1.5-km long rail crossing at Nykirke, southwest of Oslo, is part of the modernization of the railway network in Norway. Construction work is organized as a turn-key project where the contractor is responsible for engineering, planning and construction. The contract was based on target estimate and incentives for enhanced efficiency and quality.

The terrain along the proposed track was rather hilly with maximum elevation difference about 40 m. The soil conditions were dominated by outcropping bedrock and soft silty marine clay deposits. The marine clay was in some locations very sensitive and even quick. The geology and topography led to two complex constructions. At the northern end, the proposed alignment crossed a 20-m deep river valley, and required a 140-m long tunnel about halfway on the link. At the tunnel exit followed large cuts in clay and rock along most of the alignment, except for a small fill over a short distance. Work started in April 2000 and was completed in April 2001.

The paper presents the geotechnical aspects of the fill over the 20-m river valley at the northern end of the railway link.

The initial site investigations, done in a traditional manner with 54-mm dia tube sampling, gave the soil profile in Figure 1: natural water content of 25-35%, total unit weight of 19.5 kN/m^3, plasticity index between 4 and 9%, clay content between 20 and 55%. The index strength tests suggested low undrained shear strength (10-25 kPa) at depths 4 to 12 m, in part due to sampling disturbance. This shear strength profile, as measured initially, corresponded to approximately 0.2 times the *in situ* effective overburden stress.

The customer had done a preliminary design using these initial soil characteristics. The initial solution was a fill founded on concrete piles driven to bedrock. In the bidding phase, the bidders could prepare alternative solutions, provided the bidder "guaranteed" their feasibility. One of the bidders asked NGI to optimize the foundation design. The contract was made a "total enterprise" contract with incentives for all parties if savings could be achieved.

Recent projects in Norway for upgrading intercity railway lines include long stretches of railways to be built on low plasticity silty clays (I_p = 15-20%). Vane shear tests gave unreasonably low shear strengths in silty clays, especially when the clay was sensitive. In view of the difficulties NGI had experienced in the past obtaining consistent stress-strain-strength parameters from specimens recovered by traditional sampling techniques, and in view of the good results obtained with the Université de Sherbrooke block sampler, Karlsrud proposed in 1999 an approach for obtaining parameters along a railway alignment:

1. Take high quality samples with the Université de Sherbrooke block sampler at representative locations.

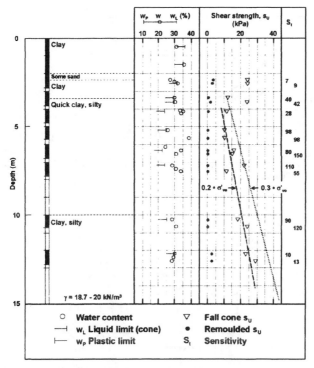

Figure 1. Soil profile, initial soil investigations, Nykirke railway link (Jernbaneverket, 2000).

2. Do oedometer tests, anisotropically consolidated triaxial compression and extension tests and simple shear tests at the same locations to determine stress-strain-strength parameters at relevant effective stresses.
3. Run piezocone penetration tests at the same locations.
4. Establish correlations between piezocone factors and the parameters determined in the laboratory on the high quality specimens.
5. Proceed with piezocone penetration tests as the prime method to determine soil parameters along the line from the correlations between cone factors and soil parameters.

Figure 2 compares the correlations between piezocone factors N_{KT}, N_{KE} and $N_{\Delta u}$ and pore pressure factor B_q, using block samples and 76-mm samples as reference (Karlsrud et al, 1996; Lunne et al, 1989; 1997). To obtain the piezocone factors, the undrained shear strength from triaxial compression tests was used.

The piezocone factors and parameter B_q are defined as:

$$N_{KT} = (q_t - \sigma_{vo})/s_u$$
$$N_{KE} = (q_t - u)/s_u$$

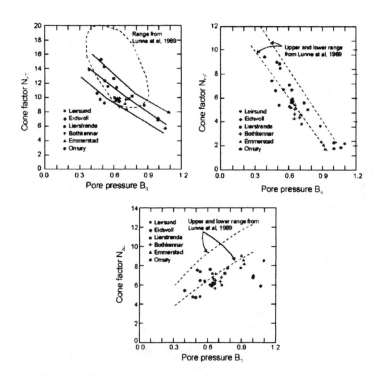

Figure 2. Piezocone factors as a function of pore pressure ratio B_q derived from anisotropically consolidated triaxial compression tests on block samples (Karlsrud et al, 1996; Lunne et al, 1997)

$$N_{\Delta u} = (u - u_o)/s_u$$
$$B_q = \Delta u/(q_t - \sigma_{vo}) = (u - u_o)/(q_c - \sigma_{vo})$$

where

q_t	=	corrected cone resistance = $q_c + (1 - a) \cdot u$
q_c	=	measured cone resistance (uncorrected)
u	=	measured pore water pressure (behind cone)
u_o	=	hydrostatic pore water pressure
Δu	=	excess pore water pressure ($u - u_o$)
a	=	net area ratio of piezocone
σ_{vo}	=	total overburden stress
s_u	=	reference undrained shear strength

The newer correlations show lower N_{KT}, N_{KE} and $N_{\Delta u}$ values. The cone factors are generally 30-40% lower using the block samples as reference. Use of newer correlations will therefore result in higher s_u-values. The undrained shear strength is obtained by inverting the above relationships. More work is presently underway with the correlations.

Figure 3 presents the derived cone factors N_{KT} and $N_{\Delta u}$ as a function of over-consolidation ratio (OCR) for six Norwegian soft clays:

$$N_{KT} = (q_T - \sigma_{vo})/s_u$$
$$N_{\Delta u} = \Delta u/s_u$$

where

q_T	=	corrected cone resistance
σ_{vo}	=	*in situ* total vertical stress
s_u	=	undrained shear strength
Δu	=	excess pore pressure

A study of the topography and geology at Nykirke suggested that erosion had probably occurred and that the clay was overconsolidated and should have higher

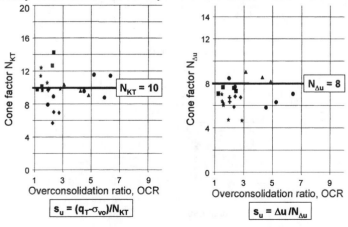

Figure 3. Reference cone factors on six soft Norwegian clays (NGI, 2000).

shear strength than assumed in the original design. With the geology that included ravines and areas with massive quick and soft clay deposits, NGI recommended cone penetration tests and large size block sampling of the soil with subsequent laboratory testing. NGI had worked over a number of years to establish relationships between cone factors and engineering parameters using block samples (Lacasse *et al*, 1985, Karlsrud *et al*, 1996 and Lunne *et al*, 1997, 1998).

Supplementary site investigations were done, including block sampling with the Sherbrooke sampler (Lefebvre and Paulin, 1979).

Figure 4 compares the stress-strain curves and the effective stress paths of two specimens from depths of 9.5 and 10.1 m. They were recovered with a 54-mm sampler and a 250-mm block sampler respectively. Figure 4 illustrates the completely different behavior, the first contractant and the second dilatant, for the 54-mm and the block sample.

The 54-mm samples suffered greatly due to sampling disturbance, and the shear strength was too low. The volume of water pressed out during consolidation in the 54-mm specimen (5.5%) is 7 times higher than for the block specimen (Fig. 2). Research projects at NGI have also tested 76-mm and 95-mm samples; they show similar trends as those shown in Figure 4, although the strength and modulus reduction is less than for the 54-mm sample.

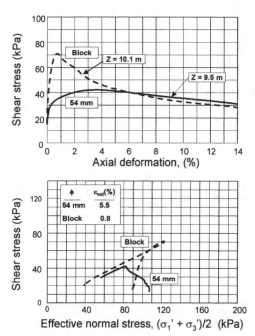

Figure 4. Comparison of triaxial compression test results on 54-mm and 250-mm samples, Nykirke railway link (NGI, 2000)

It is possible to correct for sampling disturbance based on recent research tests where disturbed samples were compared to block samples (Lunne *et al*, 1997). In this case, the shear strengths for a vertical stress path and a stress path at 1:3 were considered. The correction approach, still preliminary, is illustrated in Figure 6 for the sample at depth 11.5 m. The initially measured disturbed shear strength of 42 kPa increased to corrected values of 56 and 66 kPa for the vertical stress path assumption and the 1:3 stress path assumption respectively. The block sample gave a shear strength of 70 kPa at 10 m depth. The piezocone tests suggested a triaxial compression undrained shear strength of 72 kPa at the same depth. Since the clay is over-consolidated (Figs 4 and 7), the correction along the 1:3 stress path is probably acceptable.

Figure 5 gives the triaxial compression undrained shear strength derived with cone factors N_{KT} and $N_{\Delta u}$ of 10 and 8 respectively.

A linear s_u-profile was proposed for design. The shear strengths from triaxial compression tests on block samples verify well the proposed design profile based on the piezocone tests. For comparison purposes, an s_u-profile corresponding to a normally consolidated triaxial compression undrained shear strength of $0.3 \cdot \sigma'_{vo}$ and the triaxial test results on 54-mm samples are also shown.

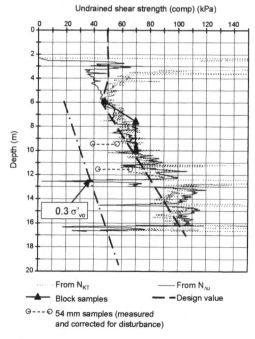

Figure 5. Undrained shear strength derived from cone penetration tests and compared with laboratory test results, Nykirke railway link (NGI, 2000).

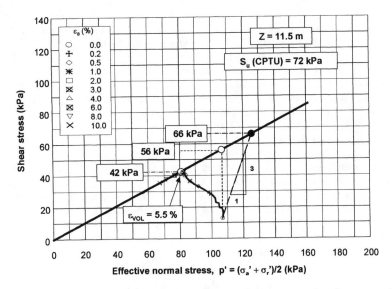

Figure 6. Correction for sampling disturbance of 54-mm sample of overconsolidated clay at 11.6 m, Nykirke railway link.

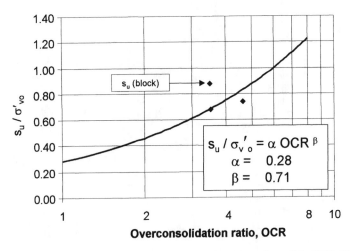

Figure 7. Normalized undrained shear strength, Nykirke railway link (NGI, 2000).

Oedometer test on specimens recovered with the block sampler suggested an OCR of 4.5 at 6-m and 3.5 at 10 m-depth. Figure 7 illustrates the normalized undrained shear strength of the block specimens as a function of overconsolidation ratio. The data suggest a normally consolidated triaxial compression strength of 0.28 (α), and the OCR-function increases with an exponent of 0.71. Both α- and β- factors are reasonable and agree with earlier published data. The relationship developed was based on Ladd *et al.*'s (1977) relationship developed between the normalized strengths of overconsolidated and normally consolidated clay and the overconsolidation ratio.

The new profile, combining high quality sampling and recent research results on the interpretation of the piezocone, demonstrated that the shear strength of the clay was significantly higher than originally believed. The higher shear strength was caused by pre-consolidation under an overburden that has since been eroded away.

The initial solution proposed to build fills founded on piles down to bedrock. The new design profile made the pile foundation unnecessary and enabled the contractor to use preloading and vertical drains to accelerate the settlements. A fill with a 3-m preloading was placed. The preloading was left in place for 6 months, then removed. Before filling, vertical drains were driven in a triangular pattern with 1.25-m spacing down to bedrock or moraine. Side berms ensured stability.

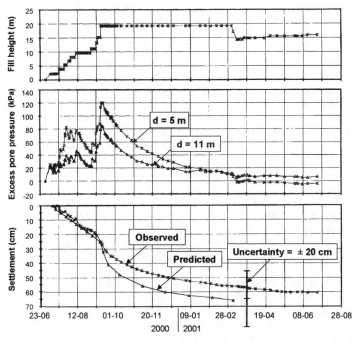

Figure 8. Fill height, pore pressures and settlements, Station 340, 7.5 m off center-line, Nykirke railway link (Hermann and Jensen, 2000; NGI files).

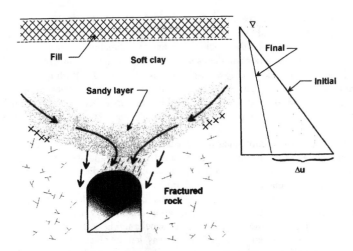

Figure 9. Pore pressure reduction and settlement due to tunnel leakage.

Piezometers were placed in the clays, and settlement indicators on top of the clay. Prediction suggested that 90% of the consolidation settlement under preloading should be completed in 5-6 months after the last fill placement. Maximum settlement was predicted as 65 cm ± 20 cm. Figure 8 presents the measured pore pressure and settlement up to August 2001. The final settlement will probably be as predicted, and the dissipation of pressure is as expected.

Over the entire 1.5-km crossing, the new strengths enabled cost reductions of USD 1.5 million, or 20% of the total project costs. The savings were made possible because the contractor agreed to run three piezocone tests and to have these interpreted with the results of recent research, and the fact that higher shear strength parameters than used in the preliminary design could be demonstrated.

2.2 Bergslia-Nydalen tunnel

For cities in coastal areas or close to large rivers, underground construction is a major challenge to geotechnical engineers. Often the tunnels and deep excavations are planned for ground that is marginally stable unless supported. Stability during construction must be assured. In urban areas, excavation-induced ground movements must be predicted and controlled so that overlying buildings, structures and services are not adversely affected.

The planning of underground constructions should be based on a good understanding of the geology. Soil and rock investigations are often not addressed properly, causing cost and time overruns. The consequences can be far-reaching, sometimes greatly detrimental to the environment. For example, when groundwater leaks into a rock tunnel overlain by compressible sediments, it can cause significant pore pressure reduction and consolidation settlements (Fig. 9). Subsidence due to leakage into rock tunnels becomes a major issue. A "good" safety factor has little significance when settlement governs. To illustrate the cost-effectiveness of additional soil investigations, an example is taken from a 1.2-km long subway tunnel section in rock between Bergslia and Nydalen in Oslo. The section is part of the subway in Oslo ("T-baneringen"). The alternative was to either design and build the tunnel based on traditional investigations or to expand the investigations to reduce some of the uncertainties. With the first option, the site investigations were inexpensive. There was also a risk of considerable leakage of groundwater through the tunnel, construction delays, unbudgeted sealing operations, compensations for damage to dwellings and other buildings, environmental damage and public complaints, as well as general discontent with the engineering profession. With the second option, the question was whether the increased site investigation costs would decrease sufficiently the risks and associated costs.

Table 1 lists the investigations carried out. The often qualitative traditional testing (rock drilling and soundings) were supplemented with more quantitative testing, including piezocone tests, piezometric measurements, sampling and enhanced seismics. Classification and oedometer tests were run in the laboratory on the recovered soil specimens.

The additional testing gave an improved knowledge base relative to what has been done earlier in the area and where leakage and damages due to settlements had been experienced. The new investigation covered a 500 to 600-m wide corridor along the tunnel alignment. They revealed that the formation overlying the tunnel included permeable soil layers along most of the alignment, sometimes as much as 30-m thick.

With the additional samples, *in situ* tests and experience from existing tunnels in the Oslo region, it was possible to:

- calculate settlements as a function of pore pressure reduction,
- estimate settlements as a function of leakage,
- identify the most critical location of the buildings and installations in each bedrock trench,
- establish maximum leakage criteria in the different tunnel sections to prevent excessive settlement on critical buildings and installations.

Table 1.   Bergslia-Nydalen Tunnel: Traditional Investigation Program and Extended Program

Test method	Traditional investigation	Additional testing
Refraction seismic profiles	3	8
Rotary pressure soundings	3	0
Rock drilling	12	0
Combined rotary and rock soundings	35	32
Piezocone penetration test	0	15
Piezometers	1	23
54-mm sampling profiles	1	4
95-mm sampling profiles	0	3
Field vane	1	0
Simple soundings	13	0

The maximum leakage criterion was 7 to 14 liters per minute per 100 m tunnel. The results showed that the maximum leakage criteria were met, and that no additional costs were incurred because of leakage. An extensive pre-grouting effort was necessary in a 50-m long tunnel section that had very poor rock conditions, a condition that had been discovered thanks to the site investigations.

The case study is an example showing that appropriate additional testing and/or sampling, aimed at solving the problem at hand and adapted to the formations *in situ*, will lead to improved engineered solutions. The example does not say that more testing is better; it says that judiciously selected testing, including methods giving higher sample quality, is cost-effective.

3   INNOVATION AND INTERACTION IN CONTRACTING

To achieve more economical solutions, the challenges facing our profession include:
- the need to use existing projects with foresight, i.e. take every opportunity to develop new solutions and to test and monitor new technology
- the need to increase interaction among owner, designer, contractor, consultant and user, e.g. with new forms of contracting and the application of value engineering, encouraging brainstorming towards optimum solutions.

A study in the building and construction industry in Norway was carried out to discern the reasons for the lack of successful collaboration in building and construction projects. The Norwegian study concluded that five factors were the main culprits:
- lack of ability and lack of an arena for collaboration;
- lack of interest in changing the "way of doing things";
- lack of competence;
- weak leadership and poor planning;
- different, and even conflicting, goals and success criteria.

These results apply to Norway, but some degree of similar situations can probably be found in many countries. Value engineering is one tool than can contribute to help remedy these aspects. Some of its concepts were used for the Nykirke project.

Value engineering's first objective is to enhance value by providing a framework of systematic procedures for conceptualization, definition, implementation and operation of a project (Fig. 10). With appropriately structured contracts, value engineering creates opportunities to manage risk and achieve positive results for owner, designer, consultant and contractor. Key elements, in addition to planning and teamwork, include (ICE, 1996; Powderham and Rutty, 1994):
- increased value to customer by elimination of unnecessary cost or improved achievement of e.g. time and quality
- evaluation of options based on required function rather than simple cost cutting
- new idea creation as formal step in project
- life cycle costing (total costs of owning and operation of a facility) as input when evaluating alternatives
- integrated team approach

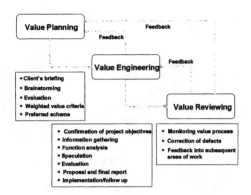

Figure 10. Value engineering (ICE, 1996)

There is a strong synergy between value engineering, risk assessment and the observational method. Risk assessment adds value by listing and quantifying options, thus minimizing the impact of all risks in the project. The observational method (Peck, 1969; 2001; Powderham, 1998) focuses on cost and time savings during construction and encourages a system to manage risk. Quoting Peck (2001), the observational method can pay-off handsomely, without more than the most elemental theory and with only qualitative predictions. In Peck's examples, there were few refinements, and no elaborate computer modeling to be "validated" by exotic remote-reading sensors. Refinements have their place, but they should not deflect attention and resources from the essence of the method.

4 SOIL INVESTIGATION AS A RISK-BASED DECISION

Soil investigations represent a risk-based decision. The complexity of a soil characterization is based on the consequences of a failure to perform as expected. Figure 11 presents an illustration of the decision process:
- a low risk project involves few hazards and has limited consequences. Simple *in situ* and laboratory testing and empirical correlations would be selected to document geotechnical feasibility.
- in a moderate risk project, there are concerns for hazards, and the consequences of non-performance are more serious than in the former case. Specific *in situ* tests and good quality soil samples are generally planned. A carefully laid soil characterization program can often lead to savings in design and execution.
- for a high-risk project involving frequent hazards and potentially risk to life or substantial material damage, high quality *in situ* and laboratory tests are required, and higher costs, and possible savings, are involved.

The decision-making process for selecting the appropriate soil investigation methods, although subconscious, is therefore risk-based. It involves consideration of requirements, uncertainties, consequences and costs.

5   SUMMARY AND CONCLUSIONS

Geotechnical engineers should be more aware of the omnipresent influence of geology. The boundaries among engineering geology, geophysics, rock mechanics, soil mechanics, hydrogeology, seismology, and a host of other disciplines are meaningless. Contribution to the solution of geotechnical problems may come from any or all of these sources. The practitioner who keeps too narrowly to one specialty is likely to overlook knowledge that could be of the greatest benefit in reaching a proper judgment. On the other hand, the geotechnical engineer must also be increasingly aware of the influence of society's needs on the goodness of the solutions provided.

A good characterization of the foundation soil, including a good understanding of the geology and geological processes, is the key to successful foundation solutions. In most cases, an investment in site characterization will lead to savings greater than the costs of the improved investigation.

The importance of interweaving good practice and research was illustrated. It is important to use the results of recent research and to offer increasingly attractive and cost-effective solutions. Our profession needs to be aware of promoting innovative solutions and to convince our clients to consider alternative solutions. For this, communication needs to be improved, and we need to thrive on change. Innovation results in new, cheaper and safer methods of construction. We should, however, be aware that it is costly to develop new approaches and new solutions.

The good function of infrastructure systems requires the expertise of the geotechnical engineer. Today's society experiences greater than ever vulnerability connected to infrastructure. Our profession, however, does not excel at telling society and policy- and decision-makers of its contributions. We have a proud profession

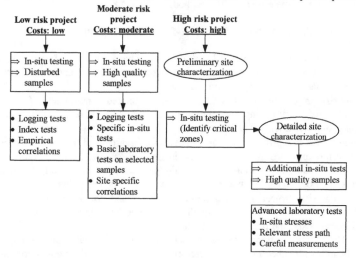

Figure 11. Risk-based soil investigation (Robertson, 1998; Lacasse and Nadim, 1998).

with many accomplishments to boast about. The engineers' voices need to be heard by politicians, policy-makers, society and the younger generation.

6 ACKNOWLEDGMENT

The authors thank their colleagues at NGI for their help in all ways. In particular, Tini van der Harst, Tim Gregory provided assistance for this paper. The authors also thank Mr. Alan Powderham, Mott MacDonald Limited, for his help with the value-engineering concept.

REFERENCES

Hermann, S., and Jensen, T.G. (2000). CPTU kombinert med blokkprøvetaking ga kostnadsbesparende løsninger for Nykirke kryssingsspor (CPTU combined with block sampling resulted in cost-saving solutions at Nykirke railway link). *Geoteknikkdagen*, Oslo. Nov 2000. pp. 32.1-32.23.

ICE (1996). *Creating value in engineering projects. Design and Practice guide,* Institution of Civil Engineers, London, UK.

Jernbaneverket Utbygging (2000). *Anbudsgrunnlag. Underbygning Parsell 5.1 Nykirke kryssingsspor*. Oslo, Norway.

Karlsrud, K. (1999). General aspects of transportation infrastructure. Invited lecture, *Proc. 12th Eur. Conf. Soil Mech. and Geot. Engg.*, Amsterdam, the Netherlands, Vol. 1, pp. 17-30.

Karlsrud, K., Lunne, T., and Brattlien, K. (1996). Improved CPTU interpretations based on block samples, *XIIth Northern Geotechnical Conference (NGM)*, Reykjavik, Vol. 1, pp 195-201.

Lacasse, S., Berre, T., and Lefebvre, G. (1985). Block sampling of sensitive clays, *Int. Conf. on Soil Mech. and Found. Eng.*, San Francisco, Vol. 2, pp. 887-892.

Lacasse, S., and Nadim, F. (1998). Risk and reliability in geotechnical engineering. State-of-the-Art paper. *Fourth International Conference on Case Histories in Geotechnical Engineering*, Proc. Intern. Symp., Paper No. 50.A.-S, St. Louis, Missouri, USA.

Ladd, C.C., Footh, R., Ishihara, K., Schlosser, F. and Poulos, H.G. (1977). Stress-Deformation and Strength Characteristics. State-of-the-Art Lecture, *Proc. Int. Conf. on Soil Mech. and Found. Eng.*, Tokyo, Vol. 4, pp. 421-494.

Lefebvre, G., and Paulin, C. (1979). A new method of sampling in sensitive clay, *Canadian Geotechnical Journal,* Vol. 16, pp. 226-233.

Lunne, T., Berre, T., and Strandvik, S. (1997). Sample disturbance effects in soft low plastic Norwegian clay. *Conf. on Recent Developments in Soil and Pavement Mechanics*, Rio de Janeiro, Proc. pp. 81-102.

Lunne, T., Berre, T., and Strandvik, S. (1998). Sample disturbance effects in deep water soil investigations. *Offshore Site Investigation and Foundation Behavior '98, SUT*, London, UK, pp. 199-220.

Lunne, T., Strandvik, S., Berre, T., and Andersen, K.H. (2000). *Deep water sampling phase 3. Analyses and recommendation report.* NGI Report 521676-9, Oslo, Norway., 359 p.

Lunne, T., Lacasse, S., and Rad, N. (1989). SPT, CPT, pressuremeter testing and recent developments on *in situ* testing of soils. General report, *12th Int. Conf. Soil Mechanics and Foundation Engineering,* Rio de Janeiro, Vol. 4, 1985, pp. 2339-2403.

Lunne, T., Robertson, P.K., and Powell, J.J.M. (1997). *Cone Penetration Testing in Geotechnical Practice*, Blackie, London, 312 p.

Mair, R. (2000). "Research and innovation in underground construction", *Ground Engineering, YGEC 2000 Supplement*, pp. 12-13.

Peck, R.B. (2001). The observational method can be simple. *Proc. Instn. Civ. Engrs., Geotech. Engng.*, Vol. GT149, Issue, pp. 71-79.

Peck, R.B. (1969). Advantages and limitations of the observational method in applied soil mechanics. 9th Rankine Lecture. *Geotechnique 19*, No. 2, pp 171-187, ICE, London.

Powderham, A.J. (1998). The observational method - application through progressive modification; *Proc. Journal ASCE/BSCE,* Vol. 13, No. 2, pp. 87-110.

Powderham, A.J., and Rutty, P.C. (1994). The observational method in value engineering. *Proc. 5th Int. Conf. Piling and Deep Foundation*, Bruges, pp. 5.7.1-5.7.12.

Robertson, P.K. (1998). Risk-Based Site Investigation. *Geotechnical News*, Sept. 1999, pp. 45-47.

ANALYZING THE EFFECTS OF GAINING AND LOSING GROUND

by

Andrew J. Whittle[1] & César Sagaseta[2]

Members ASCE

ABSTRACT

There are many problems in geotechnical engineering where the inadvertent or controlled ground loss or displacement of soil have important implications in geotechnical design. This paper shows that simplified analytical methods can provide reliable and practical solutions for estimating ground movements in two applications: Analyses of pile installation using the Shallow Strain Path Method simulates consistently the downward movement of soil beneath the advancing pile tip and the more general heave towards the stress-free ground surface. Far field movements can be estimated using simple closed-form solutions. Field monitoring data from a large piling project in East Boston provide a clear demonstration of the predictive capabilities of these analyses. Tunneling-induced ground movements can be obtained by similar approximate techniques, considering deformations around the tunnel cavity as the sum of three component mode shapes. Analytic solutions have been obtained for 2D deformations around a cylindrical tunnel cavity using assumptions of linear elastic and constant plastic flow behavior. Examples presented in the paper show that the analytical solutions are not only consistent with well established empirical methods, but also provide a more comprehensive framework for understanding the distribution of ground movements.

KEYWORDS: Displacement piles, tunnels, analytical methods, ground movements, field measurements.

---

[1] Professor, Department of Civil & Environmental Engineering, Massachusetts Institute of Technology, Cambridge, MA02139.

[2] Professor, Division de Ingenieria del Terreno, Universidad de Cantabria, 39005 Santander, Spain

INTRODUCTION

The geotechnical profession has recently enjoyed a rare moment of public adulation with the successful completion of stabilization works for the Leaning Tower of Pisa and its grand re-opening to the public in June 2001 (e.g., Boston Globe, 2001). A 10% reduction in the tilt (corresponding to approximately 170 years of accumulated movement) were achieved by the controlled extraction of soil from beneath the north (underhanging) side of the Tower (Burland et al., 1998). This elegantly simple solution (apparently first proposed by Terracina, 1962) was implemented only after extraordinarily detailed investigations of the mechanisms controlling Tower movements (Costanzo et al., 1994). The success of the scheme relies on a highly controlled drilling operation and the link between the volume of soil extracted and the settlement of the Tower foundation.

On further reflection there are many situations in geotechnical engineering where ground movements occur due to the controlled or inadvertent loss or displacement (i.e., gain) of soil. Indeed, the first author and his senior colleague at MIT, Professor Charles C. Ladd, recently collaborated in a study of excavation performance for the CA/T project in Boston. One section of this project was bedevilled by unexpectedly large ground settlements apparently caused by drilling holes for tieback anchors.

Figure 1 summarizes the ground surface settlements and lateral wall deflections measured at one particular stage of excavation for this 60m wide cut in South Boston. The lateral earth support system comprises a 0.9m thick reinforced concrete diaphragm wall that is keyed into the underlying argillite bedrock, and was designed with three levels of prestressed tiebacks inclined at 45° and anchored in the glaciomarine and rock layers. At the selected stage of the project, the excavation is only 7.6m deep (with a wide berm left in place in front of the wall) and extends through cohesive fill to the surface of an organic silt layer. Two levels of tieback anchors have been installed, tested and locked-off. The inclinometer data show that these preload forces are sufficient to pull the top of the wall away from the excavation by up to 65mm, with maximum inward bowing of less than 20mm occurring below the excavated grade. Far field surface settlements (approximately 15mm occurring more than 30m from the excavation) were attributed to effects of groundwater pumping prior to excavation, while much larger settlements (up to 100mm) occurred closer to the wall. The settlements are much larger than might be anticipated by empirical methods (e.g., Peck, 1969; Clough & O'Rourke, 1990), exceeding 1% of the excavated depth. They are also inconsistent with the deflected mode shape of the wall, implying significant compression of the soil mass. The measured wall deflections and far field settlements are very well predicted by quite sophisticated non-linear finite element analyses (Jen, 1998). However, these same calculations anticipate very small ground surface settlements closer to the wall .

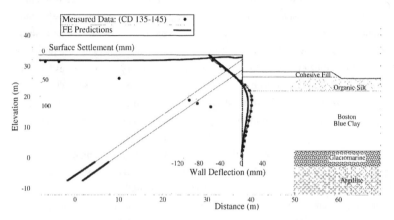

Figure 1. Ground movements of a tied-back wall, South Boston (after Jen ,1998)

Detailed correlations of the measured ground response with construction activities (Whelan, 1995) showed a close link between settlements and the installation of the tieback anchors. The latter were installed with a downhole hammer and wash water, while high air pressure was used to flush cuttings from the casing. Field observations of water and air escaping through adjacent tieback holes provided clear evidence of soil fracture and uncontrolled ground loss. Subsequent numerical analyses (Jen, 1998) have all but eliminated other possible sources of ground movement (relating to vertical displacements of the diaphragm wall, consolidation of the clay etc.).

The preceding is an example of ground movements caused by uncontrolled ground loss and arising from an unanticipated source. There are many other situations in geotechnical engineering where the source of ground loss or displacement is known, while predictions of ground movement are important parameters in design. This paper reviews recent progress in the development of simplified methods of analyses for predicting ground movements associated with two common classes of problems, 1) pile driving and 2) shallow tunneling in soft ground.

PILE INSTALLATION

Background

When piles are driven or jacked into low permeability clay, there is minimal migration of pore water within the surrounding soil mass and hence, the volume displaced during penetration must be accommodated by undrained shear deformations. It is well established that the pile penetration generates pore pressures in the surrounding clay (e.g., Roy et al., 1981; Levadoux, 1980) and causes heave at the ground surface (e.g., Hagerty & Peck, 1971; Dugan & Freed, 1984). The ground deformations associated with pile installation have a number of important practical implications on a) the performance of previously installed adjacent piles within a group, including potential loss of their end bearing capacity (Koutsoftas, 1982) or addition of structural loads (Poulos, 1994), and b) potential damage to nearby constructed facilities. The subsequent dissipation of the excess pore pressures can cause consolidation settlements that are larger than the initial heave. A well known example presented by D'Appolonia and Lambe (1971), Figure 2, shows movements of buildings caused by pile driving at four sites around the MIT campus.

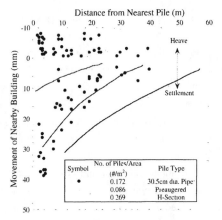

Figure 2. Movements of nearby buildings at four site on MIT campus (after D'Appolonia & Lambe, 1971)

Heave movements up to 10mm occurred during pile installation, while subsequent consolidation settlements up to 40mm were correlated with the density of the piles (i.e., the relative volumes of soil displaced).

There is a long history of research at MIT dealing with different facets of disturbance effects caused by the installation of piles and penetrometers. The original development of the piezocone by Profs. Baligh and Ladd in the 1970's was accompanied by extensive theoretical analyses to predict and interpret engineering properties of the clay from the pore pressures generated during penetration and their subsequent consolidation (Levadoux, 1980). Baligh and Levadoux (1980) recognized that theoretical analysis of penetration is particularly difficult due to the high gradients of the field variables (displacements, stresses, strains and pore pressures) around the penetrometer, the large deformations and strains in the soil, the complex constitutive behavior of soils, and non-linear penetrometer-soil interface characteristics. Experimental observations (e.g., Vesic, 1963; Robinsky & Morrison, 1964) indicate that soil deformations caused by deep penetration of rigid tools are similar in different soils, although the penetration resistance (and hence, soil stresses) can be radically different. Hence, Baligh and Levadoux (1980) proposed that soil deformations and strains caused by steady deep penetration are essentially independent of its shearing resistance and can be estimated with reasonable accuracy based only on kinematic considerations and boundary conditions. Baligh (1985) formalized this assumption in the framework of the Strain Path Method (SPM). Applications of the SPM framework for clays simulate soil deformations and strains using the steady, irrotational flow of an incompressible, inviscid fluid, while modeling of soil behavior has been accomplished using advanced effective stress models such as MIT-E3 (Whittle & Kavvadas, 1994; Whittle et al., 1994). The analyses have been applied and validated in problems ranging from sampling tube disturbance (Baligh et al., 1987), to pile set-up (Morrison, 1984, Azzouz et al., 1990; Whittle, 1992) and, most recently, the design of a tapered piezoprobe for deepwater site exploration (Whittle et al., 2001).

By considering two-dimensional deformations of soil elements, the Strain Path analyses provide a more realistic framework for describing the mechanics of deep penetration than pre-existing one-dimensional, cylindrical cavity expansion methods (CEM; Randolph et al., 1979; Yu, 2000). On the other hand, the assumptions of strain controlled behavior used in the Strain Path Method greatly simplify the penetration problem and avoid the complexity of large deformation finite element analyses (e.g., Kiousis et al., 1988; van den Berg et al., 1996).

Shallow Strain Path Method

The restriction of strain path analyses to conditions of steady, deep penetration can be acceptable for calculating strains near the tip of a pile, but has no physical meaning for far field conditions, where the presence of the ground surface is likely to affect soil deformations. For example, in the SPM analysis of pile penetration, all soil elements undergo net downward movements, whereas there are many published field observations of ground surface heave caused by pile driving.

These limitations have been addressed in the Shallow Strain Path Method (SSPM; Sagaseta et al., 1997) by explicitly including the effects of the stress free ground surface. In the SSPM method, Figure 3, the pile is represented by superimposing full-space solutions for a point source, S, and mirror image sink, S' (i.e., absorbing an equal and opposite volume to the source) at some embedment depth, h, below and above the notional ground surface, respectively. At points along the ground surface, the combined action of the source and sink will cancel out the normal stresses, but will double the shear stresses. In order to simulate a stress free surface, corrective shear tractions (Step 3, Fig. 3) are applied to the surface. This involves evaluating the shear strains due to the source and image sink and the corresponding shear stresses occurring at the ground surface (assuming a given stress-strain behavior for the soil). The strains due to these corrective shear tractions are then added to the previous fields due to the source and virtual sink. Key aspects of the SSPM approach can be summarized as follows:

1. In contrast to the source and sink solutions, the calculation of corrective shear tractions requires a specific stress-strain relation for the soil. Analytical solutions are only possible for linear, elastic behavior. In this case, the resulting deformations are inversely proportional to the soil modulus, while the corrective stresses themselves are computed using the same soil modulus. Hence, the soil modulus will cancel out and the resulting ground deformations are apparently independent of the assumed elastic properties. Nevertheless, the solution for this component of the deformation is based on the assumption of homogeneous, linear, isotropic behavior.

2. The SSPM formulation computes soil deformations by integrating the velocities (and strain rates) numerically (i.e., considering the changes in geometry) along the particle paths as the pile penetrates from the stress-free ground surface. This approach provides a good approximation of large strain conditions close to the

surface of the penetrometer (where contributions from the source and sink terms dominate).

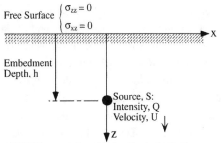

a) SSPM Representation of Shallow Penetration Problem

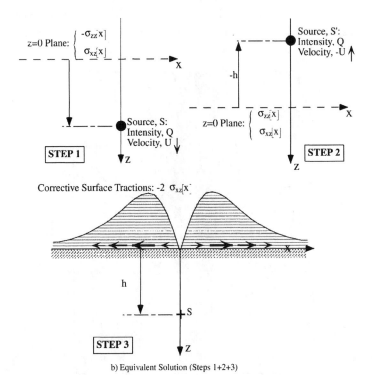

b) Equivalent Solution (Steps 1+2+3)

Figure 3. Conceptual model for the Shallow Strain Path Method (Sagaseta et al., 1997)

Figure 4 illustrates SSPM predictions of the vertical and radial displacements within the soil for a pile penetrating to an embedment depth, L/R = 10 (the rounded tip geometry was referred to as a 'simple pile' by Baligh, 1985). The figure shows the pile geometry and distortion of an initially square grid ($\Delta r/R = \Delta z/R = 0.25$) of soil elements, together with contours of selected values of the radial and vertical displacement components, δ_r/R and δ_z/R, respectively. In the vicinity of the penetrometer tip, soil deformations are very similar to the distorted grid for deep penetration (Baligh, 1985), with net downward displacements. However, settlements are restricted to a tear-shaped zone ahead of the pile tip and extend upwards along a thin annulus around the pile shaft, with a thickness of less than 10% of the pile radius. All other points in the soil mass undergo upward displacements, with maximum surface heave occurring close to the pile shaft. Sagaseta and Whittle (1997) show that the lateral spread and vertical extent of the settlement bulb are about 0.47 and 0.65 times the pile length, L.

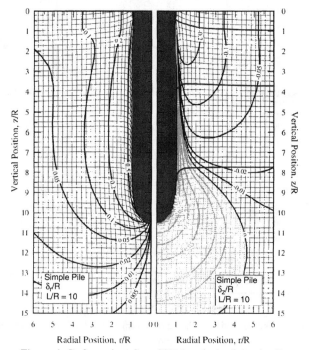

Figure 4. Deformations from SSPM analysis of simple pile
(Sagaseta & Whittle, 2001)

At some distance from the pile, the assumption of small strains becomes a good approximation, superposition techniques become valid and closed-form expressions can be derived for the ground surface displacements caused by driving of a single pile of radius, R, from the ground surface to a depth, L:

$$\delta_{rSS}(r,0) = \frac{R^2}{2} \cdot \frac{L}{r\sqrt{r^2 + L^2}} = \frac{\Omega}{2\pi} \cdot \frac{L}{r\sqrt{r^2 + L^2}}$$

$$\delta_{zSS}(r,0) = -\frac{R^2}{2} \cdot \left(\frac{1}{r} - \frac{1}{\sqrt{r^2 + L^2}} \right) = -\frac{\Omega}{2\pi} \cdot \left(\frac{1}{r} - \frac{1}{\sqrt{r^2 + L^2}} \right)$$

(1)

where the subscript 'ss' and Ω the cross-sectional area of soil displaced.

In many instances, the deleterious ground movements caused by pile driving can be mitigated by pre-augering through the upper soil layers and driving only as the pile approaches its design embedment. The effects of pre-boring to a depth L_0 can be taken into account in the SSPM analyses by simply altering the limits of integration (i.e., from L_0 to L). Using superposition of the small strain results the surface heave can then be found as:

$$\delta_{zSS}(r,0) = \frac{\Omega}{2\pi} \cdot \left(\frac{1}{\sqrt{r^2 + (\eta L)^2}} - \frac{1}{\sqrt{r^2 + L^2}} \right)$$

(2)

where $\eta = L_0/L$ is the pre-augering depth ratio.

Pre-augering is very effective in reducing the surface heave, particularly in the region close to the pile. For typical pre-boring depth ratios $\eta = 0.5 - 0.8$, equation 2 shows that the heave at a distance of half the pile length (r/L = 0.5) is reduced to just 10 to 50% of the movement that would have occurred without pre-augering.

There are many practical situations where piles are driven into a bearing layer or a rigid base is located at some finite depth below the pile tip elevations. These boundary conditions inevitably affect the displacements that occur in the overlying soil (e.g., Poulos, 1994). It is possible to incorporate the rigid base as a reflecting boundary in the proposed analyses (after Sagaseta, 1987). However, the combined action of the rigid base and the stress-free ground surface produces an infinite multiplicity of image sources and sinks. A much simpler alternative (Sagaseta & Whittle, 1996) is to evaluate the relative settlement between the surface and the rigid base, neglecting the influence of the base on the stress distribution. This is the usual approach used for incorporating a rigid bedrock in foundation settlement analyses. For a rigid base at depth H_b below the ground surface, the vertical displacement, δ_z, at a point (r, z), in the soil can be written:

$$\delta_z(r,z) = \delta_{zi}(r,z) - \delta_z(r,H_b) \tag{3}$$

where δ_{zi} is the displacement calculated for the homogeneous half-space (with no bottom boundary).

Results reported by Sagaseta and Whittle (2001) show that the influence of the rigid bedrock only becomes noticeable for H_b/L less than 1.10 (i.e., when the vertical distance between the pile tip and the base is less than 10% of the pile length). The surface heave adjacent to the end bearing pile ($H_b/L = 1$) is approximately 50% higher than that for a friction pile embedded in a deep, homogeneous clay layer ($H_b/L = \infty$). This effect decreases with the pile length ratio, and amounts to a 20% increase in heave for $L/R=100$.

Validation

Sagaseta and Whittle (2001) have evaluated the SSPM predictions using data from a variety of available sources including field measurements of i) building movements caused by installation of large pile groups, ii) uplift of a pile caused by driving of an adjacent pile within a group and iii) spatial distributions of ground movements caused by installation of a single pile.

More detailed measurements of ground displacements have also been obtained in large scale laboratory chamber tests (Gue, 1984). A series of 9mm model steel piles were driven 344mm ($L/R = 34$) into a semi-cylindrical chamber (450mm diameter and 450mm in depth) filled with Speswhite kaolin and consolidated at vertical stresses ranging from 200 to 600kPa and overconsolidation ratios from 1 to 10. Ground surface heave was measured during the driving process at several distances from the pile axis, while displacements within the soil mass were evaluated from optical measurements of a grid of lead shot markers embedded in the clay. Figure 5 summarizes the final radial and vertical displacement components measured in the soil mass at the end of pile driving. These data are presented as radial profiles at three elevations: 1) near the ground surface (at $z_m = 3R\pm2R$), 2) at mid-pile depth ($z_m = L/2\pm2R$) and 3) near to the tip elevation (at $z_m = L\pm2R$).

There is reasonable agreement between SSPM predictions and measured radial displacements at all three elevations (Figs. 5a, c and e). Small differences between computed and measured radial displacements are of the order of the scatter in the measurements and can be explained, in large part, by the proximity of the rigid lateral boundary in the laboratory tests (r/R = 28). For the vertical displacements, there is good agreement near the tip elevation (where settlements extend to r/R = 18) and at mid-depth (Fig. 5d). However, near the ground surface (Fig. 5b) the SSPM analyses underpredict the measured surface heave by a factor of approximately two (for r_0/R = 6-10). On further investigation, Sagaseta and Whittle (2001) suggest that this discrepancy may be related to the occurrence of radial cracks observed at the ground surface during pile installation and is consistent with tensile horizontal strains computed in the SSPM analyses.

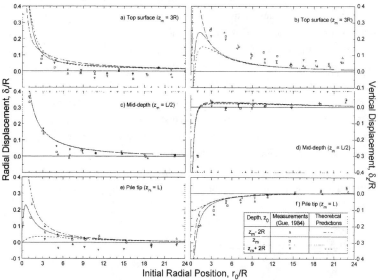

Figure 5. Comparison of predictions with measured displacements within soil mass from calibration chamber tests (Sagaseta & Whittle, 2001)

Case Study, East Boston

Payiatakis and Davie (1998) recently published a very interesting study of ground response to pile driving during construction of a pile foundation for a new egress ramp at the I-90 Logan airport interchange (part of the CA/T project) in East Boston. Figure 6b shows a plan of the site, with locations of the individual piles, covering an area of approximately 30m by 75m (sub-divided into an approach slab and the foundation for the east abutment), and their relative proximity to the Logan Hilton hotel. The foundation comprises a total of 353, 0.4m square, prestressed,

precast concrete piles (each tipped with a 1.5m H-pile stinger) spaced at intervals of
1.37m to 2.44m. The piles were driven 48m to 58m through the deep layer of
marine clay (BBC; Fig. 6a) to act in end-bearing in the underlying glaciofluvial and
glaciolacustrine deposits. The hotel complex includes high rise (four concrete
buildings with 5-12 stories) and low rise (1-2 stories) sections enclosing a courtyard
and pool (Fig. 6b). Surface settlements were monitored extensively through a
network of deformation monitoring points (DMP) located along the hotel frontage
and adjacent areas. During the first two month phase (I in Fig. 6b) piles were driven
on the far side of the Approach Slab (with a minimum horizontal separation of 27m
from the structure). These activities generated surface heave movements in the
range 40mm – 44mm at the hotel, causing minor cracking of the low rise structures.
It is also interesting to note that 25% of the piles had to be re-driven when they were
heaved more than 19mm by installation of the neighboring piles. A second one
month phase (II, Fig. 6a), proceeded after a delay of one month during which time
ground movement mitigation measures were agreed with the contractor (Payiatakis
& Davie, 1998). Construction during the second phase of pile driving used: 1) pre-
augering (0.3m diameter) to depths of 24.4m for all piles, 2) a line of wick drains
(4in diameter, 38.1m deep) were installed at 0.9m centers, parallel to the approach
slab one week after the start of driving, and 3) the sequence of driving was carefully
controlled based on field monitoring of deformations and pore pressures.

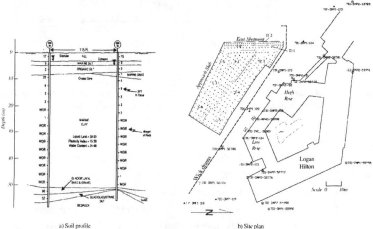

a) Soil profile   b) Site plan

Figure 6. Pile driving project for egress ramp in East Boston
(after Payiatakis & Davie, 1998)

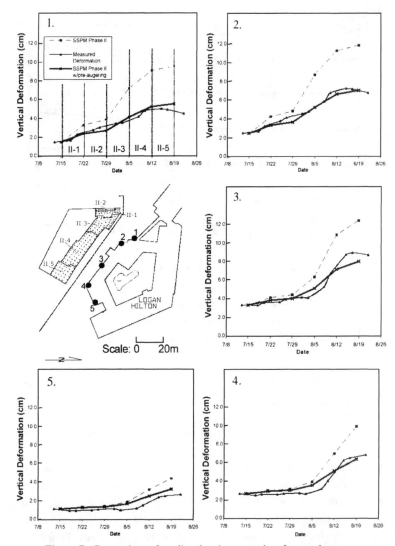

Figure 7. Comparison of predicted and measured surface settlements

Figure 7 summarizes SSPM predictions of surface heave movements at five locations along the hotel frontage, relative to conditions at the start of Phase II

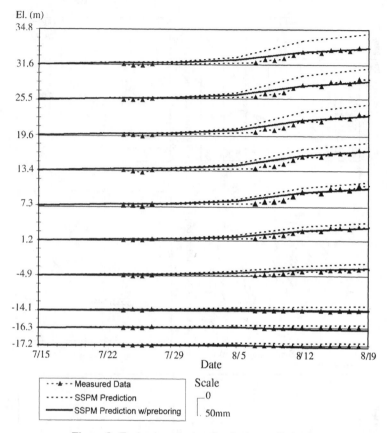

Figure 8. Evaluation of subsurface heave predictions

(Rehkopf, 2001). The predictions are based on simple superposition of the far field deformations (eqns. 1 and 2) using the detailed plan of pile locations, assuming an average length L = 53.3m, both with and without the effects of pre-augering. The analyses do not consider the effects of the underlying rigid base (which are expected to be minor for such slender piles; L/R = 233, for an equivalent radius, R = 0.23m), nor do they account for the presence of the hotel buildings or partial drainage occurring during the timeframe of Phase II. The results confirm that pre-augering has a major effect in reducing the maximum heave by 25mm – 50mm along the hotel frontage (points 1-4; Fig. 7). Excellent agreement can also be seen between predictions with pre-augering and the measured data at all five of these monitoring points. Figure 8 shows further comparisons between SSPM predictions (solved numerically) and vertical displacements measured at selected elevations in the soil profile using a multi-point heave gauge (MPHG; location in Fig. 6b). Again the agreement is quite remarkable suggesting that the proposed SSPM analysis can provide an excellent first approximation for the distribution of ground movements. More detailed comparisons presented by Rehkopf (2001) suggest that the SSPM analyses do tend to overestimate the measured vertical displacements (notably in Phase I), and can give a good first order estimate of horizontal movements in the upper soil layers (data at depth are constrained by the underlying rock layer).

SHALLOW TUNNELS IN SOFT GROUND
Introduction

The prediction and mitigation of damage caused by construction-induced ground movements represents a major factor in the design of tunnels in congested urban environments. This is an especially important problem for shallow tunnels excavated in soft soils, where expensive remedial measures such as compensation grouting or structural underpinning must be considered prior to construction.

Ground movements inevitably arise from changes in soil stresses around the tunnel face and overexcavation of the final tunnel cavity. Sources of movements are closely related to the method of tunnel construction ranging from a) closed-face systems such as tunnel boring machines (with earth pressure or slurry shields), where overcutting occurs around the face and shield ('tail void') but local ground loss is constrained by grout injected between the soil and precast lining system; to b) open-face systems (such as the New Austrian Tunneling Method, NATM) where ground loss around the heading is minimized by expeditious installation of lining systems in contact with the soil (typically steel rib or lattice girder and shotcrete) with additional face support provided by a shield or other mechanical reinforcement (soil nails, sub-horizontal jet grouting etc.). In all cases, it is easy to appreciate the complexity of the mechanisms causing ground movement and their close relationship with construction details, especially given the non-linear, time dependent mechanical properties of soils, and their linkage to groundwater flows. Indeed, this complexity has encouraged the widespread use of numerical analyses, particularly non-linear finite element methods, over a period of more than 30 years (reviewed recently by Gioda and Swoboda, 1999). Although powerful numerical analyses undoubtedly provide the most comprehensive framework for modeling tunneling processes, their predictive accuracy is closely tied to the knowledge of in situ conditions and the modeling of soil behavior. The great majority of published analyses consider 2-D (plane strain) conditions around the final tunnel lining. In these situations, predictions of ground movements are controlled by artificial parameters used to account for the sequence of tunnel excavation, face support and lining installation (e.g., through convergence-confinement methods; Panet & Guenot, 1982). Displacements at the tunnel cavity or stress reduction factors for these calculations are most commonly estimated from back analyses of similar previous projects.

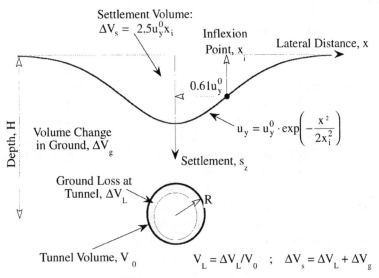

Figure 9. Empirical function for transversal surface settlement trough
(after Peck, 1969)

In practice, most predictions of ground settlements are based on empirical methods first proposed by Peck (1969) and Schmidt (1969). The transversal surface settlement trough is characterized by a Gaussian distribution curve (Fig. 9):

$$u_y\left(x, \ y = 0\right) = u_y^0 \cdot \exp\left(-\frac{x^2}{2x_i^2}\right) \qquad (4)$$

Hence, the settlement distribution is defined by two parameters 1) x_i the location of the inflexion point in the curve, and 2) u_y^0 the settlement above the crown of the tunnel.

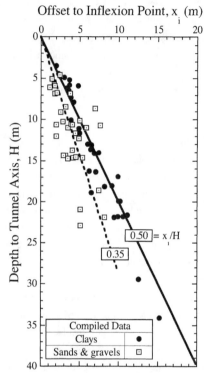

Figure 10. Empirical estimation of inflexion point for surface settlement trough caused by soft ground tunneling (after Mair & Taylor, 1997)

There is considerable scatter in empirical estimates of the parameter x_i that controls the settlement distribution as shown in Figure 10. However, there is general agreement that the parameter x_i is related to the tunnel depth, H. The recent data compiled by Mair and Taylor (1997) suggest average values, $x_i/H = 0.35$ and 0.50 for tunnels in sands and clays, respectively.

Assuming there is no change of volumetric strain within the soil mass, the volume contained by the surface settlement trough, ΔV_s (integral of eqn. 4) can also be equated with the ground loss around the tunnel cavity ΔV_L (Fig. 9) such that:

$$\Delta V_L = \sqrt{2\pi} u_y^0 x_i \approx 2.5 u_y^0 x_i \tag{5}$$

Analytical (closed-form) solutions can offer an attractive alternative to the empirical method. These analyses make gross approximations of real soil behavior (i.e., constitutive equations) but otherwise fulfill the principles of continuum mechanics. Analytical solutions can be obtained for shallow tunnels in terms of a relatively small number of physically meaningful input parameters. Hence, these solutions can provide a complete framework for understanding the relationships between the distribution of far-field deformations, construction methods and ground conditions.

Analytical Solutions for Shallow Tunnels

The analytical solutions described in this section relate ground displacements to a prescribed set of displacements around the tunnel cavity. Figure 11 shows that these displacement boundary conditions around a circular tunnel section can be expressed as the summation of three basic modes; 1) uniform convergence (u_ε) ; 2) pure distortion (ovalization, u_δ, with no change in volume of the cavity), and 3) vertical translation (Δu_y). The convergence parameter u_ε is clearly related to the change in volume of the tunnel cavity (per unit length), $2u_\varepsilon/R = \Delta V_L/V_0$, where V_0 is the initial tunnel volume.

Verruijt (1997) has presented complete elastic solutions for the uniform convergence mode using the complex formulation of planar elasticity. The physical domain is mapped into an annular space and displacements are solved as series solutions of Goursat functions (after Muskhelishvili, 1963). Pinto (1999) has subsequently extended this method to solve the ovalization mode. Vertical translation components are solved from both of these preceding calculations by prescribing zero vertical displacements far from the tunnel (i.e., the vertical translation is a function of the deformations at the tunnel cavity, u_ε and u_δ, as well as the model input parameters, R/H, v). The principal advantage of these elastic solutions is that they do represent the physical dimension of the tunnel cavity, R. However, the results are inconveniently derived in an infinite series form and therefore most easily accessed in graphical form.

An alternative simplified formulation proposed by Sagaseta (1987), and later refined by Verruijt and Booker (1996) and Pinto (1999), is based on the superposition of singularity solutions to represent the effects of ground loss and ovalization in an elastic half-plane. This formulation can also address partially the

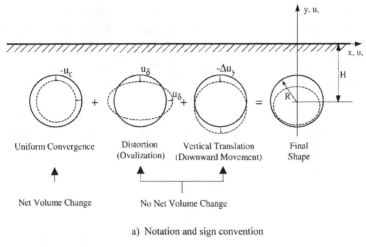

Uniform Convergence · Distortion (Ovalization) · Vertical Translation (Downward Movement) · Final Shape

Net Volume Change · No Net Volume Change

a) Notation and sign convention

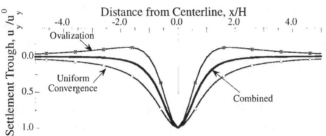

b) Component contributions to surface settlement trough

Figure 11. Deformation modes around tunnel cavity

effects of soil plasticity and dilation (Sagaseta & González, 1999; Pinto & Whittle, 2001) and is readily extended to 3-D conditions (Pinto, 1999). The superposition principle is identical to that used in the SSPM formulation for pile installation. Ground loss (i.e., uniform convergence) in the half-plane is represented by combining fundamental solutions for: 1) a line sink in a full plane, 2) an image source of equal strength located equidistant (depth, H) above the plane of the ground surface (which cancels the normal stress component along this surface), and 3) a distribution of corrective shear tractions along the bisecting line (simulating a traction-free ground surface). Similarly, the ovalization mode is represented by superimposing fundamental solutions for distortion of a circular cavity in a prestressed elastic full plane (after Kirsch, 1898; this represents effects of stress deviation due to K_0 conditions). Appendix I summarizes the complete solutions for ground movements (u_x, u_y; in an elastic half-plane). It should be noted that both the uniform convergence and ovalization modes induce vertical translation of the tunnel cavity (Δu_y; Fig. 11a, eqns. I.1c and I.2c). Further solutions have also been reported by Chatzigiannelis and Whittle (2001) for cross-anisotropic, elastic soils.

Figure 11b illustrates the distribution of surface settlements for the uniform convergence and ovalization modes. It is clear that the resulting trough shape is controlled primarily by the ratio of the two modes, $\rho = -u_\delta/u_\varepsilon$, which is subsequently referred to as the 'relative distortion' parameter. For deep tunnels in an elastic medium, the parameter ρ can be written:

$$\rho = \frac{1-K_0}{1+K_0+2\cdot r_u} \cdot \frac{3-4\cdot v}{1-r_p} \tag{6}$$

where $r_p = p_i/p_0$ is the internal pressure ratio, defined as the uniform pressure that the tunnel lining transmits to the soil in the radial direction divided by the hydrostatic component of the total stress (acting at the tunnel springline) prior to tunnel excavation; and $r_u = p_w/\sigma'_{v0}$ is the ratio of the in-situ pore pressure to the vertical effective stress. The far field pressure is:

$$p_0 = \sigma'_{v0}\left(\frac{1+K_0}{2}\right)+p_w \tag{6a}$$

By estimating typical values of the parameters in equations 6, Pinto and Whittle (2001) suggest that the typical range for the relative distortion is $-0.5 \le \rho \le 3$. The high end of this range can be expected in soft, normally consolidated deposits, while highly overconsolidated soils will have low (even negative) values of ρ.

Figure 12 compares analytical solutions of ground movements (horizontal, u_x/u_ε; and vertical, u_y/u_ε) around a very shallow tunnel with R/H = 0.45 (and assumed values $\rho = 0.5$ and $v = 0.25$) using the exact (complex variable) and approximate (superposition of line sources) methods. The agreement is surprisingly close between these two solutions (differences are within 10% above the crown, and indistinguishable in much of the soil mass), especially considering this is such a shallow tunnel. Hence, the results confirm the efficacy of using approximate superposition methods for shallow tunnels.

One of the key limitations of the proposed analysis is the assumption that the soil behavior can be approximated by linear elasticity. In practice, the zone of soil close to the tunnel cavity may experience significant shearing (full mobilization of

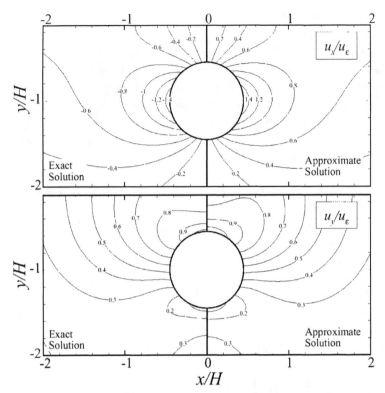

Figure 12. Comparison of elastic solutions for shallow tunnel
$(R/H = 0.45, \rho = -0.5, \nu = 0.25)$

shear strength) which may be accompanied by irrecoverable volume expansion (dilation due to shearing). The effects of this local failure zone have been evaluated using analytical solutions derived by Yu and Rowe (1998) (see also Yu, 2000) for the expansion (or contraction) of a cylindrical cavity in a full plane assuming Mohr-Coulomb yield of the soil. The results show two key features: 1) soil plasticity affects predictions of ground displacements within a zone up to one radius beyond the tunnel wall; and 2) within the failure zone the elasto-plastic solutions predict ground displacements that are larger than corresponding elastic analyses. These results imply that far field (i.e., r/R ≥ 2) ground deformations can be adequately modeled by elastic theory, while deformations measured at the tunnel wall will always be underestimated by the elastic analyses. This behavior is clearly important if far field ground movements are to be related to deformations measured at (or close to) the tunnel cavity. Pinto (1999) suggests that the tunnel cavity (or near field) deformations can be reduced to an equivalent elastic behavior and has presented a set of reduction factors based on specified shear stiffness and strength parameters for the soil mass (G/p'_0, c'/p'_0, ϕ'). Estimation of these parameters can be made using correlations presented by Ladd et al. (1977) and Mesri and Abdel-Ghaffar (1993).

Apart from linear elasticity, analytical solutions for ground movements can also be obtained for simple soil plasticity models under the assumption of a constant flow rule:

$$\varepsilon_{vol} = -\sin \psi \cdot \gamma_{max} \tag{7}$$

where ε_{vol} is the volumetric strain, γ_{max} the maximum shear strain, and ψ an average angle of dilation.

Under this assumption (and neglecting elastic strains), Sagaseta (1987) shows that the radial displacements around a line source attenuate according to $u_r \sim 1/r^\beta$, where $\beta = (1+\sin\psi)/(1-\sin\psi)$. For soils where there is no dilation (e.g. undrained shearing of clays), $\beta = 1$ and the displacement field, $u_r \sim 1/r$ (identical to elastic case with $\nu = 0.5$). Sagaseta and González (1999) and Pinto (1999) have subsequently presented analytical solutions for the uniform convergence and ovalization deformation modes in terms of $\alpha = (1+\beta)/2 = 1/(1-\sin\psi)$. Assuming a maximum dilation angle, $\psi = 30°$, then $1 \leq \alpha \leq 2.0$.

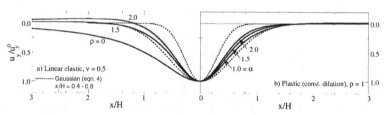

Figure 13. Effect of input parameters on surface settlement distribution

Figures 13a and b compare the surface settlement distributions (u_y/u_y^0) obtained analytically by assumptions of linear elastic (ρ, ν) and constant dilation plastic (ρ, α) material behavior, respectively, with the empirical relations (eqn. 5) assuming $x_i/H = 0.4 - 0.8$. It is clear that the parameters ρ and α control the predicted settlement trough shape (ν plays only a secondary role) and can match a large fraction of the empirically observed settlement distributions (cf. Fig. 10).

Effects of More Complex Soil Behavior

Further understanding of the role of soil behavior in predicting the settlement trough shape can be achieved by comparing the analytical solutions with numerical calculations using more realistic constitutive models for soils. For example, Pestana and Whittle (1999) have presented a generalized elasto-plastic model, referred to as MIT-S1, which is capable of predicting the rate independent, effective stress-strain-strength behavior of uncemented soils over a wide range of confining pressures and densities. Figure 14 illustrates typical MIT-S1 predictions for a series of five standard, drained triaxial compression shear tests (with parameters derived for a reference cohesionless soil, Toyoura sand) performed at the same confining pressure but a range of initial void ratios (very dense, $e_0 = 0.6$, to very loose, $e_0 = 0.95$). The predicted stress-strain behavior is consistent with measured data, with non-linear stiffness properties clearly seen at axial strains $\varepsilon_a < 0.01\%$, and dilation occurring at $\varepsilon_a = 0.5\%$ for dense samples at high mobilized friction angles.

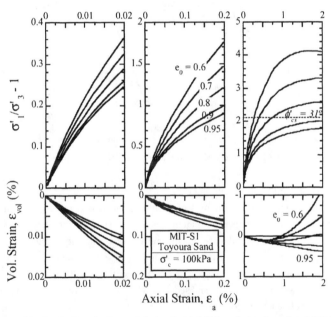

Figure 14. Typical predictions of drained triaxial shear behavior using MIT-S1 model

Figure 15 compares the distributions of surface settlements (u_y/u_y^0) for a tunnel with R/H = 0.45 in a cohesionless, medium-dense Toyoura sand $(e_0 = 0.75)$ in the uniform convergence and ovalization modes of deformation (Hsieh, 2001). The distribution of surface settlements predicted by MIT-S1 is very similar to the analytical solutions (linear, isotropic soil) for very small values of u_ε/R (0.002, 0.02%, Fig.1 5a). However, as the ground loss increases to $u_\varepsilon/R = 0.2\%$, there is a very substantial narrowing of the settlement trough associated with an incipient failure mechanism predicted by MIT-S1. In contrast, Fig. 15b shows surprisingly close agreement between the non-linear numerical solutions and analytical solutions for ovalization up to $u_\delta/R = 0.2\%$. These results represent a first step towards a more comprehensive understanding linking soil properties to tunnel-induced ground movements.

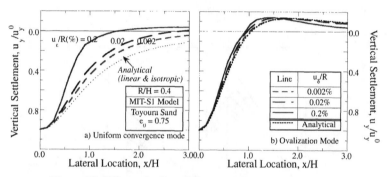

Figure 15. Effects of soil modeling on surface settlement trough

Interpretation of Field Monitoring Data

The analytical predictions of ground deformations for a given tunnel are controlled by three independent parameters 1) v, ρ, and u_ε for the linear elastic soil; or 2) α, ρ and u_ε for the case of constant dilation plasticity. These three input parameters can be derived from a set of three independent field measurements. In practice, it is common to measure the surface settlement above the crown of the tunnel. However, the location and scope of other field data varies significantly from one project to another. In order to illustrate the design capability and to evaluate the analyses it is necessary to assume some reference measuring system. Pinto and Whittle (2001) recommend using three pre-defined measurements to interpret the model parameters as shown in Figure 16: 1) the centerline surface settlement, u_y^0; 2) the surface settlement offset at a distance H from the centerline of the tunnel, u_y^1; and 3) the horizontal displacement measured at the springline elevation (i.e., at depth H) in an inclinometer located 2R from the center of the tunnel, u_x^0. Figures 17a, b and c illustrate the design charts for a tunnel with R/H = 0.2. The standard interpretation procedure first uses the measured ratios, u_x^0/u_y^0 and u_y^1/u_y^0 to estimate parameter sets (ρ, ν) or (α, ρ) (e.g., Fig. 17a). these parameters are then used to find the ground loss ratio u_y^0/u_ε from either Fig 17b or 17c (under the preferred assumption of soil model).

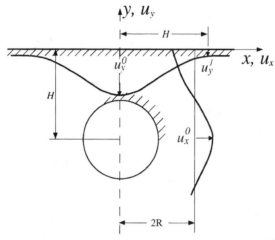

Figure 16. Reference field measurements for proposed design method (Pinto, 1999)

Clough et al. (1983) describe measurements of ground movements associated with construction of the N-2 contract tunnel for the San Francisco clean water project. This was the first usage of an Earth Pressure Balance (EPB) tunnel boring machine in the US. Figure 18 shows a typical cross-section of the tunnel with radius, R = 1.78m located at a depth, H = 9.6m (R/H = 0.185). The soil profile comprises 6.6m of rubble fill underlain by a 7.1m layer of Recent Bay Mud, containing the tunnel and underlain by colluvium and residual sandy clay. The current analyses consider the settlements and lateral deflections measured at one line of instrumentation (line 4; Clough et al., 1983) 15 days after the passage of the tunnel shield. The surface settlements were only measured over a width -0.5 < x/H < 0.5, from which u_y^0 = -30.6mm is well defined while u_y^1 is not measured. An

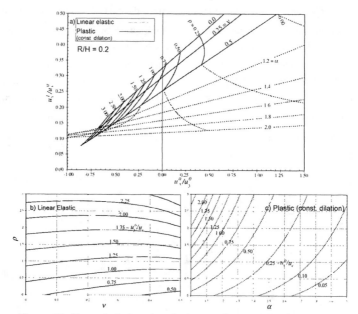

Figure 17. Example design chart for interpreting tunnel-induced ground
displacements (after Pinto & Whittle, 2001)

$u_x^0/u_y^0 = -0.77$	R/H (from design charts)		
$\nu = 0.5$	0.15	0.2	**R/H = 0.18**
ρ	1.7	2.5	**2.15**
u_y^0/u_ε	1.30	2.30	**1.86**
u_ε [mm]	-24	-13	**-19**
$(\Delta V_L/V_0 \times 100\%)$			**(2.15%)**
u_δ [mm]	--	--	**41**
Δu_y [mm]	Eqn. I.1c:	Eqn. I.2c:	**-9**
	$\Delta u_y/u_\varepsilon = 0.175$	$\Delta u_y/u_\delta = -0.170$	
	$\Delta u_y = -3.5\text{mm}$	$\Delta u_y = -5.6\text{mm}$	
Crown settlement [mm]: $u_c = u_\varepsilon - u_\delta + \Delta u_y$			**-69**

Note: Sign convention for displacements given in Figure 11a

Table 1. Interpretation of tunnel deformation parameters, N-2 water tunnel, San
Francisco

inclinometer installed at x = 3.6m (x/R ≈ 2.0), enables direct evaluation of u_x^0 = 23.5mm. Given that the soil is a soft clay, the analyses can be applied under the assumption of incompressible behavior (v = 0.5). Table 1 summarizes the model parameters derived from these data (based on design charts at R/H = 0.15, 0.20):

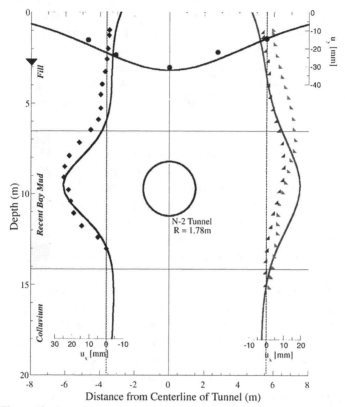

Figure 18. Comparison of computed and measured deformations, N-2 tunnel project (predictions – solid lines)

The proposed analyses identify a relatively large distortion parameter, ρ = 2.15, consistent with the low K_0 conditions anticipated in Recent Bay Mud. Figure 18 shows that the predicted surface settlement trough is in very good agreement with the measured data. It should be noted that these same data were also well matched by Clough et al. (1983) using the empirical Gaussian curve with x_i/H = 0.42. The analyses also give very reasonable estimates of the outward lateral deflections measured in inclinometers located 3.6m and 5.6m (data for two inclinometers are combined in the figure) from the tunnel centerline, Figure 18. The analyses predict maximum lateral movements at the springline elevation, while the measured maxima occur between the crown and springline. Table 1 also details model predictions of the vertical displacements at the crown of the tunnel, u_c = -69mm (Table 1). This corresponds to approximately 90% of the 76mm tail void for this particular tunnel shield (assuming no pitching of the machine). Although the tail void was infilled with a sand-bentonite-cement grout (injected at a pressure of 210-315kPa), Clough et al. (1983) hypothesized that the Bay Mud did indeed fill much of the tail void prior to grout injection. The current analyses are consistent with this hypothesis.

DISCUSSION

Current geotechnical practice for estimating ground movements caused by pile driving and shallow tunneling relies almost exclusively on empirical experience. This paper shows that simplified analytical models can achieve predictions of these movements that are not only consistent with empirical data but also provide a more comprehensive framework for understanding these processes. The preceding examples have shown the ability to assess the effectiveness of pre-augering for driven piles and the consistency of lateral soil displacement predictions around a shield tunnel.

There are many direct extensions of the current analyses that are of great potential. For example, Pinto (1999) has derived analytical solutions for 3-D tunnel headings that can relate ground movements to the distribution of ground loss around the tunnel face, shield and lining (and hence, enable comparisons for different methods of tunneling). The solutions also offer a very convenient framework for modeling effects of related processes such as compaction grouting.

The results in this paper also suggest that certain classes of geotechnical problem can be solved with adequate accuracy using minimal knowledge of soil properties. This confirms the importance of the underlying kinematic constraints, such as the incompressibility condition for undrained shearing of clay (basis of cavity expansion and strain path methods; Bishop et al., 1945; Hill, 1963; Baligh, 1985 and Yu, 2000), and the ability to use linear, elastic theory to represent stress-free conditions at the ground surface. Although the same theories can be applied to near field problems (e.g., effects of PV drain installation; Saye, 2001), the modeling of soil behavior becomes a dominant factor in these predictions. Generalized effective stress models have proved instrumental in achieving reliable predictions of friction pile set-up in clays (e.g., Whittle, 1992; Whittle & Sutabutr, 1999) and have even been used to quantify the effects of installation disturbance on measurements made by pressuremeter devices (Aubeny et al., 1998).

In light of the experience reported in Figure 1, it is interesting to review the potential applications of simplified analytical methods for predicting ground movements around excavations. Hashash and Whittle (1994, 1996) have shown, through a series of parametric finite element studies, that non-linear and inelastic soil behavior controls the distribution of excavation-induced surface settlements around braced excavations in clay; while simple elastic models generate wholly unrealistic trough shapes (compared to empirical data). However, other sources of ground movements such as wall installation (either driving of sheet pile sections or slurry trench methods) are clearly amenable to simplified methods of analysis. For example, Sagaseta and Whittle (2001) show good agreement between SSPM predictions of ground deformations caused by sheet pile wall installation and measured field data reported by Finno et al. (1988). These solutions provide an important first step towards modeling of wall installation in subsequent analyses of excavation performance. Although ground losses associated with tieback installation (such as Fig. 1) were unexpected, the simplified analyses clearly provide a framework for explaining and assessing their relation to movements in the overlying soil.

CONCLUSIONS

This paper has demonstrated that simplified analytical methods can provide reliable predictions of ground movements caused by pile driving (well-defined displacement of soil volume, 'gain of ground') and shallow tunneling (ill-defined loss of ground). The analyses (especially the convenient closed-form solutions) are practical to use, provide more complete insight on factors controlling the distributions of ground movements than existing empirical methods, and require minimal information on soil properties. Further validation of these methods can be found in related papers by Sagaseta and Whittle (2001) and Pinto and Whittle (2001).

ACKNOWLEDGMENTS

The title of this paper is an obscure pun on our colleague Chuck Ladd's enthusiasm for gridiron football. Research on the study of ground movements associated with pile driving was initially sponsored by the offshore oil industry through a Joint Industry Project on the design of suction caisson anchors, while work on shallow tunneling has been carried out in conjunction with underground construction of the Tren Urbano in Río Piedras. The Authors would like to make special acknowledgment of Prof. Mohsen Baligh who initiated the use of strain path analyses in geotechnical engineering. David Druss of PBQD generously supplied data on ground movements in East Boston. Finally, our thanks to current and recent graduate students who have contributed to this work, notably Federico Pinto, Jon Rehkopf, Yannis Chatzigiannelis and Yo-Ming Hsieh.

REFERENCES

Aubeny, C.P., Whittle, A.J. & Ladd, C.C. (1998) "Effects of disturbance on undrained strengths interpreted from pressuremeter tests," *ASCE Journal of Geotechnical and Geoenvironmental Engineering*, 126(12), 1133-1145

Azzouz, A.S., Baligh, M.M. & Whittle, A.J. (1990) "Shaft resistance of friction piles in clay," *ASCE Journal of Geotechnical Engineering*, 116(2), 205-221.

Baligh, M.M. (1985). "Strain Path Method." *ASCE Journal of Geotechnical Engineering*, 111(9), 1108-1136.

Baligh, M.M. & Levadoux, J-N. (1980). "Pore pressure dissipation after cone penetration." *Research Report R80-11*, Dept. of Civil Engineering, MIT, Cambridge, MA.

Baligh, M.M., Azzouz, A.S. & Chin, C.T. (1987). "Disturbances due to 'Ideal' tube sampling." *ASCE Journal of Geotechnical Engineering*, 113(7), 739-757.

Bishop, R.F., Hill, R. & Mott, N.F. (1945) "The theory of indentation and hardness tests," *Proc. Phys. Soc.,* London, 57, 149-159.

Boston Globe (2001) "In a political and physical feat, tower set to re-open," Kevin McCullen, June 15.

Burland, J.B., Jamiolkowski, M. & Viggiani, C. (1998) "Stabilizing the Leaning Tower of Pisa," *Bull. Eng. Geol. Env.*, 57(1), 91-99

Chatzigiannelis, Y. & Whittle, A.J. (2001) "Analysis of ground deformations caused by shallow tunneling in cross-anisotropic soil," *submitted for publication.*

Clough, G.W., Sweeney, B.P., & Finno, R.J. (1983) "Measured Soil Response to EPB Shield Tunneling," *ASCE Journal of Geotechnical Engineering*, 109(2), 131-149.

Clough, G.W. & O'Rourke, T.D. (1990). "Construction Induced Movements of Insitu Walls." *Proc. ASCE Specialty Conf. on Design and Performance of Earth Retaining Structures,* Ithaca, NY., 439-470.

Costanzo, D., Jamiolkoswki, M., Lancellotta, R. & Pepe, M.C. (1994) "Leaning Tower of Pisa: Description of the behavior," *Proc. Settlement'94, ASCE,* Special Volume – Banquet Lecture, College Station, TX, 55p.

D'Appolonia, D.J., & Lambe T.W. (1971) "Performance of four foundations on end-bearing piles," *J. Soil Mechs. & Found. Engrg. Div., ASCE,* 97(1), 77-93.

Dugan, J.P. & Freed, D.L. 91984) "Ground heave due to pile driving," *Proc. 1st. Int. Conf. on Case Histories in Geot. Engrg.,* S. Prakash, Ed., St. Louis, MO., 1, 117-122..

Finno, R.J., Atmatzidis, D.K. & Nerby, S.M. (1988) "Ground response to sheet pile installation in clay," *Proc. 2nd. Int. Conf. on Case Histories in Geot. Engrg.,* S. Prakash, Ed., St. Louis, MO., 1, 1297-1301.

Gioda G., & Swoboda G. (1999) "Developments and applications of the numerical analysis of tunnels in continuous media," *International Journal for Numerical and Analytical Methods in Geomechanics,* 23, 1393-1405.

Gue, S.S. (1984) *Ground heave around driven piles in clay,* PhD. Thesis, University of Oxford, UK.

Hagerty, D.J. & Peck, R.B. (1971) "Heave and lateral movements due to pile driving." *J. Soil Mech. and Found. Div., ASCE,* 97(11), 1513-1532.

Hashash, Y.M.A & Whittle, A.J. (1996). Ground movement prediction for deep excavations in soft clay. *ASCE J. Geotechnical Engrg.,* 111(3), 302-318.

Hill, R. (1963) "A general method of analysis of metal-working processes," *J. Mech. Phys. Solids,* 11, 305-326.

Hsieh, Y.M. (2001) Ph.D Thesis in progress, MIT, Cambridge, MA.

Jen, L. (1998) *The design and performance of deep excavations in clays,* Ph.D Thesis, Dept. of Civil & Environmental Engineering, MIT, Cambridge, MA.

Kiousis, P.D., Voyiadis, G.Z. & Tumay, M.T. (1988) "A large strain theory and its application in the analysis of the cone penetration mechanism *Intl. Journal for Numerical and Analytical Methods in Geomechanics,* 12(1), 45-60.

Koutsoftas, D.C. (1982). "H-pile heave: a field test." *ASCE Journal of Geotechnical Engineering,* 108(8), 999-1016.

Ladd, C.C., Foott, R., Ishihara, K., Schlosser, F., and Poulos, H.G. (1977) "Stress-deformation and strength characteristics," *Proc. 9th Intl. Conf. Soil Mechs. & Found. Engrg.,* Tokyo, 2, 421-494.

Levadoux, J-N. (1980) *Pore pressures in clays due to cone penetration*, Ph.D Thesis, Dept. of Civil Engineering, MIT, Cambridge, MA.

Mair, R.J. & Taylor, R.N. (1997) "Bored tunnelling in the urban environment," *Proc. 14th Intl. Conf. Soil Mechs. & Found. Engrg*, Hamburg, (4), 2353-2385.

Mesri, G. & Abdel-Ghaffar, M.E.M. (1993) "Cohesion intercept in effective stress stability analysis," *ASCE Journal of Geotechnical Engineering*, 119(8), 1229-1249.

Morrison, M.J. (1984) *In situ measurements on a model pile in clay*, Ph.D Thesis, Dept. of Civil Engineering, MIT, Cambridge, MA.

Muskhelishvili, N. I. (1963) *Some Basic Problems of the Mathematical Theory of Elasticity*, P. Noordhoff Ltd., Groningen, The Netherlands.

Panet, M. & Guenot, A. (1982) "Analysis of convergence behind face of a tunnel," *Proc. Tunneling'82*, 197-204.

Payiatakis, S. & Davie, J. (1998) "Structure heave and settlement due to pile driving – A case history," *Effects of Construction on Structures*, ASCE GSP 84, 16-29.

Peck, R.B. (1969) "Deep Excavations and Tunnels in Soft Ground". *Proc. 7th Intl. Conf. Soil Mechs. & Found. Engrg*, Mexico City, State of the Art Volume, 225-290.

Pestana, J.M. & Whittle, A.J. (1999) "Formulation of a unified constitutive model for clays and sands," *Intl. Journal for Numerical and Analytical Methods in Geomechanics*, 23, 1215-1243.

Pinto, F. (1999) *Analytical methods to interpret ground deformations due to soft ground tunneling*, S.M Thesis, Dept. of Civil and Environmental Engineering, MIT, Cambridge, MA.

Pinto, F. & Whittle, A.J. (2001) "Evaluation of analytical solutions for predicting deformations around shallow tunnels in soft ground," *submitted for publication*.

Poulos, H.G. (1994) "Effect of pile driving on adjacent piles in clay," *Canadian Geotechnical Journal*, 31(6), 856-867.

Randolph, M.F., Carter, J.P. & Wroth, C.P. (1979), "Driven piles in clay: Effects of installation and subsequent consolidation," *Géotechnique*, 29(4), 361-393.

Rehkopf, J.C. (2001) *Prediction and measurement of ground movements due to pile driving in clay: A case study in East Boston*, M.Eng Thesis, Dept. of Civil and Environmental Engineering, MIT, Cambridge, MA.

Robinsky, E.I. & Morrison, C.F. (1964) "Sand displacement and compaction around friction piles," *Canadian Geotechnical Journal*, 1, 81-93.

Roy, M. Blanchet, R., Tavenas, F. & La Rochelle, P. (1981). "Behavior of a sensitive clay during pile driving," *Canadian Geotechnical J.*, 18, 67-86.

Sagaseta, C. (1987) "Analysis of Undrained Soil Deformation due to Ground Loss," *Géotechnique* 37(3), 301-320.

Sagaseta, C. (1998) On the role of analytical solutions for the evaluation of soil deformations around tunnels. *Application of Numerical Methods to Geotechnical Problems* (A. Cividini, ed.), CISM Courses and Lectures No. 397, 3-24.

Sagaseta, C. & González, C. (2000) "Predicción teórica de subsidencias," Chapter 6.2 Aspectors Geotécnicos de la Ampliación del Metro de Madrid.

Sagaseta, C. & Whittle, A.J. (1996). Discussion to 'Effect of pile driving on adjacent piles in clay', *Canadian Geotechnical Journal*, 33(3), 525-527.

Sagaseta, C. & Whittle, A.J. (1997). "Heave and settlements around piles driven in clay." *Computational Plasticity. Fundamentals and Applications, Proc. Complas V*, D.R.J. Owen, E. Oñate and E. Hinton, Eds., 2, 1722-1728.

Sagaseta, C., Whittle, A.J. & Santagata, M. (1997). "Deformation analysis of shallow penetration in clays.", *Int. J. Num. and Anal. Meth. in Geomech.*, 21, 687-719.

Sagaseta, C. & Whittle, A.J. (2001). "Ground movements due to pile driving in clay," *ASCE Journal of Geotechnical & Geoenvironmental Engineering*, 127(1), 55-66.

Saye, S.R. (2001) "Assessment of soil disturbance by the installation of displacement sand drains and prefabricated vertical drains," Soil Behavior & Soft Ground Construction, Ladd Symposium

Schmidt, B. (1969) *Settlements and ground movements associated with tunnelling in soil*, Ph.D. Thesis, Dept. of Civil Engineering, University of Illinois, Urbana-Champaign.

Terracina, F. (1962) "Foundation of the Tower of Pisa," *Géotechnique*, 12(4), 336-339.

Timoshenko, S. P., & Goodier, J. N. (1970) *Theory of elasticity*, Third Edition, McGraw Hill Kogakusha, Ltd.

van den Berg, P., de Borst, R. and Huetink, H. (1996) "An Eulerian finite element model for penetration in layered soil," *International Journal for Numerical and Analytical Methods in Geomechanics*, 20, 865-886

Verruijt, A. (1997) "A Complex Variable Solution for a Deforming Tunnel in an Elastic Half-Plane," *International Journal for Numerical and Analytical Methods in Geomechanics*, 21, 77-89.

Verruijt, A. and Booker, J.R. (1996) "Surface settlements due to deformation of a tunnel in an elastic half plane," *Géotechnique,* 46(4), 753-756.

Vesic, A.S. (1963) 'Bearing capacity of deep foundations to sand', *Stresses in Soils and Layered Systems, Highway Research Board*, No. 39, 112-153.

Whelan, M.P. (1995) *Performance of deep excavations in South Boston for the Central Artery/Third Harbor Tunnel project*, S.M. Thesis, Dept. of Civil and Environmental Engineering, MIT, Cambridge, MA.

Whittle, A.J. (1992) "Assessment of an effective stress analysis for predicting the performance of driven piles in clays," *Advances in Underwater Technology, Ocean Science and Offshore Engineering Volume 28*, Offshore Site Investigation and Foundation Behavior, Society for Underwater Technology, London, 607-643.

Whittle, A.J. & Hashash, Y.M.A. (1994) "Soil modeling and prediction of deep excavation behavior," *Proc. Intl. Symp. on Pre-Failure Deformation Characteristics of Geo-Materials* (IS-Hokkaido'94), Sapporo, 1, 589-595.

Whittle, A.J. & Kavvadas, M.J. (1994) "Formulation of the MIT-E3 constitutive model for overconsolidated clays," *ASCE Journal of Geotechnical Engineering*, 120(1), 173-199.

Whittle, A.J. & Sutabutr, T. (1999) "Prediction of pile set-up in clay," *Transportation Research Record*, 1663, pp. 33-41

Whittle, A.J., DeGroot, D.J., Ladd, C.C., & Seah, T-H. (1994) "Model prediction of the anisotropic behavior of Boston Blue Clay," *ASCE Journal of Geotechnical Engineering*, 120(1), 1994, 199-225.

Whittle, A.J., Sutabutr, T., Germaine, J.T. & Varney, A. (2001) "Prediction and interpretation of pore pressure dissipation for a tapered piezoprobe," *Géotechnique*, 51(7), 601-617.

Yu, H.S. (2000) *Cavity expansion methods in geomechanics*, Kluwer Academic Publishers, Dordrecht, The Netherlands.

Yu, H.S., & Rowe, R.K. (1998) "Plasticity solutions for soil behavior around contracting tunnels and cavities," *International Journal for Numerical and Analytical Methods in Geomechanics*, 23, 1245-1279.

APPENDIX I. ANALYTICAL SOLUTIONS FOR GROUND DEFORMATIONS IN AN ELASTIC HALF-PLANE

1. Uniform Convergence Mode

$$\frac{u_x}{u_\varepsilon} = x \cdot R \cdot \left\{ \frac{1}{x^2 + (y+H)^2} - \frac{1}{x^2 + (y-H)^2} + \frac{4 \cdot (1-v)}{x^2 + (y-H)^2} - \frac{4 \cdot (y-H) \cdot y}{\left[x^2 + (y-H)^2 \right]^2} \right\} \tag{I.1a}$$

$$\frac{u_y}{u_\varepsilon} = R \cdot \left\{ \begin{array}{l} \dfrac{(y+H)}{x^2 + (y+H)^2} - \dfrac{(y-H)}{x^2 + (y-H)^2} + \dfrac{4 \cdot (y-H) \cdot x^2 + 2 \cdot H \cdot \left[x^2 - (y-H)^2 \right]}{\left[x^2 + (y-H)^2 \right]^2} - \\[3mm] \ldots - \dfrac{4 \cdot (1-v) \cdot (y-H)}{x^2 + (y-H)^2} \end{array} \right\} \tag{I.1b}$$

Associated vertical translation of tunnel springline

$$\frac{\Delta u_y}{u_\varepsilon} = 4 \cdot \frac{R}{H} \cdot \frac{8 \cdot (1-v) - (1-2 \cdot v) \left(\dfrac{R}{H} \right)^2}{\left[4 + \left(\dfrac{R}{H} \right)^2 \right]^2} \tag{I.1c}$$

2. Ovalization Mode

$$\frac{u_x}{u_\delta} = \frac{R \cdot x}{3 - 4 \cdot v} \cdot \left\{ \begin{array}{l} \dfrac{(3 - 4 \cdot v) \cdot \left[x^2 + (y+H)^2 \right]^2 - \left[3 \cdot (y+H)^2 - x^2 \right] \cdot \left[x^2 + (y+H)^2 - R^2 \right]}{\left[x^2 + (y+H)^2 \right]^3} - \\[4mm] \ldots - \dfrac{(3 - 4 \cdot v) \cdot \left[x^2 + (y-H)^2 \right]^2 - \left[3 \cdot (y-H)^2 - x^2 \right] \cdot \left[x^2 + (y-H)^2 - R^2 \right]}{\left[x^2 + (y-H)^2 \right]^3} + \\[4mm] \ldots + \dfrac{x^2 + y^2 - H^2}{\left[x^2 + (y-H)^2 \right]^2} \cdot 8 \cdot (1-v) - 8 \cdot y \cdot \dfrac{y \cdot (x^2 + y^2) + 2 \cdot H \cdot (H^2 - x^2) - 3 \cdot y \cdot H^2}{\left[x^2 + (y-H)^2 \right]^3} \end{array} \right\} \tag{I.2a}$$

$$\frac{u_y}{u_\delta} = \frac{R}{3 - 4 \cdot v} \cdot \left\{ \begin{array}{l} (y - H) \cdot \dfrac{(3 - 4 \cdot v) \cdot \left[x^2 + (y - H)^2 \right]^2 - \left[3 \cdot x^2 - (y - H)^2 \right] \cdot \left[x^2 + (y - H)^2 - R^2 \right]}{\left[x^2 + (y - H)^2 \right]^3} - \\[6mm] \ldots - (y + H) \cdot \dfrac{(3 - 4 \cdot v) \cdot \left[x^2 + (y + H)^2 \right]^2 - \left[3 \cdot x^2 - (y + H)^2 \right] \cdot \left[x^2 + (y + H)^2 - R^2 \right]}{\left[x^2 + (y + H)^2 \right]^3} + \\[6mm] \ldots + \dfrac{x^2 \cdot (2 \cdot H - y) - y \cdot (y - H)^2}{\left[x^2 + (y - H)^2 \right]^2} \cdot 8 \cdot (1 - v) - \\[6mm] \ldots - \dfrac{8 \cdot (y - H) \cdot \left\{ H \cdot y \cdot (y - H)^2 - x^2 \cdot \left[\left(x^2 + y^2 \right) + H \cdot (y + H) \right] \right\}}{\left[x^2 + (y - H)^2 \right]^3} \end{array} \right\} \quad \text{(I.2b)}$$

Associated vertical translation of tunnel springline:

$$\frac{\Delta u_y}{u_\delta} = \frac{2}{3 - 4.v} \cdot \frac{R}{H} \cdot \frac{(1 - 8.v) \cdot \left(\dfrac{R}{H} \right)^4 - (11 - 8.v).4 \left(\dfrac{R}{H} \right)^2 - 32}{\left[4 + \left(\dfrac{R}{H} \right)^2 \right]^3} \quad \text{(I.2c)}$$

Post-Preload Settlements of a Soft Bay Mud Site

ABSTRACT

The paper presents a case history involving a soft bay mud site that had been reclaimed for commercial development by filling over old marshlands along the margins of San Francisco Bay. Prefabricated vertical drains spaced 1.22 m on center and surcharge were used to accelerate the consolidation of the soft and compressible bay mud. Slower than anticipated settlements and construction schedule limitations resulted in premature removal of the surcharge, resulting in significantly higher post-construction settlements than anticipated. A large single-story retail building was constructed before the full extent of potential post-construction settlements was realized. Post-preload settlements were so severe that settlements up to 41 mm were recorded within the two-month construction period for the building. An extensive investigation was undertaken to evaluate the severity of the settlement problems and to develop a plan of monitoring and implementing remedial measures to keep the building functional. The investigations included additional explorations and testing, reanalysis of the preload data, installation of piezometers and groundwater monitoring wells at four instrumentation stations, and monitoring the settlements at each of the columns and along the exterior walls. The building experienced post-construction settlements ranging between 140 mm and 376 mm over a period of 9½ years (October 1991 – February 2001) and differential settlements up to 100 mm, with 20% of the column bays experiencing differential settlements in excess of the tolerable limit of 38 mm (over a span of 13.72 m). Two of the four walls of the building experienced severe differential settlements which resulted in extensive cracking that required significant strengthening measures. Many of the columns had to be jacked up and shimmed to keep the differential settlements within tolerable limits.

The paper reviews the design of the preload program and the settlements recorded during the preload period. It reviews the post-construction investigations, summarizes the geotechnical characteristics of the bay mud, and evaluates the performance of the preload. It summarizes the results of the monitoring program and reviews the lessons learned regarding the use of vertical drains in combination with preloading to accelerate consolidation of soft clay soils.

[1] Principal, URS Corporation, 221 Main St., Suite 600, San Francisco, CA 94105; phone (415) 243-3840.

INTRODUCTION

A large shopping center was developed along the northwestern margins of the San Francisco Bay by filling over an area that had been previously occupied by marshlands. Fills up to 6 m thick were required to raise the site grades to the final design elevations and to accommodate up to 2.4 m of settlement resulting from consolidation of the soft bay mud deposit that underlies the site. Vertical drains spaced 1.22 m on center were installed to accelerate consolidation of the bay mud. The design recognized that post-construction settlements in the range of 100 mm to 150 mm would develop due to secondary compression of the bay mud, but it was estimated that the differential settlements would remain within tolerable limits. The design of the vertical drains relied on empirical correlations to estimate the coefficients of horizontal consolidation (C_h) based on coefficients of vertical consolidation (C_v) measured from a limited number of consolidation tests. The design was based on the expectation that primary consolidation would be completed in less than six months, and that surcharging the site would reduce the rates of secondary compression. The design of the surcharge was based on the assumption that the preload would be removed when the measured settlements reached or exceeded the theoretically calculated settlements to achieve 100% consolidation of the bay mud under the final stresses imposed by the fill and the superstructure. No pore water pressure measurements were made until the very end of the preload period, at which time schedule considerations required removal of the surcharge.

A combination of factors led to premature removal of the surcharge: (1) the rate of consolidation was significantly slower than anticipated because the assumed values of C_h were several times higher than the actual; (2) the last 3 m of fill were placed very slowly and thus the full weight of the fill was not effective until five months after installation of the vertical drains; and (3) although the time of removal of the surcharge was postponed by about 4½ months, no active measures were taken to accelerate the consolidation process, and in the end time ran out and the surcharge had to be removed to avoid jeopardizing the project's construction schedule. A dispute between the owner of the building and the developer responsible for the reclamation of the site regarding the suitability of the site for building construction led to an independent investigation to review the available data and advise the building owner as to the expected future settlements. After a brief review of the available information, the owner of the building was advised that settlements over the life of the building would be at least twice the values that the design was based on. This resulted in a short interruption of construction. Eventually the building owner and the developer agreed upon financial and legal terms to allow construction of the building to proceed, on the assumption that post-construction settlements would be within the design limits.

Settlement measurements over a period of two months during building construction began to reinforce the conclusion drawn by the independent investigation that the settlements would be significantly higher than initially estimated by the designers. At that point, the scope of the independent investigation was expanded to provide a more comprehensive evaluation of the extent of the settlement problems, and assist in developing a program to mitigate the impact of the settlements on the building. After a two-month investigation, it was concluded that the settlement monitoring program would have to continue for the life of the structure, and remedial measures would need to be implemented in due course as required. The results of these investigations as well as the results of the monitoring over a period of 9½ years are summarized in this paper, followed by the conclusions drawn from the study regarding the design of vertical drains in combination with preloading to reclaim soft bay mud sites.

HISTORY OF SITE DEVELOPMENT AND PRELOAD DESIGN

Based on available records, it appears that the design of the surcharge and vertical drains were based on the following assumptions: (1) the bay mud was essentially normally consolidated although it was known that there was a crust layer several feet thick near the surface; (2) a compression ratio of 0.4 was used to estimate consolidation settlements; (3) a coefficient of horizontal consolidation of 3.7 m²/yr was used to design the vertical drains; (4) a surcharge of 1.22 m corresponding approximately to 20% of the fill thickness would be used; (5) the bay mud thicknesses were estimated from the contours shown on Fig. 1, which indicate

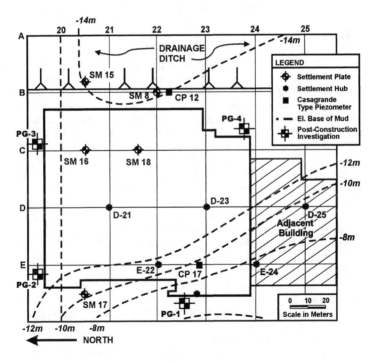

**Figure 1. Project Site Plan, Instrumentation, and
Elevation Contours of the Bottom of the Bay Mud**

variations in excess of 6 m; and (6) a coefficient of secondary compression of 2% was assumed to estimate secondary consolidation. A series of graphs was produced presenting estimated consolidation settlements as a function of bay mud and fill thicknesses, which were used to design the preload and to aid in the interpretation of the settlement measurements. The performance of the preload is summarized below.

Initial filling of the site took place in the summer and fall of 1989 when 3 m to 4 m of fill were placed over the site. The site was regraded in late April 1990, and prefabricated drains consisting of a plastic core 100 mm wide with a filter fabric jacket were installed between April 26 and May 16, 1990. Horizontal strip drains, overlapping the vertical drains, were placed on top of the fill to provide drainage for the water discharging from the drains, so that it could flow towards an adjacent drainage ditch located east of the building site. The site was graded at a slope of 2½ % towards the drainage ditch to promote flow of water through the strip drains to the ditch. Shortly after installation of the drains, filling of the site resumed but progressed rather slowly, continuing until November 1990. Fig. 1 shows a plan of the site, the instrumentation for monitoring the consolidation of the site, and elevation contours of the base of the bay mud as presented in the geotechnical report. Information regarding the elevation of the top of the bay mud is less clear, but it is surmised that the top of the bay mud was close to the surface, which varied from elevation 0 to –0.6 m.

All of the settlement markers except SM8 were installed after a significant amount of fill had been placed, and more importantly, approximately two months after the drains had been installed. Consequently, the early settlements resulting from consolidation after drain installation were not recorded. Settlement marker SM8 had been installed and monitored from the beginning of the initial filling in August 1989. However, SM8 had been installed close to the top of the slope of the drainage ditch, and filling in this area was deliberately slow to avoid slope failure. Even though the east slope of the embankment adjacent to the drainage ditch was quite flat, inclinometer measurements showed lateral deformations up to 305 mm. Therefore, the concern for slope stability was justified. The two piezometers shown on Fig. 1 were installed in late April 1991, shortly before removal of the surcharge (see Fig. 11 for details).

Fig. 2 shows the history of filling and the installation of the vertical drains, and settlements measured at the five settlement markers shown on Fig. 1. Monitoring at four of the five settlement markers was not started until late June 1990, by which time a substantial portion of the fill in the area under the footprint of the building had been placed, and almost one month after the drains had been installed. The settlement records of SM15 to SM18 show distinct changes in the rate of settlement between August and November 1990, consistent with records that show that new fill was being added during this period. Fig. 3 shows typical plots of settlements versus the logarithm of time. It also notes distinct periods when additional fill was being placed in the vicinity of these markers. The placement of significant amounts of fill as late as December 1990, approximately 7 months after installation of the wick drains, is evident as are the corresponding changes in the

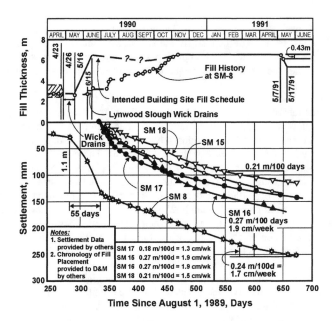

Figure 2. Fill History and Embankment Settlements

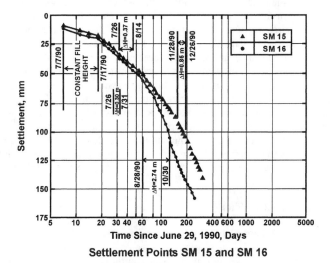

Settlement Points SM 15 and SM 16

**Figure 3. Semilogarithmic Plots of Settlements
Measured During the Preload Period**

rates of settlement. The amount of surcharge removed at SM8 in May 1991 was only 0.43 m, less than 7% of the total fill thickness. How much surcharge had been removed from elsewhere within the building site is not clear, but it is likely that higher-than-anticipated settlements might have resulted in the actual surcharge being less than planned.

All five settlement markers show that at the time of removal of the surcharge, the rate of settlement ranged between 13 mm to 19 mm per week, and the logarithmic plots on Fig. 3 show no indication of leveling off. The decision to remove the surcharge was based exclusively on the settlement data from marker SM8, which was located 9 m east of the building site. The data show a dramatic increase in settlements at the location of SM8 over a period of 55 days, starting immediately after installation of the drains, at an essentially constant fill height. Thereafter, the rate of settlement decreased drastically and remained essentially constant over a period of approximately 250 days. The settlement data at SM8 do not show any signs of increased rate of settlement as new fill was being placed, similar to what is seen on the records of the four other markers. Simple analyses show that the development of 1.13 m of settlement over a period of 55 days under the initially placed fill (about 3 m) would imply approximately 70% consolidation and a C_h value of 3.7 m^2/yr. If C_h had actually been that high, it is unlikely that the rate of settlement would have decreased so dramatically after the first 55 days, as new fill was being placed. These observations cast doubts as to whether the settlements measured at SM8 were representative of the consolidation behavior of the bay mud. It is likely that the settlements at SM8 had been affected by lateral deformations of the adjacent slopes and/or the settlement plate might have been disturbed during installation of the wick drains. In spite of the apparent anomalies of the data, the surcharge was removed on the premise that the settlements at SM8 had exceeded the theoretical values corresponding to full consolidation of the bay mud under the permanent fill and superstructure loads. It appears that no adjustment was made to the settlement measurements to account for undrained deformations, in spite of the large lateral deformations measured with the inclinometers. The surcharge was removed in spite of the significant rates at which settlement was still taking place, and in spite of the excess pore pressures measured at the two piezometer locations (see Fig. 1), which were installed shortly before removal of the surcharge. Subsequent analysis of the measurements made at these two piezometers indicated excess pore pressures of 0.6 m to 1.2 m of water head a few feet below the top of the bay mud, and 3.05 m to 3.66 m of water head near the mid-depth of the bay mud.

SITE CHARACTERIZATION

After the building was completed, a comprehensive investigation was undertaken to better understand the conditions causing the large post-construction settlements and to assess potential future settlements to plan appropriate mitigation measures, if necessary. As part of the investigation program, four boreholes were drilled near the four corners of the building, at locations PG-1 to PG-4 shown on Fig. 1. At each location, pneumatic piezometers were installed to measure in-situ pore water pressures and their variation with depth. Because the tops of the drains were at least 3 m below existing grades, it was not feasible to excavate pits to expose the drains, and the boreholes were located without the benefit of knowing the position of the piezometers relative to the vertical drains. For consistency, the piezometers were installed in holes that were spaced 1.22 m apart with the intent that all of the piezometers would be approximately at the same distance from the drains. Conceptually, this layout of the piezometers was thought to be essential to allow rational interpretation of the measurements. It was hoped that at one or two of the groups the piezometers might be located at a sufficient distance from the drains to provide a reasonable representation of the average pore pressures between the drains. Undisturbed piston samples of the bay mud were obtained for index, consolidation, and strength testing. The results of the exploration program are summarized below.

Fig. 4 summarizes Atterberg limit data on the plasticity chart. It shows data from the north bay site, a site at the San Francisco Airport where similar mudflat conditions prevail (UAL site), and typical data from areas in downtown San Francisco that had been reclaimed by filling over the bay, i.e., from areas outboard of the old shoreline, beyond the old mudflats. It is evident that the bay mud in the mudflat areas has significantly different plasticity characteristics from bay mud at sites beyond the old shoreline, from which much of the empirical data for compressibility characteristics of the bay mud had been developed. Liquid limits range from 80% to 110% with Plasticity Index values in excess of 40%. The Atterberg limits lie consistently below the A-line on the plasticity chart, indicating that the bay mud at this site was organic.

Fig. 5 shows the variation in moisture content data with depth. Most of the data were generated from post-preload investigations; however, at boreholes M3 and

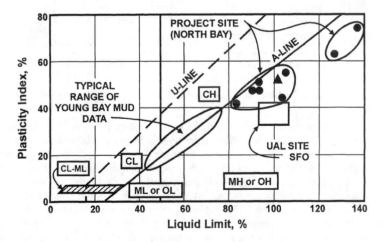

Figure 4. Atterberg Limit Data

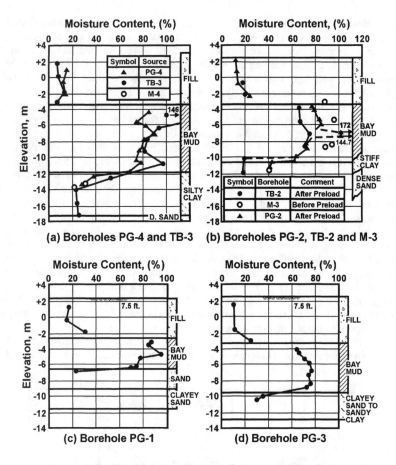

Figure 5. In-Situ Moisture Contents Before and After Preload

M4, data were available from before preloading. The data show very high moisture content values, even after preloading that resulted in large consolidation settlements. Moisture content reductions on the order of 20% to 30% resulting from consolidation are evident. The explorations revealed a layer of very stiff clay below the bay mud that had been classified in the original investigation as bay mud. The data also showed the presence of thin layers of clay rich with organics, hereinafter referred to as "highly organic" clay, which is much more compressible than the bay mud. Because of the presence of the stiff clay layer below the bay mud, the contours of the bottom of the bay mud and the corresponding bay mud thicknesses had to be reinterpreted. The results of the reevaluation are shown on Fig. 6.

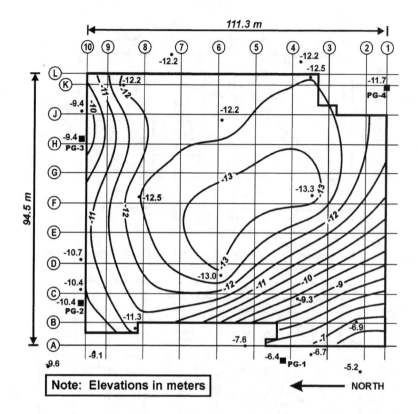

Figure 6. Contour Lines of Bottom of Bay Mud Elevation

Approximate bay mud thicknesses can be obtained from Fig. 6 by assuming that the top of the bay mud is approximately at elevation −3.05 m, which was the elevation of the top of the bay mud determined from boreholes drilled as part of the post-construction investigations.

The results of consolidation tests are summarized on Figs. 7 and 8. Fig. 7a shows compression ratios (CR) as a function of moisture content. The compression ratio is defined as the incremental strain in the virgin compression zone for a tenfold increase in effective stress. The results indicate compression ratios between 0.3 and 0.4 for the bay mud, values between 0.45 and 0.5 for the highly organic clay, and close to 0.2 for the stiff clay. Given that the moisture contents before preloading were 20% to 30% higher than the values shown on Fig. 7a, a value of CR of 0.4 appears reasonable as an average. However, the high CR values of the organic clay were an indication that significant and perhaps random variations in settlements could develop.

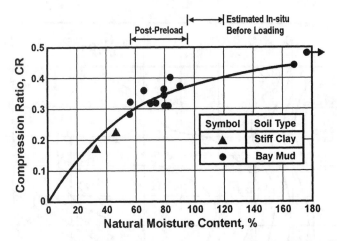

(a) Compression Ratios From Post-Preload Consolidation Tests

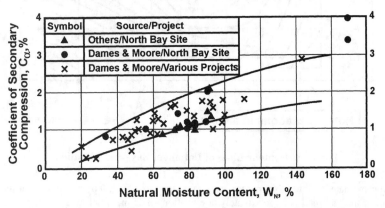

(b) Correlation Between C_α and Moisture Content for Bay Mud

**Figure 7. Compression Ratios and Coefficients
of Secondary Compression of Bay Mud**

Fig. 7b summarizes coefficients of secondary compression for normally consolidated bay mud as a function of moisture content. Data are shown from the north bay site and from various other project sites. Most of the data for the north bay site indicate values between 1% and 2%, with values in the range of 3.5% to 4% for the highly organic clay layers.

Fig. 8 summarizes C_v values, including data from the north bay site and from the UAL site at the airport for comparison. As expected given the high liquid limits

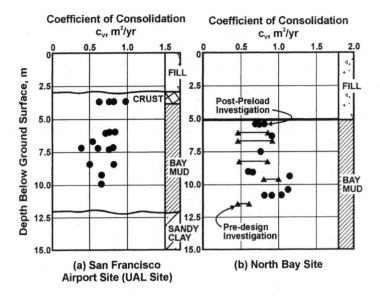

Figure 8. Coefficients of Consolidation of Bay Mud

of these soils, the C_v values are quite low, ranging from 0.46 m²/yr to 1.11 m²/yr. Values of C_v determined from tests on bay mud samples taken from many sites in downtown San Francisco typically range from 1.0 to 1.5 m²/yr. The value of C_h of 3.7 m²/yr selected for the design of the vertical drains and preload were 4 to 8 times the C_v values.

The results of the post-preload investigations lead to the following conclusions: (1) the thickness of the bay mud was much more variable than initially assumed, leading to concerns that differential settlements would be much more severe than anticipated earlier; (2) the bay mud had very low coefficients of consolidation, which explain the slow consolidation, but also suggest that significant additional consolidation settlements might develop; (3) the thicknesses of the bay mud were generally less than assumed in the design, because of misclassification of a stiff clay layer that was present below the bay mud (this meant that consolidation settlements might be less than estimated, but the rate of consolidation might be slower because of poor drainage at the base of the bay mud); and (4) the presence of random layers of bay mud rich in organics made prediction of total and differential settlements very difficult. Recent investigations (Koutsoftas, 2001) at a nearby site in similar geologic terrain suggest that the bay mud might have been lightly overconsolidated with overconsolidation ratios (OCR) in the range of 1.6 to 2.0, and confirmed the presence of a thin crust, both of which had been ignored in the design.

In summary, the soil conditions at the north bay site were much more complicated than initially anticipated. Simply by review of the geotechnical characteristics of the bay mud, it is evident that the accuracy of any theoretical estimates of consolidation settlements was subject to significant compensating errors, and any success in predicting the actual total settlements would probably be fortuitous. It emphasizes the need for comprehensive observations of pore pressures in addition to settlements to decide when to remove the surcharge.

POST-PRELOAD INVESTIGATIONS

Post-preload investigations included a brief independent investigation conducted shortly after removal of the surcharge, and additional post-construction investigations. The results of these investigations are summarized below.

1.   Initial Investigation to Evaluate the Effectiveness of the Surcharge. Some of the measurements of settlements made during the preload period were presented in Fig. 2, and showed that the rates of settlements were still quite high when the surcharge was removed. Several of the records of settlements were analyzed using the semi-empirical method proposed by Asaoka (1978) and discussed in detail by Jamiolkowski et al. (1985). Plots from typical records are shown on Fig. 9. The data indicate projected additional settlements to complete primary consolidation ranging from 0.21 m to 0.28 m. These results, however, were of little practical value because the ultimate settlements were not known — the settlements that had occurred prior to July 1990 were not recorded. However, if one assumes that the ultimate consolidation settlements estimated from consolidation analyses, in the range of 2.1 m to 2.4 m, were reasonably representative of actual settlements, the degree of consolidation of the bay mud might have been close to 90%.

The data shown in Fig. 9 as well as the data from other locations were analyzed following the method described by Jamiolkowski et al. (1985) to obtain approximate estimates of C_h. The results of the analyses are summarized in Table 1

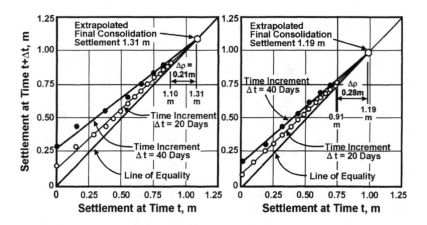

**Figure 9. Extrapolated Final Consolidation Settlements
Using Asaoka's Procedure – Hub D21, D23, and E22**

$C_h = -de^2 F_n \ln\beta/8\Delta t$; de = 1.34 m, F_n = 2.29			
MARKER	β	Δt (year)	C_h (m^2/yr)
SM-8	0.904 0.743	0.1 0.2	0.52 0.76 (0.64)
SM-15	0.935 0.860	0.05 0.10	0.69 0.77 (0.73)
SM-16	0.901 0.836	0.05 0.10	1.07 0.92 (0.99)
SM-17	0.913 0.837	0.05 0.10	0.94 0.92 (0.93)
SM-18	0.946 0.892	0.05 0.10	0.58 0.59 (0.59)
D-21 D-23 D-25	0.74 0.82 0.86	0.11 0.11 0.11	1.37 0.90 (0.99) 0.71
PAD V	0.80	0.14	0.84
Post-Preload Measurements 12 Columns	0.743	0.273	0.58
Average During Preload Period			0.82

Table 1. Estimates of C_h from Settlement Data (Based on Asaoka's Method)

and indicate values of C_h ranging between 0.52 m^2/yr and 1.37 m^2/yr, with an average value of 0.82 m^2/yr. Supplemental analyses were performed using the settlements from twelve columns that are considered representative of the range of conditions within the building. The results of the analyses are included in Table 1 and indicate an average value of C_h of 0.57 m^2/yr. Because the method of analysis ignores the effects of vertical drainage, the estimated C_h values are higher than the actual values. Given the available data, a C_h value of 0.75 m^2/yr would appear reasonable for design purposes. This value is about the same as the average of the C_v values shown on Fig. 8.

The low values of C_h are in part due to the organic nature of the bay mud at this site, and to a significant degree due to smear and remolding caused by the installation of the wick drains using a mandrel. It is important to recognize that the average values back-calculated from the analysis of the settlement data do not represent fundamental properties of the bay mud, but rather average values over the entire thickness of the mud, including the effects of smear due to drain installation and changes in C_h due to changes in vertical effective stress during consolidation. However, these values are very useful from a practical point of view in estimating rates of consolidation.

The most important information regarding the potential for future settlements was obtained from the two hydraulic piezometers that were installed in late April 1991, near the end of the surcharge period. Measurements between May 1 and June 14, 1991 are presented in Fig. 10a. They show that it took approximately 16 days for the pore pressures to reach equilibrium in the instruments. On May 17, 1991, coincident with the beginning of surcharge removal, measurements at

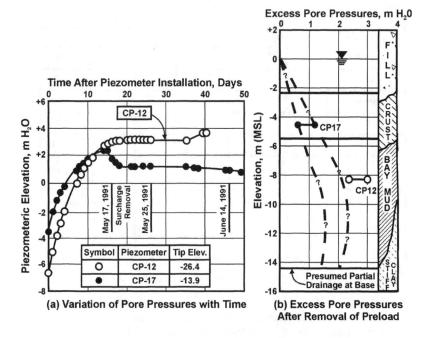

Figure 10. Excess Pore Pressures at Time of Preload Removal

piezometer CP-17 indicated a 1.22 m drop in pore water pressures. The pore pressures at CP-17 remained constant until June 5, when a slight additional drop in pore pressure was recorded. The pore pressures at CP-12 remained constant for about 20 days and then showed a 0.61 m increase. Static water levels were assumed to be at elevation 0, but as will be seen later, actually groundwater levels fluctuate with time. The initial interpretation was based on an assumed static water table elevation that might vary between 0 and –0.6 m.

The excess pore pressures estimated from these two piezometers are shown on Fig. 10b, together with the interpretation of probable excess pore pressure profiles. Because the fill was very clayey, the excess pore pressures were assumed to be 0 at the level of the groundwater, not at the top of the bay mud. Likewise, the presence of stiff clay below the bay mud led to the conclusion that the bottom of the bay mud might not be fully drained. This assessment was based on prior experience from the detailed studies of a test fill for the Hong Kong Airport project (Foott et al., 1987; Koutsoftas and Cheung, 1994). The interpreted profiles were tempered by the understanding that piezometer CP-12 was located outside the footprint of the building, and might have been affected by the late placement of fill which might not be representative of the conditions within the footprint of the building.

Based on the excess pore pressure profiles shown on Fig. 10b, future consolidation settlements in the range of 127 mm to 178 mm were estimated. In addition, it was assumed, based entirely on prior experience, that primary consolidation would be completed in 2 to 3 years, and on that basis secondary consolidation settlements in the range of 100 mm to 125 mm were estimated over the useful life of the building (30 years). Thus, total post-preload settlements of 230 mm to 305 mm were estimated. These values are twice the values assumed in the design.

Post-preload settlement measurements were made at five locations within or very close to the footprint of the building over a period of 64 days, between the time of the removal of surcharge on May 21, and July 16, 1991, as construction of the building foundations was getting underway. The settlements ranged between 21 mm and 40 mm, or the equivalent of 2.5 to 4.3 mm/week. Simple analysis of these measurements led to the conclusion that secondary consolidation settlements over this period could not be more than 6.5 mm, and therefore the measured settlements could not possibly be due to secondary consolidation, but rather due to continued dissipation of excess pore pressures. The rather large values of recorded settlements within such a short period of time raised concerns that the settlements estimated based on the approximate excess pore pressures shown on Fig. 10b might be too low, since the two piezometer measurements might not be representative of average conditions. Based on these considerations, it was concluded that the settlements calculated based on the excess pore pressure profiles on Fig. 10b would probably be minimum values, and that actual settlements could be significantly greater. Construction of the building was stopped shortly after these predictions were published.

At that point, various options were evaluated for mitigating the impacts of the settlements, the most promising of which was to remove 1.5 m to 3 m of the existing fill (which had total unit weights in the range of 18.9 kN/m^3 to 22.0 kN/m^3) and replace it with lightweight fill, to essentially eliminate the excess pore pressures. After a short interruption, construction of the building resumed, without implementation of any mitigation measures, and the building was substantially completed by October 10, 1991. As part of the decision to resume construction, a more comprehensive settlement monitoring program was instituted starting on August 8, 1991, with weekly measurements being made at approximately one-third of the building columns. Typical plots of settlements versus time from this program are shown on Fig. 11. They indicate fairly steady rates of settlement over a period of 155 days between August 8, 1991 and January 9, 1992. Settlements measured at thirty-eight column locations between August 8 and October 10, 1991, the time of substantial completion of the building, ranged between 15 mm and 40 mm, with an average value of 28 mm. Linear extrapolation of the early measured settlements backwards over a 78-day period between May 21 and August 8, 1991, the period between surcharge removal and the beginning of settlement measurements, indicates a probable range of additional settlements between 18 mm and 53 mm that would have developed between the time of removal of surcharge and the beginning of building construction. Therefore, estimated total settlements, between the time of removal of the surcharge and the end of building construction, ranged between 33 mm and 94 mm. On average, the settlements between the time of removal of surcharge and the end of building construction (141 days) were more than one-half of the total post-construction settlement of 100 mm to 150 mm estimated to occur over the 30-year life of the building that the original building design was based upon.

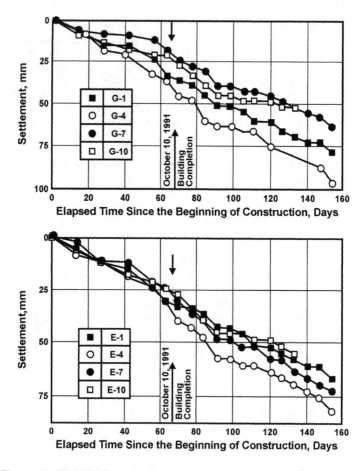

Figure 11. Early Building Settlements: August 8, 1991 to January 9, 1992

2. *Estimates of Future Settlements from Pore Pressure Measurements.* The pore pressure measurements made at four instrumentation stations within the first two weeks after installation of the piezometers are summarized in Fig. 12. At three of the four locations, the excess pore pressures (the difference between the hydrostatic line and the measured pore pressures) were small, typically less than 1.52 m of pressure head. The data at station PG-1 are significant because even though the bay mud was only 3.65 m thick, with apparent top and bottom drainage, the data still indicate small positive excess pore pressures. Data measured in the clayey sand layer below the bay mud at station PG-3 provide a check on the hydrostatic pressures. The data at station PG-4 not only show very significant excess pore pressures, but also suggest only partial drainage at the base of the bay mud. Fig. 12 shows data obtained from two separate investigations that were not measured at the same locations, but were sufficiently close to allow meaningful comparisons. After installation of the piezometers, it was found that the drains at PG-4 were spaced 1.52 m apart rather than 1.22 m apart. The data shown with solid circles were measured from a nearby location where the drains were spaced 1.22 m apart; these data show values that are consistent with the values measured at PG-4. The important fact revealed from the data on Fig. 12 was that two independent investigations led to the same conclusion that positive excess pore pressures remained after removal of the surcharge, suggesting that primary consolidation had not been completed. Visual examination of the exposed tips of the horizontal strip drains in the adjacent drainage ditch revealed evidence of seepage from the drains, which was interpreted as further evidence that consolidation was still ongoing.

The pore pressure measurements shown on Fig. 12 were used to estimate future consolidation settlements. Secondary consolidation settlements, over a period of 30 years after building construction, were estimated using a C_α value of 2%. The estimated settlements are summarized in Fig. 13. Primary and secondary consolidation settlements are shown together with the total estimated settlements. The consolidation settlements were estimated based on pore pressure measurements made on December 30, 1991. In order to estimate the total settlements after building construction, settlements measured between October 10, 1991, and January 9, 1992 (the measurements closest to December 30, 1991), at columns close to the four stations were added to the estimated values. These added settlements ranged between 25 mm and 38 mm. The predictions shown on Fig. 13 were used only as an approximate assessment of future settlements, recognizing the inherent limitations of the pore pressure measurements that arise from the fact that the location of the piezometers relative to the vertical drains was not known. However, the estimated values were considered as the minimum settlements that should be expected over the life of the structure.

SECOND ROUND OF SETTLEMENT PREDICTIONS

Approximately two years after the building had been substantially completed, as part of the building owner's planning to mitigate his risks, he requested a re-evaluation of the settlements expected to develop over the 30-year life of the building. A theoretical analysis was not considered fruitful because of the many variables and uncertainties regarding the subsurface and loading conditions cited earlier. An approximate approach was chosen for settlement prediction as illustrated by the typical example in Fig. 14. Settlements were estimated at twelve columns located near the four corners of the building, where the thicknesses of the bay mud were reasonably well known. Based on the author's personal experience and considerations of fundamental consolidation behavior, and in view of the measurements made up to that point, the following assumptions were made to facilitate the predictions.

1.   Settlements would continue for at least one more year (beyond the latest measurements) at the same rate as observed during the preceding twelve months.

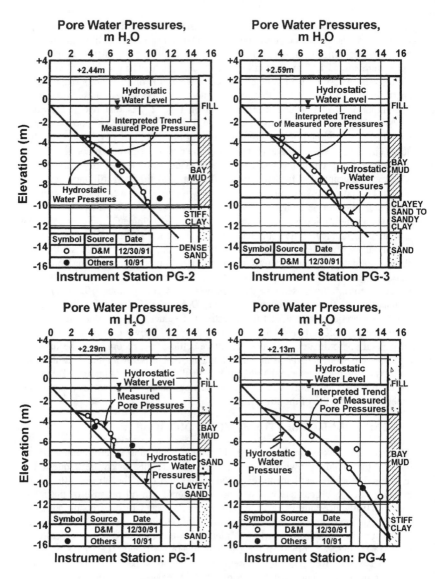

Figure 12. Pore Pressures Measured 7 Months After Removal of Surcharge

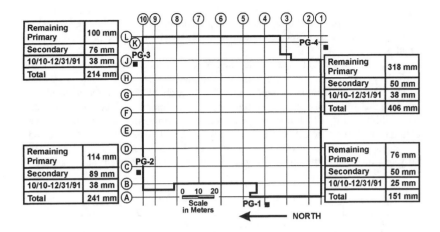

Figure 13. Estimated Post-Construction Settlements (January 1992)

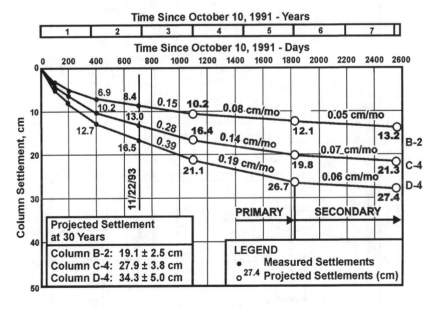

Figure 14. Typical Example of Projected Settlements Vs. Time

2.   Over the subsequent 2-year period, settlements would develop at a rate equal to one-half of the settlement rates observed during the previous twelve months.

3.   Subsequent settlements (i.e., beyond October 1996) would be due to secondary compression and were estimated assuming a C_α of 2%.

The study included three columns near each of the four corners of the building that were considered to be representative of the range of conditions within the building.   Fig. 14 shows the predicted settlements for three of the twelve columns that were selected for this study.   Fig. 14 also shows the predicted settlements over a period of 7 years as well as the predicted settlements over the 30-year life of the building.   Fig. 15 compares the estimated and measured

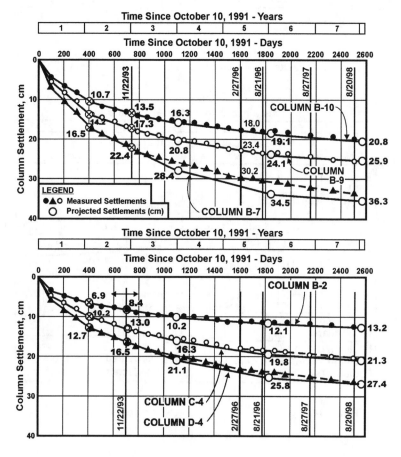

Figure 15. Comparison of Predicted and Measured Settlements (7-Year Record)

settlements for six of the twelve columns, which are typical of the rest of the columns where predictions were made. The measured settlements shown on Fig. 15 are reasonably close to the predicted settlements and are considered a good indication that the 30-year estimates should provide a reasonable approximation of the expected settlements. The estimated total settlements over the 30-year life of the building are presented in Table 2. The measured settlements between October 10, 1991 and January 18, 2001 are also included in this table for comparison. The measured settlements as of January 18, 2001 range from 75% to 84% of the 30-year estimates predicted in November 1993.

Area	Column	30-Year Estimates: Jan. 1992 Predictions (mm)	30-Year Estimates: Nov. 1993 Predictions (mm)	Measurements up to January 18, 2001 (9½ years) (mm)
Southwest Corner	B-2	152 at PG-1, near column B-2	190 ± 25	142
	C-4		280 ± 38	226
	D-4		343 ± 50	288
Northwest Corner	B-7	241 at PG-2, near column B-10	445 ± 50	372
	B-9		343 ± 38	287
	B-10		292 ± 38	221
Southeast Corner	G-4	406 at PG-4, near column J-4	445 ± 38	349
	H-4		445 ± 38	350
	J-4		445 ± 38	336
Northeast Corner	J-7	216 at PG-3. Near column G-9	318 ± 38	247
	H-8		318 ± 38	244
	G-9		267 ± 38	197

**Table 2. Settlement Predictions Versus Measurements
at Selected Column Locations**

POST-CONSTRUCTION SETTLEMENT MEASUREMENTS

Fig. 16a shows contours of settlements measured between October 10, 1991 and January 18, 2001. Measured values range between 140 and 368 mm. The high density of the settlement contours within the western half of the building is an indication of high differential settlements within this area. The pattern of settlements mirrors to some extent the contours of bay mud thicknesses shown in Fig. 6. Fig. 16b shows contours of settlements over the last year of monitoring. After 9½ years, settlements are still progressing at 3.5 to 11.5 mm per year.

Fig. 17 shows annual settlement rates of six columns representative of the overall behavior of the building. The annual rates of settlement decreased dramatically during the first 3 years and fluctuated somewhat over the next 3 years, but appear to have remained fairly constant over the last 3 years.

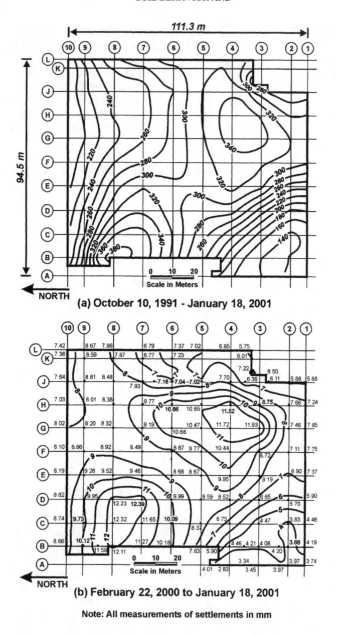

(a) October 10, 1991 - January 18, 2001

(b) February 22, 2000 to January 18, 2001

Note: All measurements of settlements in mm

Figure 16. Measured Post-Construction Settlements

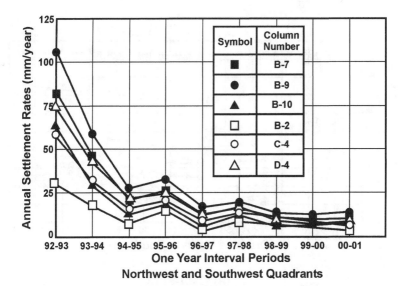

Figure 17. Typical Annual Settlement Increments

Fig. 18 shows settlement profiles along the east and west walls of the building, based on measurements made at the top of the exterior walls, which extend above the roof of the building. The east wall settled fairly uniformly except for the 18.2 m near the north end, where differential settlements of 50 mm over 18.2 m have developed. The settlements of the north wall of the building are similar to the profiles of the east wall. The settlements of the west wall are far more variable and indicate large differential settlements between columns B-5 and B-10. The settlement depression between columns B-7 and B-8.5 reflects the settlement trough shown in Fig. 16a. The settlement profiles along the south wall of the building are similar to the profiles of the west wall, with a significant depression and corresponding high differential settlements near the east end of the wall.

The segment of the wall north of column B-5 had experienced severe diagonal cracking. The cracks typically opened to a maximum of 3 mm to 5 mm, and then remained stable. As the differential settlements increased, new cracks appeared on the wall. Spacing of the cracks varies from 100 mm to more than 1,000 mm. Extensive cracking also developed on the south wall, while, as expected, the north and east walls experienced only minor cracking. The south and west walls had to be strengthened to maintain their structural integrity.

Fig. 19 illustrates the development of differential settlements with time at nine column bays that are considered typical of the behavior of the building. The major portion of the differential settlements developed within the first 800 to 1,200 days. The differential settlements after the first 1,000 days were relatively small over most of the building, although some cases involving significant differential settlements are evident from Fig. 19.

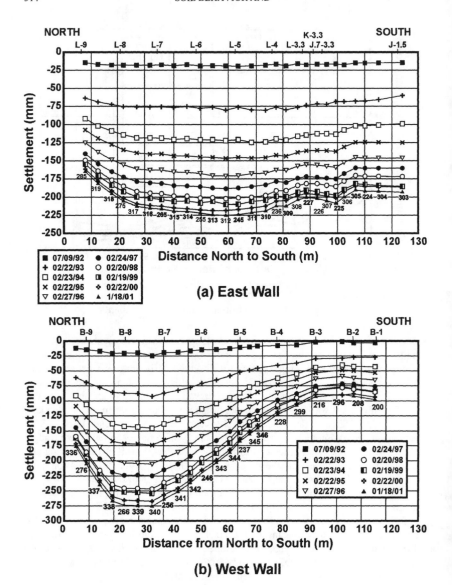

**Figure 18. Development of Settlement with Time: East and West Walls
(April 23, 1992 and January 18, 2001)**

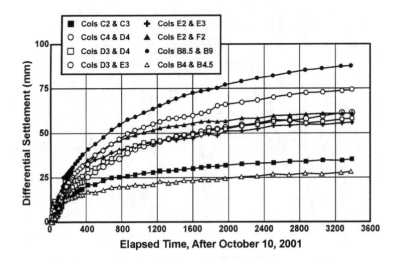

**Figure 19. Typical Differential Settlements Versus Time
(October 10, 1991 and January 18, 2001)**

A statistical analysis of the total and differential settlements measured between October 10, 1991 and January 18, 2001 reveals the following: (1) total settlements range between 132 mm and 386 mm, with a median value of slightly more than 254 mm; and (2) approximately 20% of the differential settlements exceed 38 mm, which is the established tolerable limit for the building (equivalent to 25 mm in 9.1 m, or 38 mm in 13.7 m). Many of the columns had to be jacked and shimmed, some of them repeatedly, to keep the differential settlements within tolerable limits. The maximum differential settlement is close to 102 mm over a span of 13.7 m.

Review of the measured groundwater levels and pore water pressures is instructive in understanding the settlement behavior of the site, and in particular the fluctuations in annual rates of settlements shown in Fig. 17. Groundwater levels measured at three locations (PG-1, PG-2, and PG-3) are summarized in Fig. 20. The results show significant and unexpected long-term fluctuations in groundwater levels. The cause of these fluctuations is not known, but they are surprising in view of the clayey nature of the fill. They may be related to some hydraulic connection to the drainage ditch (or other underground water sources) provided by the horizontal strip drains that were installed during filling to allow discharge of water from the vertical drains to flow to the ditch.

Fig. 21 summarizes fluctuations in annual settlement rates and groundwater levels. It appears that the settlement rates correlate well with the observed groundwater fluctuations. The large reductions in settlement rates between 1993-1994 and 1994-1995 appear to coincide with periods of high groundwater levels.

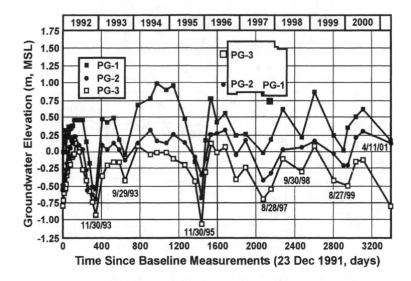

Figure 20. Groundwater Level Fluctuations

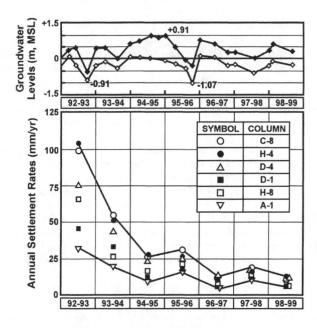

Figure 21. Variations in Annual Settlement Rates and Groundwater Levels

Also, the increased settlement rates in 1995-1996 coincide with a sharp drop in groundwater levels during the same period. The rather steady settlement rates during the last 3 years of records shown on Fig. 21 coincide with rather steady water levels, as shown on Fig. 20. Fig. 22 attempts to provide an explanation for the observed fluctuation in annual settlement rates. As the groundwater levels change, the boundary conditions along the vertical drains also change, causing changes in the hydraulic gradient between the drains and therefore affecting the rate of dissipation of excess pore pressures. Also, even if there were no excess pore pressures, the changes in hydrostatic pressures in the drains change the effective stress field around the drains and perhaps slows down the rate of secondary consolidation, because C_α is a function of stress level.

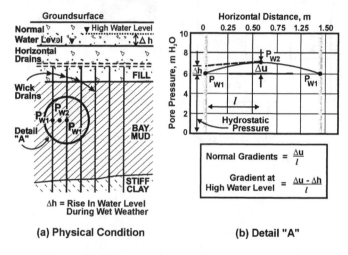

(a) Physical Condition   (b) Detail "A"

**Figure 22. Effects of Increased Groundwater Levels
on Consolidation of the Bay Mud**

ANALYSIS OF SETTLEMENT DATA

Fig. 23 presents semi-logarithmic plots of settlements measured at twelve column locations typical of the settlement behavior of the building. The data show fairly linear trends during the first 2 years of post-construction monitoring, followed by a gradual reduction in the rates of settlement over a period of one year or so, before beginning to exhibit reasonably linear trends consistent with secondary consolidation behavior. It appears from these plots that the beginning of secondary consolidation occurred after June 30, 1994, approximately three years after building construction was completed. However, this assessment is by necessity somewhat subjective and therefore it can only be considered as very approximate. In an effort to better understand the settlement characteristics, measured settlements were

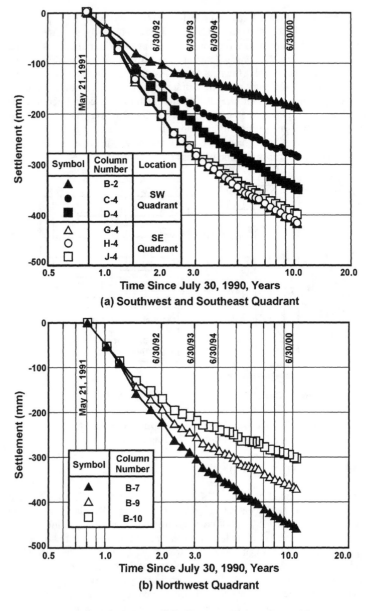

Figure 23. Post-Preload Settlements Versus Logarithm
of Time at Selected Columns

analyzed using Asaoka's (1978) procedure. Fig. 24 shows two of the plots that are representative of the range of characteristic behavior of the settlements of most of the columns. The data show approximately bilinear behavior, with the first part of the line being flatter than the line of equality and the second part following approximately along the line of equality. As pointed out by Jamiolkowski et al. (1985), the point of intersection of the data with the line of equality is assumed to represent approximately the end of primary consolidation. In both cases shown on Fig. 24, the end of primary consolidation occurs in August 1994, which is close to the subjective evaluation made on the basis of the semi-logarithmic plots shown on Fig. 23. A more rigorous analysis was undertaken by calculating the secondary consolidation settlements over different periods of time and comparing the calculated settlements, based on a C_α of 2.0%, with the actual measurements. The results of the analyses are summarized on Fig. 25. Fig. 25a compares the calculated and measured settlements over a 2-year period between August 24, 1994 and August 21, 1996. Even though there is some scatter in the data, eleven of the twelve data points indicate C_α values in the range of 2.0% to 3.5%. These values of C_α are significantly higher than the laboratory data would suggest. Fig. 25b compares calculated settlements with measurements over an approximately five-year period between February 27, 1996 and January 18, 2001. The data indicate C_α values that range between 1.6% and 3.0%, with most of the values between 2.0% and 2.4%.

The values of C_α indicated in Fig. 25 represent average values over the full thickness of the bay mud at each column location. The actual values of C_α are probably more variable than indicated in Fig. 25, given that layers of highly organic clay are likely to be present at many locations. Another important limitation of the data shown in Fig. 25 is that the thickness of the bay mud is not known accurately at each column location, and therefore the calculated settlements are somewhat approximate. However, overall, the data appear to provide a reasonable representation of the range of average C_α values over the full depth of the bay mud at each column location.

The analysis of the data presented in Fig. 25 underscores the difficulties in determining the time when primary consolidation is completed and secondary consolidation begins. The data on Fig. 25 are interpreted to suggest that secondary consolidation probably had not started until early 1996, approximately 4½ years after removal of the surcharge. This experience points out how slow the last 5% to 10% of primary consolidation can actually be, even with closely spaced vertical drains. One of the major challenges in designing vertical drains in combination with surcharge is to control post-construction settlements within tolerable limits.

FUTURE SETTLEMENTS

The analyses of settlement data presented earlier indicate that the site experienced secondary consolidation settlements for the past five to seven years, and provided back-calculated values of C_α (as well as the product of C_α times the layer thickness, H) that can be used to estimate future settlements over the remaining useful life of the building. The estimated settlements are summarized below and compared to the estimates made in 1993. The calculation of future settlements did not use individual C_α or H values, but

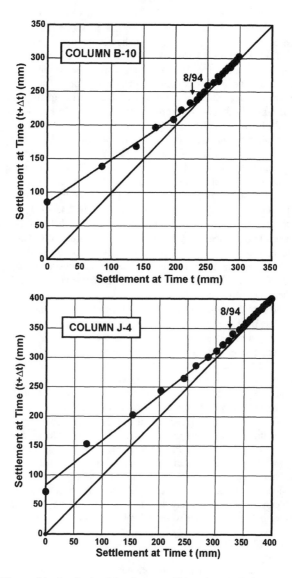

Figure 24. Analysis of Settlements Using Asaoka's Method

the product of $C_\alpha H$ back-calculated from the analysis of the data shown on Fig. 25. Thus, errors in C_α and/or H values compensate for each other, and do not affect the results of the estimated future settlements.

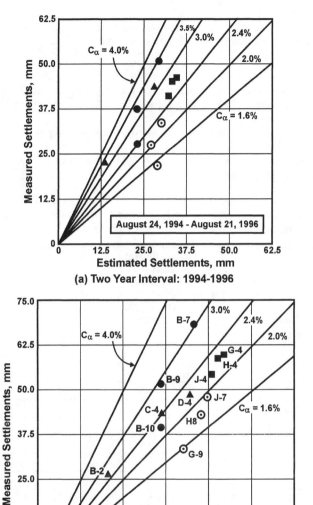

Figure 25. Analysis of Secondary Consolidation Settlements

Using the $C_\alpha H$ values back-calculated from the information shown in Fig. 25b, future settlements can be estimated following the expression:

$$\Delta_S = C_\alpha \, H \log T_2/T_1$$

where Δ_S is the incremental settlement expected to develop between times T_2 and T_1, with January 18, 2001 corresponding to T_1 of 3,825 days and October 10, 2021 corresponding to T_2 of 11,390 days;

C_α is the average coefficient of secondary consolidation, which is different for each column;

H is the bay mud thickness at the corresponding column location; and

T_1 and T_2 are times in days.

Table 3 summarizes the settlements measured up to January 18, 2001, the expected incremental settlements, the total estimated settlements over the 30-year life of the building, and the estimates made in November 1993.

Area	Column	Measured Settlements Oct. 10, 1991 to Jan. 18, 2001 (mm)	Estimated Incremental Settlement Jan. 18, 2001 to Oct. 10, 2021 (mm)	2001 Estimates of 30-Year Settlements (mm)	1993 Estimates (mm)
Southwest Corner	B-2	142	46	188	190 ± 25
	C-4	226	76	302	280 ± 38
	D-4	288	85	373	343 ± 50
Northeast Corner	B-7	372	120	492	445 ± 50
	B-9	287	91	378	343 ± 38
	B-10	221	67	288	292 ± 38
Southeast Corner	G-4	349	105	454	445 ± 38
	H-4	350	104	454	445 ± 38
	J-4	336	95	431	445 ± 38
Northeast Corner	J-7	247	85	332	318 ± 38
	H-8	244	76	320	318 ± 38
	G-9	197	59	256	267 ± 38

Table 3. Future Settlements

The predictions made in 1993 based on rather limited data are very close to the predictions made in 2001, after having the benefit of approximately 7½ years of additional measurements. It is anticipated that the actual total settlements over the 30-year life of the building would probably be within ±15 mm of the indicated values.

CONCLUSIONS

The development of a soft bay mud site using vertical drains in combination with surcharge experienced major difficulties because of a variety of factors including: (1) highly variable subsurface conditions over the footprint of a single building which were not detected by the exploration program and were not accounted for in the design; (2) overly optimistic assessment of the drainage characteristics of the bay mud; (3) reliance on published correlations between C_v and C_h, which were not appropriate for the bay mud encountered at this site; (4) insufficient instrumentation and reliance exclusively on settlements to determine the degree of consolidation achieved, on the basis of which the time for removal of the surcharge was determined; (5) poorly controlled placement of the fill; (6) insufficient flexibility in the project schedule to account for uncertainties in terms of the duration of consolidation; and (7) inability of the parties involved, or unavailability of an appropriate mechanism, to facilitate timely resolution of the issues involved and to take appropriate corrective measures.

The review of the design, the settlement measurements made during the surcharge period, extensive post-construction investigations, and a comprehensive settlement monitoring program over a period of 9½ years led to the following conclusions:

1.   The design of vertical drains in combination with surcharge requires a very careful and comprehensive evaluation of the compressibility characteristics and stress history of the bay mud, with particular attention to potential variations both spatially and with depth. The presence of random layers of highly organic clay tends to slow down the dissipation of excess pore pressures and causes larger than anticipated settlements due to secondary compression. The above factors are of paramount importance to the design of the preload and must be clearly understood in order to plan an effective preload program.

2.   Estimating C_h based on empirical correlations in terms of ratios of C_h/C_v must be viewed with some skepticism and conservatism unless local experience is available that is directly relevant to a particular project. At the north bay site, both the magnitude of C_v and the ratio of C_h/C_v had been overestimated, leading to unrealistically low predictions of the time required to achieve the specified degree of consolidation. When the effects of smearing and disturbance caused by drain installation are not explicitly incorporated in the analysis of the rates of consolidation, the selection of C_h must consider these effects. C_h/C_v ratios higher than 1.0 are difficult to justify without the benefit of a test fill, or other specific local experience to base the selection of C_h/C_v.

3.   Settlement measurements during the surcharge period are helpful in evaluating the progress of consolidation and the rate of settlements, but are not sufficient by themselves to make a reliable determination of the degree of consolidation at any given time and to determine the time for the removal of surcharge. Pore pressure measurements are indispensable. The instrumentation program must include both settlement and pore pressure measurements with appropriate redundancy and should be combined with reliable information about the subsurface conditions at each instrumentation station. Settlements and pore pressure

measurements are of little value if the subsurface and drainage boundary conditions are not well understood.

4.   The interpretation of pore pressure measurements is critically dependent on accurate knowledge of the position of the piezometers relative to the drains, the top and bottom drainage conditions, and understanding the groundwater conditions and potential fluctuations with time. Nevertheless, if a sufficient number of piezometers are installed and the data are carefully evaluated, the observational approach can be used to determine when the surcharge can be removed and situations where the surcharge is removed prematurely can be avoided.

5.   Successful application of vertical drains in combination with surcharge requires conservatism, especially when the ground conditions are variable, so that the desired results can be achieved uniformly over the site. The disturbance caused by the installation of the drains causes significant changes in the compressibility and drainage characteristics of the bay mud that are difficult to evaluate during the design stage. The design of vertical drains must include an allowance in the schedule to prolong the consolidation period, if required, and/or make allowances for remedial measures, such as: (a) increasing the surcharge height to cause further consolidation within the available time; and/or (b) removal of portion of the fill and replacement with lightweight fill to reduce excess pore pressures.

6.   One of the difficulties with the design of the surcharge is the fact that often the amount of consolidation settlement can be seriously underestimated. The greater-than-anticipated settlements can either reduce the amount of surcharge fill that can be removed, or may result in the need to add extra fill late in the surcharge period, and therefore reduce the effectiveness of the surcharge.

REFERENCES

Asaoka, A. (1978). "Observational Procedure for Settlement Prediction," Soils and Foundations, Vol. 18, No. 4, December 1978, pp. 87-101.

Foott, R., Koutsoftas, D.C., and Handfelt, L. (1987). "Test Fill at Chek Lap Kok, Hong Kong," Proc. ASCE, Journal of the Geotechnical Engineering Division, Vol. 113, No. 2, February 1987, pp. 106-126.

Jamiolkowski, M., Ladd C.C., Germaine, J.T., and Lencellotis, R. (1985). "New Developments in Field and Laboratory Testing of Soils," State-of-the-Art Report, 11th ICSMFE, Vol. I, pp. 57-153.

Koutsoftas, D.C. (2001), Discussion: "Factors of Safety and Reliability in Geotechnical Engineering" by J.M. Duncan, ASCE Journal of Geotechnical and Geoenvironmental Engineering, Vol. 127, No. 8, pp. 706-710.

Koutsoftas, D.C. and Cheung, R.K.H. (1994). "Consolidation Settlements and Pore Pressure Dissipation," Proc. ASCE, Specialty Conference on Vertical and Horizontal Deformations of Foundations and Embankments, Geotechnical Special Publication No. 40, Vol. 2, pp. 1100-1110.

Assessment of Soil Disturbance by the Installation of Displacement Sand Drains and Prefabricated Vertical Drains

By Steven R. Saye[1], Member, ASCE

Abstract

Prefabricated vertical (PV) drains are commonly used to decrease the drainage paths within soft soils in order to accelerate the time for primary consolidation. PV drains are full displacement drains of small volume that are thought to exhibit considerably less disturbance to the soil mass than displacement sand drains. Experience at three sites where excessive disturbance due to installation of PV drains is thought to have occurred, combined with available published data regarding the performance of displacement sand drains and PV drains, provides the basis for an empirical approach to assess the excessive installation disturbance effects imposed by both PV drains and displacement sand drains. The size of both the installation mandrel and the anchor are important factors in the disturbance effects. The disturbance caused by PV drains is shown to be similar to displacement sand drains when the drain spacing ratio is based on the effective diameter of the mandrel/anchor combination calculated using the mandrel/anchor combination perimeter, not the end area. With this approach, a modified drain spacing ratio (effective drain spacing / effective mandrel diameter) greater than 7 to 10 is considered necessary to reduce the excessive disturbance effects of PV drain installation.

Introduction

The significant disturbance associated with the installation of displacement sand drains was identified by Casagrande and Poulos (1969) with the suggestion that full displacement type sand drains frequently had no benefit, and in some instances were harmful in comparison with no treatment. Displacement sand drains were used extensively in highway construction up to the 1960's then were replaced, first by non-displacement sand drains(augered and jetted), and later by prefabricated

---

[1]Senior Engineer, Geotechnical Services, Inc., 7050 South 110th Street, Omaha Nebraska 68128. ssaye@gsinetwork.com

vertical (PV) drains and fabric-encased sand drains installed with a mandrel. PV drains are described by Rixner et al. (1986) as full displacement drains of small volume that are thought to impose considerably less disturbance to the soil mass than displacement sand drains due to the small size of the mandrel and drain in comparison to the large pipe (often about 0.46 m diameter) commonly used to install displacement sand drains. The concept is presented in this paper that PV drains develop disturbance effects similar to, or greater than, the disturbance of displacement sand drains and that at close spacing (less than about 2 m) installation of PV drains causes progressively more disturbance that substantially reduces the rate of consolidation of the foundation soils, as noted by Casagrande and Poulos (1969) for displacement sand drains.

Installation Disturbance Effects
Design efforts to assess the disturbance effects for displacement drains largely rely on the Barron (1948) and Hansbo (1979) theoretical treatment described in Equation 1 that assumes an incompressible smear zone of low-permeability soil around the displacement drain that retards the movement of water to the drain.

$$t = \frac{d_e^{\,2}}{8c_h} F \ln\left(\frac{1}{1 - \overline{U}_h}\right) \tag{1}$$

Where:

t = the time to achieve average degree of consolidation, $\overline{U}_h$
d_e = the equivalent drain spacing (1.05 times the triangular spacing S)
c_h = the horizontal coefficient of consolidation for undisturbed soil outside of the smear zone
$F = F_n + F_s$
F_n = drain spacing factor [ln (n) - 0.75], with n = d_e/d_w and d_w = equivalent drain diameter
F_s = the disturbance factor [$(k_h/k_s - 1) \ln (d_s/d_w)$] with k_h = the horizontal permeability for the undisturbed soil, k_s = the permeability within the smear (remolded) zone and d_s = the diameter of the smear (remolded) zone

Evaluation of the actual performance of PV drains and displacement sand drains at different spacings through evaluation of settlement and piezometer data can be used to gain an understanding of the installation disturbance effects. Values of F_s from Equation 4 are summarized in Table 3 and plotted with respect to the average d_e in Figure 1 using the assessments of $c_h(e)$, the effective field c_h, with instrumentation data summarized in Tables 1 and 2 for displacement sand drains and PV drains.
Based on Equation 1, Ladd (1989) showed that $c_h(e)$ for vertical drains

having a homogeneous smear zone can be expressed as:

$$c_h(e) = c_h \frac{F_n}{F_n + F_s} \quad \text{or} \quad \frac{c_h(e)}{c_h} = \frac{1}{\left(1 + \dfrac{F_s}{F_n}\right)} \tag{2}$$

Where: $c_h(e)$ is the back-calculated c_h from field settlement and piezometer data using $F=F_n$.

For identical drains with the same smear zone characteristics installed at different spacings in soil of constant c_h, Equation 2 can be transformed to back-calculate F_s as:

$$F_s = \frac{c_h(e_1) - c_h(e_2)}{\left(\dfrac{c_h(e_2)}{F_{n2}}\right) - \left(\dfrac{c_h(e_1)}{F_{n1}}\right)} \tag{3}$$

The back-calculated F_s values are generally negative, which can have no physical significance. The negative values of F_s occur because F_s is not constant and F_s changes at decreasing spacing raising the concern that the Barron (1948) – Hansbo (1979) theoretical treatment is not valid for drain spacings used in practice.

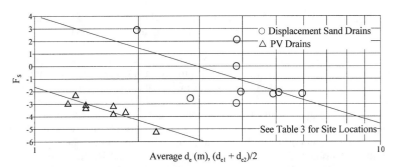

Figure 1. Calculated Disturbance Factor, F_s , For Displacement
Sand Drains and PV Drains Based on Performance Records

Onoue et al. (1991) presented the results of laboratory experimental studies to evaluate variations in permeability alongside a circular displacement drain installed in a uniform clay with $k_h = k_v$. Figure 2 shows that two zones of reduced permeability developed along this displacement drain (circular mandrel): a remolded zone where the soil structure is destroyed with a major reduction in permeability that extended for a distance about 1.6 times the displacement drain/mandrel radius; and a disturbed zone where the soil structure is modified with a moderate decrease in permeability that extended for a distance about 6.5 times the displacement drain/mandrel radius. At closer drain spacing, the disturbed zones (Zone II in Figure 2) interact, further decreasing the permeability of the soil. The observed disturbed zone agrees with the suggestion of Jamiolkowski et al. (1981) that the disturbance zone extends 5 to 6 times the effective mandrel diameter.

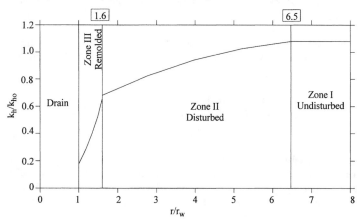

Figure 2. Expected Variation in Horizontal Permeability Following Installation of Displacement Drains (from Onoue et al. 1991)

Figure 3 illustrates 3 types of installation mandrel/anchors that were used respectively for the Storz Expressway project in Omaha, Nebraska; for the reconstruction of Interstate 15 in Salt Lake City, Utah; and the rhombic mandrel that originated in Europe and was used for the Woodrow Wilson Bridge test fill in Alexandria, Virginia. The I-15 and Storz Expressway mandrels consist of a channel, stiffening rib, and reinforcing bar anchor that results in a relatively large surface area in proportion to the circular sand drain mandrel described by Barron (1948) - Hansbo (1979). Jamiolkowski et al. (1981) and Rixner et al. (1986) convert the mandrel into an effective diameter d_m, as the diameter of a circle with an area equal to the mandrel's greatest cross sectional area, or the cross sectional area of the anchor or tip, whichever is greater. For the circular displacement drains and the rhombic mandrel without a large protruding anchor, this calculation is considered reasonable; however for the irregularly-shaped mandrels with a reinforcing bar anchor used at Storz Expressway and Interstate 15 this calculation is thought to significantly underestimate the effective size of the mandrel and the associated disturbance. Furthermore, when bottom anchorage of the PV drains does not occur successfully in the field, the contractor will either increase the size of the bar anchor or select a sheet metal plate anchor, such as shown in Figure 3b for the I-15 project. This further increases the size of the mandrel/anchor combination and increases the disturbance. For the plate anchor, the disturbance effects are dependant on the degree of collapse of the plate anchor around the mandrel.

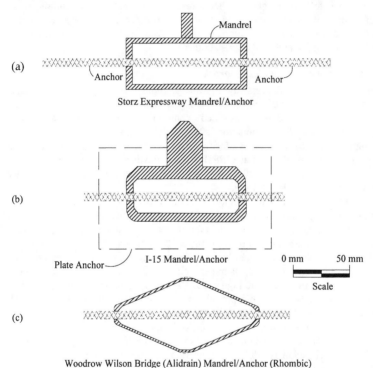

Figure 3. Examples of Installation Mandrels and Anchors (Plan View)

Evaluation of Effective c_h from Geotechnical Instruments

The effectiveness of PV drains (or conversely the adverse effects of installation disturbance) require the determination of the effective horizontal coefficient of consolidation $c_h(e)$, from field instrumentation data. Proper evaluation of the instrumentation data requires knowledge of the loading conditions, the internal drainage conditions, and the rate of consolidation. In the writer's opinion, instrumentation data for soft ground projects should be evaluated during periods of constant load by determining $c_h(e)$. Evaluation of $c_h(e)$ is preferably done using piezometer data and the Orleach (1983) method, i.e., plotting the slope of log excess pore water pressure vs. time. Settlement data is preferably evaluated using the Asaoka (1978) analysis of interval settlement data over specific intervals and using surface settlement data. It is important to obtain settlement and piezometer data away from highly permeable drainage boundaries. Evaluation of $c_h(e)$ in this manner is expected to yield a calculated c_h slightly higher than the actual c_h.

The following three sections discuss the writer's evaluation of the performance of PV drains installed at different spacings at three sites. Subsequent sections describe the evaluation of some data from reports regarding the performance of PV and sand drains at other sites.

Storz Expressway - Omaha, Nebraska

The Arthur C. Storz Expressway highway alignment extends along the Missouri River floodplain in Omaha, Nebraska; crossing thick deposits of soft highly plastic clay in a recent, abandoned, clay-filled oxbow that was named "Florence Lake". Index properties, preconsolidation stress values, and uncorrected field vane strengths presented in Figure 4 illustrate the typical geotechnical properties of the Florence Lake clay. Saye et al. (1988) describe the design and performance of the precompression program with staged loading for the embankment which was constructed during 1984 to 1989.

PV drains were placed in a triangular pattern spaced at 1.0 meter intervals for two-stage construction, at 1.6 meter intervals for single-stage construction, and at a 1.7-meter interval beneath the stability berms. The contractor fabricated a cable-driven mandrel formed by a 51 mm x 127 mm box with a 13 mm x 25 mm stiffening rib on one face. The PV drain was anchored in the underlying sand with a 0.3 meter-long, 10 mm diameter reinforcing bar. The mandrel/anchor combination is illustrated in Figure 3a. The project specifications limited the mandrel area to 64.5 cm², but provided no restrictions on the anchor size. The stiffener rib was added after work began and was required to limit bending of the mandrel during installation of the drains to depths of 12 to 16 m through the soft clay. The actual area of the mandrel/anchor was 84.5 cm². The effective diameter of the mandrel/anchor is calculated to be 10.5 cm based on the projected end area and 24 cm based on the perimeter.

Geotechnical instrumentation provided measurements of $c_h(e)$ from piezometer data using the Orleach (1983) method and from interval settlement data using the Asaoka (1978) method. The field evaluations of $c_h(e)$ for the Florence Lake clay with the piezometer data are summarized in Figure 5 showing significant variations in $c_h(e)$ between areas where PV drains were installed at a 1.0 meter spacing and where PV drains were installed at 1.6 and 1.7 meter spacings. The lowest $c_h(e)$ values of about 0.001 m²/day were measured in the most-plastic clay near depths of 3 to 7 meters where PV drains were installed at a 1.0-meter spacing. The lowest $c_h(e)$ values were about 40 percent of the design c_h value and required an unscheduled extension of the precompression time. The

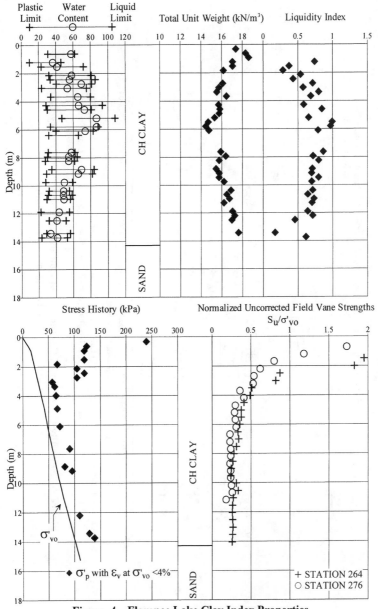

Figure 4. Florence Lake Clay Index Properties

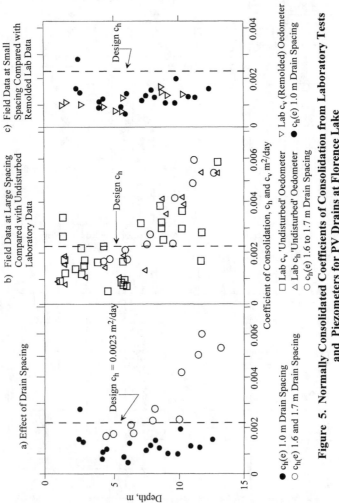

Figure 5. Normally Consolidated Coefficients of Consolidation from Laboratory Tests and Piezometers for PV Drains at Florence Lake

oedometer tests on undisturbed specimens gave a c_v of about 0.014 m^2/day using the square root of time calculation method. Oedometer tests run in the horizontal orientation showed c_h/c_v values to be approximately unity consistent with the recommendations of $c_h/c_v = 1.2+/- 0.2$ for clays with no evidence of layering by Rixner et al. (1986) and similar to $c_h(e)$ data where PV drains were installed at the 1.6 and 1.7 meter spacings. Figure 5 also shows that $c_h(e)$ data from locations where PV drains were installed at 1.0 meter spacing approximately matched laboratory c_v data of remolded samples, indicating that installation of the PV drains at the 1 meter spacing disturbed a significant zone of soil around the drain, not just smearing a zone of soil immediately adjacent to the drain. The significant disturbance is thought to result from the combined effect of the large geometry of the installation mandrel/anchor and the close spacing of the PV drains.

An important practical lesson regarding the detrimental effects of excessive PV drain installation effects is seen in Figures 6 and 7. Figure 6 shows that the disturbance associated with installation of PV drains at a 1.0 m spacing resulted in about 0.25 m additional ground surface settlement due to the weight of the sand blanket and increased the settlement by about 0.45 m at the end of Stage IA compared to the area treated with PV drains at a 1.6 m spacing. Figure 7 presents the field compressibility curves based on the applied stress and the Asaoka (1978) end-of-primary surface settlements and strains (calculated as the surface settlement / clay thickness ratio) for adjoining instrument locations (separated by about 185 m). Where PV drains were installed at a modified drain spacing ratio (perimeter calculation) of 9 (1.6 m triangular spacing), a preconsolidation stress about 15 to 20 kPa above the effective overburden stress was observed. Where PV drains were installed at a modified drain spacing ratio (perimeter calculation) of 5.5 (1.0 m spacing), normally consolidated stress conditions were observed suggesting that PV drain installation at the 1.0 m spacing disturbed the soil structure sufficiently to eliminate the small overconsolidation stress. The increased disturbance and settlement increased the total fill quantities needed to establish the design fill elevation, increased the final effective embankment stress, reduced the rate of consolidation, and reduced the embankment stability.

Using the assessment of the remolded and disturbed zones for a circular mandrel in Figure 2 suggested by Onoue et al. (1991), the following disturbance zones would be expected for the Storz Expressway mandrel/anchor based on two different assessments of the effective diameter of the mandrel:

Mandrel dia. (cm) calculation method	Dia. of remolded zone cm	Dia. of disturbed zone cm
Area Basis (11)	17	70
Perimeter Basis (24)	38	155

With the perimeter calculations of the effective diameter of an irregularly-shaped

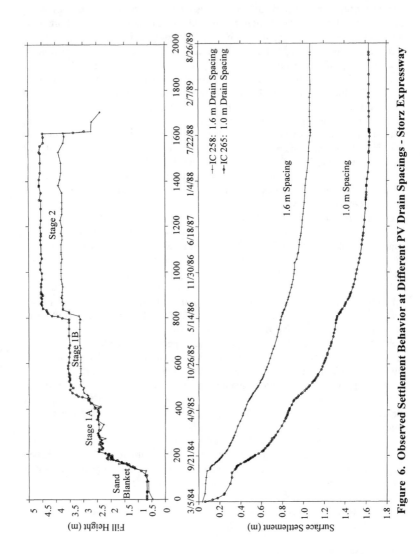

Figure 6. Observed Settlement Behavior at Different PV Drain Spacings - Storz Expressway

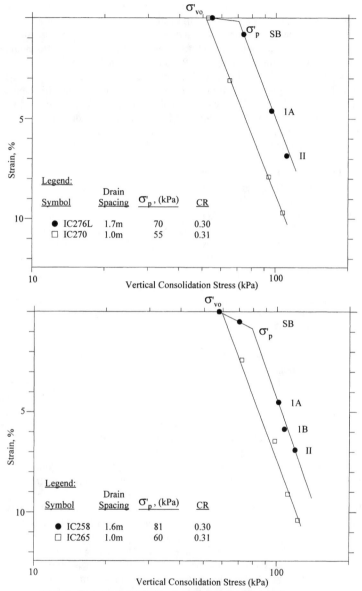

Figure 7. Effect of PV Drain Installation Disturbance on Field Compression Curves of Florence Lake Clay

mandrel, extensive interaction of the remolded and disturbed zones would be expected at the 1.0 m spacing, but a significantly reduced interaction of adjoining drains would be expected at drain spacings of 1.6 m and 1.7 m. These assessments of installation disturbance and the field behavior at Storz Expressway are thought to demonstrate the practical concerns for PV drains installed at close spacings and to support calculation of the effective diameter of the irregularly-shaped mandrel/anchor using the perimeter calculations, and to support identification of both remolded and disturbed zones around the mandrel. Additional evidence supporting the calculation of the effective diameter of the mandrel/anchor using the perimeter is presented in Figure 10 where field performance data from PV drains and displacement sand drains are available from the same site.

Interstate 15 In Salt Lake City, Utah

The reconstruction of Interstate 15 in Salt Lake City, Utah required the installation of about 7.4 million meters of PV drains to accelerate the consolidation of the thick deposits of layered silts and clays deposited in the last two stages of Glacial Lake Bonneville. Saye et al. (2001) describes the design of the foundation improvements for I-15. The roadway was completed in 2001. The typical soil conditions for a portion of the 600S design segment, near downtown Salt Lake City, are presented in Figures 8 and 9.

The initial construction of the I-15 embankments in the early 1960's included sections where displacement sand drains were installed at spacings from 3.1 to 4.9 m in a square pattern to support embankments up to 12 m high. The results of Asaoka (1978) analyses of surface settlement data for sand drain areas obtained by the Utah Department of Transportation are summarized in Table 2 and Figure 10. Figure 10 shows a consistent (although scattered) decrease in $c_h(e)$/lab c_v with decreasing drain spacing ratio.

Concerns of the writer and Dr. Charles C. Ladd regarding excessive installation disturbance associated with PV drains at close spacing resulted in the selection of a minimum design spacing of 1.5-m and the specification requirement to limit the area of the mandrel/anchor combination to 65 cm². Following negotiations with the design-build team, the installation mandrel used by the contractor for the I-15 work in the 600S area was a stepped mandrel with a lower section that had an area of 75.7 cm² with options to use a 20 cm long anchor bar, a 23 cm long anchor bar, or a 10 cm x 18 cm anchor plate as anchoring conditions became progressively more difficult. A vibrator was required to install most of the PV drains to the design depths, but its use was limited to penetration of fill and dense sand layers. The lower section of the mandrel, the small reinforcing bar anchor, and the plate anchor are illustrated in Figure 3b.

The specification restrictions on mandrel and anchor size and the minimum drain spacing were questioned by the some members of the design-build team. To check these restrictions, a test section was established in the North Temple area of Design Section 600S during the initial phase of construction to assess the impact of installation disturbance. PV drains were installed in a triangular pattern at spacings

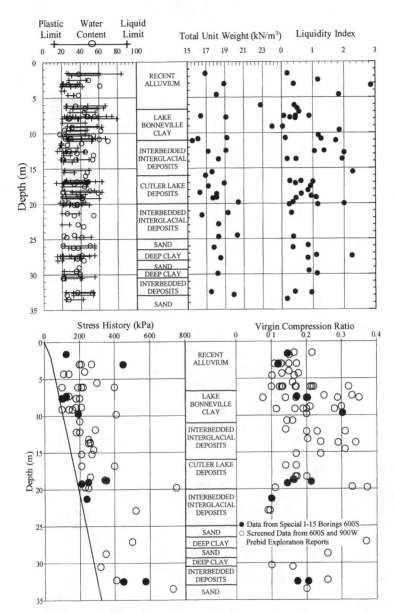

Figure 8. Lake Bonneville Clay Index Properties - I-15 Salt Lake City, Utah

SOIL BEHAVIOR AND

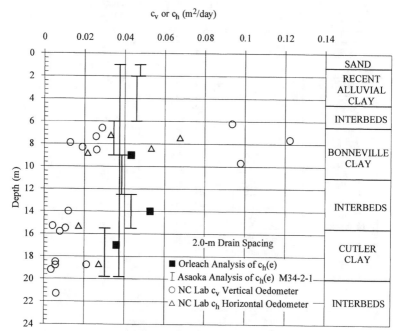

Figure 9. Laboratory and Field Values of c_h and c_v
Interstate 15 Salt Lake City, Utah

of 1.0 m, 1.5 m, and 2.0 m with a 23 cm long, 1.3 cm diameter, reinforcing bar anchor; and PV drains were installed at one location at a 1.5 m triangular spacing with a plate anchor (102 mm by 178 mm by 0.8 mm). The mandrel/short reinforcing bar anchor combination results in an effective diameter of 10.5 cm using the area calculation and 17.8 cm using the perimeter calculation. The mandrel/anchor plate combination, assuming a partially "crushed" plate anchor based on a limited number of field measurements, results in an effective diameter of 15.2 cm with the area calculation and 17.2 cm with the perimeter calculation. Asaoka (1978) analyses of surface settlement data, interval settlement data; and Orleach (1983) analyses of piezometer data yield calculations of $c_h(e)$/lab c_v for the different drain spacings. These data are referenced to a lab $c_v = 0.008$ m^2/day in the Cutler clay. Figure 9 also illustrates the reasonable agreement in $c_h(e)$ observed between Asaoka (1978) analyses of settlement data and Orleach (1983) analyses of piezometer data at the test section where PV drains were installed at a 2.0 m spacing and disturbance is thought to have been minimized. Figure 10 presents the $c_h(e)$/lab c_v data for displacement sand drains and PV drains with respect to the drain spacing ratio defined as d_e/d_m, where d_e is the effective drain spacing (S x 1.05 for a triangular pattern) and d_m = the effective diameter of the installation mandrel. The performance

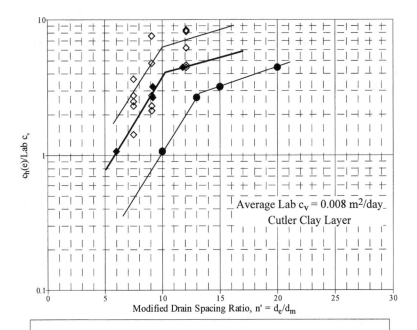

Figure 10. Normalized $c_h(e)$ from Surface Settlement Data vs. Modified Drain Spacing Ratio (d_e/d_m) for PV Drains and Sand Drains, Interstate 15 Salt Lake City, Utah

data for the PV drains in Figure 10 are presented for d_m calculated using both the perimeter and area of the mandrel/anchor combination. The data in Figure 10 show that the PV drains performed similarly to displacement sand drains if the perimeter calculation of the mandrel/anchor diameter is used supporting the use of the mandrel/anchor perimeter to evaluate installation disturbance effects. Figure 11 shows the calculated time for 95 percent consolidation, t_{95}, at each drain spacing assuming a constant $c_h(e)$ equal to the 2.0-m drain spacing value (where disturbance is thought to be least) from the Asaoka (1978) assessments of t_{95}. At drain spacings less than 1.75 m, limited decreases in the time for consolidation are observed. The higher number of PV drains and the cost for the ground treatment increases, and increased settlement due to higher disturbance effects occurs, increasing the amount of fill/embankment material. The disturbance effects approximately offset or negate the theoretical benefit of decreased drainage distance. Following the results of the test section, the remaining drains in the 600S area were installed at a 1.75 m spacing in a triangular pattern.

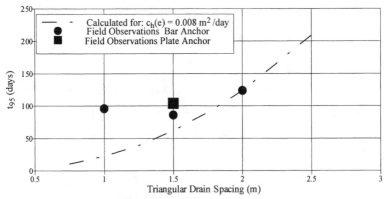

Figure 11. Effect of Installation Disturbance on Consolidation Times with PV Drains at I-15, Salt Lake City, Utah - (Saye et al. 2001)

Woodrow Wilson Bridge Test Embankment

Reconstruction of I-95 at the US 1 Intersection adjacent to the Woodrow Wilson Bridge in Alexandria, Virginia requires precompression of about 6 m of organic, highly plastic, alluvial clay described locally as the A-1 clay. The original embankment had been treated with displacement sand drains and surcharge in the 1960's. The new construction will widen each side of the existing embankment. A combination of PV drains and surcharge with deep soil mixing are planned to treat the A-1 clays (Haley and Aldrich, Inc. 2001). The site conditions are illustrated in Figure 12. The preliminary geotechnical study made by URS Greiner Woodward Clyde (1999) addressed the design of the PV drains and a series of Rowe cell tests and oedometer tests were used to help assess the normally consolidated value of c_h and the ratio of laboratory c_h/c_v. These data are presented in Figure 12, showing c_h/c_v values ranging from about 3 to 12. The very high c_h/c_v values were considered questionable. Based on the experiences at Storz Expressway and I-15, the intent of the PV drain design was to select a mandrel/anchor combination to minimize the installation disturbance effects and to take advantage of the higher c_h/c_v values expected in the A-1 soils. In this instance the limited thickness and depth of the A-1 soils was expected to permit use of a "small" mandrel.

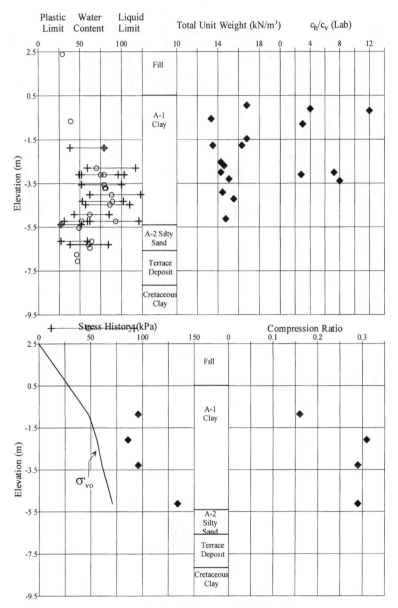

Figure 12. Woodrow Wilson Bridge Soil Properties

The preliminary geotechnical study identified uncertainties associated with the performance of the PV drains in the A-1 clays, especially the value of $c_h(e)$, and a test section was recommended such that actual performance data could be obtained to select construction schedules for staged and phased embankment construction. The test section was implemented in 2000 and PV drains were installed in a triangular pattern at spacings of 1.2 m and 1.5 m using the rhombic mandrel (11 mm by 76 mm) and 19 cm long, 9.5 mm diameter, reinforcing bar anchor shown in Figure 3c. The mandrel/anchor has an effective diameter of 7.1 cm (area basis) and 13.8 cm (perimeter basis) to minimize the disturbance of the A-1 soils. Figure 13 presents the assessments of $c_h(e)$ based on Asaoka (1978) analyses of ground surface and interval settlement data from magnet reed switch instruments suggesting a $c_h(e)$/lab c_v of 6 for the 1.5 m drain spacing and a $c_h(e)$/lab c_v of 3.5 for the 1.2 m drain spacing, consistent with the observations at other sites listed in Table 2. The actual settlement time for 95 percent consolidation at the 1.2-m spacing was similar to the settlement time observed at the 1.5 m. A design spacing of 1.4 m in a triangular pattern was recommended for the production PV drains.

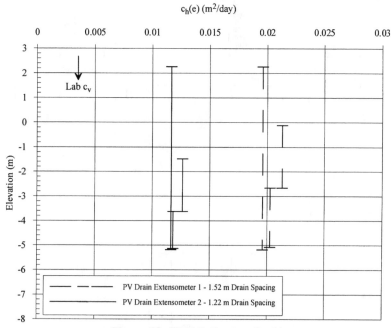

Figure 13. Field Estimates of $c_h(e)$
Woodrow Wilson Bridge Test Embankment

Other Sites Treated with PV Drains

A limited number of published case histories, listed in Table 2, were available where PV drains had been installed and values of $c_h(e)$ and c_v were reported. These data are summarized in Figure 14. Long et al. (1994) report observations for PV drains installed in Connecticut Valley varved clays near South Windsor, CT. The natural deposits exhibit c_h/c_v values that significantly improved the rate of consolidation when the PV drain spacing was relatively large and disturbance minimized. Data for a single spacing of 3.7 m showed an average $c_h(e)$/lab c_v value

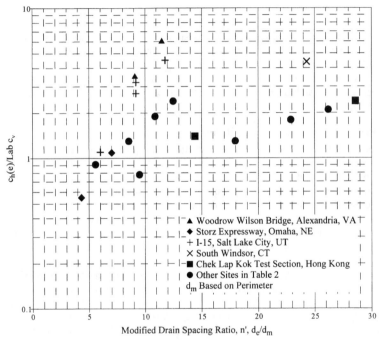

Figure 14. Normalized $c_h(e)$ vs. Modified Drain Spacing Ratio For PV Drains

of 3.9. A 102 mm by 150 mm mandrel was used to install the drains, but the anchor was not described. A modified drain spacing ratio of 24 is estimated by the writer. No significant installation disturbance appears to have occurred at this large drain spacing, consistent with the data in Figure 14.

Foott et al. (1987) report performance data from the Hong Kong airport test fill for PV drains installed at 1.5 and 3 m spacings and 0.5 m displacement sand drains installed at a 3 m spacing. Lo and Mesri (1994) describe the PV drain installation mandrel as a diamond-shape with external dimensions of 75 mm and

166 mm yielding an estimated effective diameter of 11.6 cm (perimeter basis). Koutsoftas (2001) describes the anchor as a 100 mm by 100 mm plate with a hook to attach the PV drain. The performance data are summarized below:

Drain Type and Spacing (m)	Modified Drain Spacing Ratio (d_e/d_m)	$c_h(e)$/lab c_v
PV drain, S=1.5	13.6	1.4
PV drain, S=3.0	27	2.4
Sand drain, S=3.0	6.3	0.7

These PV drain performance data are also plotted in Figure 14. (Note: the displacement sand drain performance is presented in Figure 15 and shows evidence of significant disturbance with $c_h(e)$/lab c_v of 0.7). Figure 14 suggests that two relationships for $c_h(e)$/lab c_v versus drain spacing occur, an initial slight decrease at wide drain spacings, and initiation of "excessive" disturbance at close spacings.

Sites with Displacement Sand Drains

Aboshi and Inoue (1984) present observations of $c_h(e)$ for three land reclamation projects in Japan (West Hiroshima, East Hiroshima and Fukuyama) treated with displacement sand drains. The relationship of $c_h(e)$/lab c_v vs. the drain spacing ratio is presented in Figure 15 and the data are presented in Table 1, showing $c_h(e)$/lab c_v values of about 4 +/- 1 at drain spacing ratios of 10 to 12 and a value less than 0.5 at drain spacing ratio of 5. Aboshi and Inoue (1984) conclude that the undisturbed c_h is 4 to 6 times lab c_v for these deposits, but recommend that in normal design calculations $c_h(e)$ should be assumed equal to the lab c_v to allow for disturbance effects.

Data from other sites where displacement sand drains were installed are also presented in Figure 15 and Table 1. The largest drain spacing ratio is about 13 with corresponding $c_h(e)$/lab c_v values ranging from 2 to 5. At decreasing drain spacing ratios, the value of $c_h(e)$/lab c_v decreases to about 0.5 at a drain spacing ratio of about 5. Importantly $c_h(e)$/lab c_v drops below 1 for a drain spacing ratio less than 6 to 7.

Figure 16 reproduces data from Gould (1976) for the post-glacial organic silts in the Potomac and Hudson estuaries (including the A-1 soils at the Woodrow Wilson Bridge site) where sand drains had been installed. These data were re-plotted to reflect the drain spacing ratio with an assumed drain diameter of 0.46 m. Although the corresponding laboratory c_v values were not reported, a consistent decrease in $c_h(e)$ is observed with decreasing drain spacing similar to the observations for displacement sand drains and PV drains at other sites. Gould (1976) also reports the results of jetted drains at Battery Park City at a drain spacing of 3.7 m (estimated drain spacing ratio of 9) that shows the $c_h(e)$ with limited installation disturbance to be 0.02 to 0.03 m^2/day, significantly greater than

the $c_h(e)$ for displacement drains at this spacing and similar to the reported $c_h(e)$
value of the displacement drains at a spacing of 5.5 m, the largest spacing reported,
where disturbance would be least.

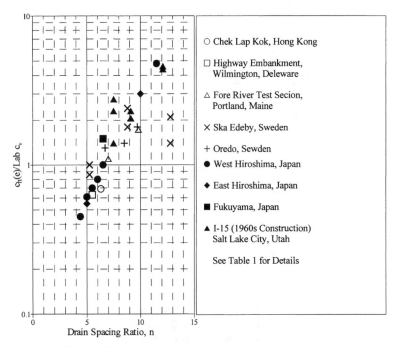

**Figure 15. Normalized $c_h(e)$ vs. Drain Spacing Ratio for
Displacement Sand Drains**

Sites with Non-Displacement Sand Drains

A limited number of settlement records were obtained regarding $c_h(e)$ for jetted
and augered sand drains where installation disturbance effects should be relatively
small to negligible. These data are summarized in Table 4. As noted above,
Gould (1976) showed that for the jetted sand drains at Battery Park City, $c_h(e)$ was
about 3 times the $c_h(e)$ indicated by the trendline for displacement sand drains at
the same spacing. Aldrich and Johnson (1972) showed that both jetted drains and
augered drains had $c_h(e)$ values from 1.2 to 2 times that of displacement drains at
the Fore River and Back Cove tests sections in Portland, Maine. These data
confirm that c_h/c_v ratios higher than unity can be achieved in natural soils with
vertical drains that minimize installation disturbance effects, and hence approach
the natural c_h.

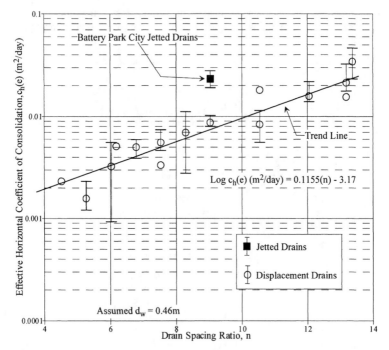

Figure 16. $c_h(e)$ vs. Drain Spacing for Organic Silt in Potomac and Hudson Estuaries from Gould (1976) Based on Data from A.A. Seymore - Jones

Empirical Prediction of Disturbance Effects

Figure 17 combines the performance data for displacement sand drains and PV drains described in Tables 1 and 2. The drain spacing ratio for the PV drains is defined as d_e/d_m (perimeter basis). Little difference is seen between the performance of PV drains and displacement sand drains when the modified drain spacing ratio is less than about 8 to 10. At close spacings, Figure 17 suggests that the disturbance effects can be approximated as:

$$c_h(e) \Big/ {}_{lab\, c_v} = 0.066\, e^{\, 0.44\, (n')} \qquad (4)$$

where: $c_h(e)$ = the effective horizontal coefficient of consolidation, lab c_v = the laboratory vertical coefficient of consolidation, e is the natural log, and n' = the modified drain spacing ratio, d_e/d_m, where d_e = the effective drain spacing and d_m is the effective diameter of the mandrel/anchor combination calculated using the perimeter.

The data for sites where PV drains were installed at a wide spacing, such as the Hong Kong airport (Foott et al. 1987) show a modest reduction in $c_h(e)$/lab c_v as the drain spacing ratio decreases to 13.5. A series of flat slopes are expected with different soils having different k_h/k_v values, as seen in Figure 17, resulting in a moderate decrease in $c_h(e)$/lab c_v at closer drain spacings. Figure 17 suggests that "excessive" disturbance effects would begin to occur at a drain spacing ratio of about 7 and the corresponding c_h/c_v in the "undisturbed" soil is near 1. Disturbance effects begin to occur at a drain spacing ratio of about 10 where c_h/c_v in the undisturbed soil is high.

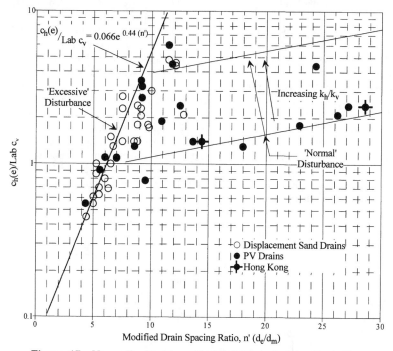

Figure 17. Normalized $c_h(e)$ vs. Modified Drain Spacing Ratio For Displacement Sand Drains and PV Drains

This empirical relationship is normalized to the laboratory c_v value. Natural variations in the value of c_v and variations in the testing and analysis methods will be factors in the actual behavior of soils treated with displacement drains using this empirical relationship.

The schedule demands for many soft ground projects seems to require increasingly shorter settlement periods, thus making closer and closer PV drain spacings appealing to design teams. To illustrate the false security of using very close design PV drain spacings, the empirical Equation 5 for installation disturbance and data in Figure 17 are applied to the Salt Lake City I-15 PV drain tests section data. The calculated times for 95% consolidation are based on a constant $c_h(e)$ =0.008 m²/day observed for the 2.0-m drain spacing (which is thought to reflect the limited disturbance effects) giving the predicted consolidation times in Figure 18 (same line as in Figure 11). In this case, the additional cost associated with PV drains installed closer than about 1.75-m provides limited, if any, benefit to the schedule, similar to the opinions expressed by Casagrande and Poulos (1969) and Gould (1976) for displacement sand drains at close spacings.

The actual value of c_h/c_v for the "undisturbed" soil and the "excessive" disturbance effects (defined by the steep-sloped trendline in Figure 17) caused by the selected design spacing and installation mandrel/anchor result in a minimum consolidation time that cannot be significantly improved by placing drains at a closer spacing. The disturbance effects appear to be greatest in layered soils with high values of c_h/c_v. Additional empirical evidence is needed to address the disturbance effects of different mandrels and anchors at spacings where "excessive" disturbance does not occur and to optimize the size of mandrels and anchors.   Furthermore, the experience at Storz Expressway suggests that "excessive" disturbance effects associated with a 1.0 m drain spacing and an "unfavorable" mandrel/anchor combination destroyed the natural soil structure sufficiently to significantly increase the compressibility of the foundation soil and likely decreased the soil strength.

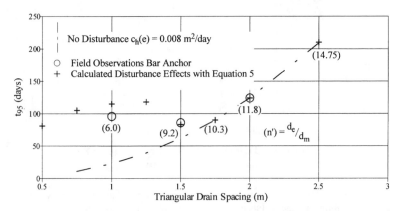

Figure 18. Calculated Disturbance Effects Interstate 15

Conclusions

Based on the writer's experience with the performance of prefabricated vertical (PV) drains at the Storz Expressway project in Omaha, Nebraska; the reconstruction of Interstate 15 in Salt Lake City, Utah; and the test embankment for the Woodrow Wilson Bridge in Alexandria, Virginia; combined with review of available published performance data for PV drains, displacement sand drains, and non-displacement sand drains at various sites; and with consultation and guidance from Dr. Charles C. Ladd; the following conclusions are presented with regards to the effects of installation disturbance with PV drains and to guide the design of PV drains:

- PV drains should be considered as displacement drains containing the same limitations identified by Casagrande and Poulos (1969) for disturbance caused by displacement sand drains. Disturbance will occur at all drain spacings, with the effect increasing as the drain spacing decreases.

- To assess the effects of "excessive" installation disturbance on the effective horizontal coefficient of consolidation, $c_h(e)$ with PV drains, the writer proposes that a modified drain spacing ratio be determined using the effective diameter of the installation mandrel/anchor combination based on the perimeter of the mandrel/anchor and the spacing of the PV drains. The assessments of the settlement data for the Lake Bonneville soils at I-15 presented in Figure 10 clearly show that disturbance effects of displacement sand drains and PV drains are similar when the effective diameter of the mandrel/anchor combination for the PV drains is calculated using the mandrel/anchor perimeter. Figure 17 is proposed for preliminary design to approximate the disturbance effects of PV drain installation based on available performance data that are considered reliable. Two types of disturbance effects are shown in Figure 17: 1) a series of flat-sloped lines varying as k_h/k_v changes in the natural soil where disturbance is likely controlled by the installation of single drains and 2) a steep-sloped line where large disturbance effects occur at close drain spacing and the disturbance effects of individual drains significantly interact. Development of additional test sections are recommended to refine the assessment of installation disturbance effects, to confirm the assessment of the effective mandrel/anchor diameter using the perimeter calculations, and to evaluate optimal mandrel/anchor geometries.

Variations in the geometry of individual installation mandrel/anchor combinations are expected to change the disturbance effects and the rate of consolidation at individual sites. Figure 17 shows that "excessive" disturbance effects are expected to begin with displacement drains at a modified drain spacing ratio of about 10, when c_h/c_v in the undisturbed clay is high, and at a modified drain spacing ratio of about 7 when c_h/c_v in the undisturbed clay is near 1. The effective horizontal coefficient of consolidation, $c_h(e)$, becomes less than the laboratory vertical coefficient of consolidation at a modified drain pacing ratio less than about 7.

- Special attention to the selection of the installation mandrel/anchor combination is considered an important part of ground improvement projects using PV drains. Designers should recognize that difficult installation conditions occur that will require large mandrels and have difficult PV drain anchorage conditions in the assessment of disturbance effects and the selection of drain types. The writer prefers use of the rhombic mandrel shape (Alidrain) in Figure 3c with a minimum length of reinforcing bar anchor. The design and field engineers should be cognizant to the damaging effects of excessive disturbance imposed by use of a larger mandrel/anchor and the potential increase in the time for construction imposed by indiscriminate selection of mandrel/anchor systems.

- Disturbance effects in displacement sand drains and PV drains create a condition that requires a minimum settlement period that can not be reduced by reducing the drain spacing. The value of $c_h(e)$ decreases due to disturbance and at close spacings this decrease can offset the theoretical effects of closer drain spacing. In the writer's opinion, design conditions may exist where the higher costs associated with use of non-displacement sand drains with minimal installation disturbance are required to achieve aggressive schedules and consolidation times.

- The assessment of c_h/c_v for the design of PV drains is difficult and large variations in both c_v and c_h/c_v often occur. Careful and extensive laboratory testing is considered important, but installation of test embankments utilizing different drain spacings in the design phase of large projects is considered the best approach to determine the effectiveness of PV drains. This approach is expected to be especially important for those projects where the available construction time is limited. For smaller projects, laboratory measurements of c_h/c_v using oedometer tests should be considered in conjunction with the guidance from Rixner et al. (1986).

Acknowledgements

The writer gratefully acknowledges the past support of Woodward-Clyde Consultants (now URS Corporation). The writer also gratefully acknowledges Wasatch Constructors, Inc. and Sverdrup/De Leuw for permission to present the Interstate 15 Reconstruction project; for the Nebraska Department of Roads and HDR Engineering, Inc. for their past support in conjunction with the Storz Expressway project; and the Virginia Department of Transportation and URS Corporation for access to the Woodrow Wilson Bridge test embankment data. The writer thanks Jennifer Williams and Jorge Santos for their assistance in preparation of the manuscript. Dr. Charles C. Ladd was a consultant on the Storz Expressway, I-15, and Woodrow Wilson Bridge projects providing invaluable guidance to the writer and other team members. Dr. Charles C. Ladd also made valuable suggestions to improve the manuscript prior to its submission.

Appendix: References

Aboshi, H. and Inoue, T. 1984. "Construction of a Reclaimed Land on a Very Soft Clay in Hiroshima - Soil Improvement and Construction Control. Seminar on Soil Improvement and Construction Techniques in Soft Ground, Singapore. January. pp. 85-102.

Aldrich, H.P., Jr., and Johnson, E.G., 1972. "Embankment Test Sections to Evaluate Field Performance of Vertical Sand Drains at I-95, Portland, Maine". Highway Research Record 405, pp 60 – 74.

Asaoka, A. 1978. "Observational Procedures or Settlement Prediction." Soils and Foundation, Volume 18, No. 4, 1978, pp 87-100.

Barron, R. A. 1948. Consolidation of Fine-Grained Soils by Drain Wells. (Trans. American Society of Civil Engineers, 113, 1948), p. 718.

Casagrande, L. and Poulos, S. 1969. "On the Effectiveness of Sand Drains." Canadian Geotechnical Journal, Volume 6, No. 3. 1969. pp. 286-326.

Eriksson, L. and A. Ekstrom. 1983. "The Efficiency of Three Types of Vertical Drain - Results from a Full-Scale Test." Proc. VIII, ECSMFE. Helsinki. pp. 605-610.

Foott, R., Koutsoftas, D.C. and Handfelt, L.D. 1987. "Test Fill at Chek Lap Kok, Hong Kong." ASCE GT. February 1987. pp 106-126.

Garassino, A., Jamiolkowski, M., Lancellotta, R, and Tonghini, M. 1979. "Behavior of Pre-Loading Embankments on Different Vertical Drains with Reference to Soil Consolidation Characteristics." VII ECSMFE, Vol. 3. Brighton, England. pp. 213-218.

Gould, J.P. 1976. "Determination of Soil Properties from Observations of Prototype Performance". National Capitol Section Annual Geotechnical Seminar, ASCE, Washington D.C.

Haley and Aldrich, Inc. 1970 "Test Sections to Evaluate Field performance of Vertical Sand Drains". Report 2, January.

Haley and Aldrich, Inc. 2001. "Engineering Report on Ground Improvement for Embankments I-95 Route 1 Interchange" VDOT Project No. 0095-96A-06, PE-1-1, Alexandria, Virginia, June.

Hansbo, S. 1979. "Consolidation of Clay by Band Shaped Prefabricated Drains". *Ground Engineering* Vol. 12, No. 5. 21-25.

Hansbo, S. 1982 Discussion Vertical Drains Thomas Telford Ltd. London

Hansbo, S. and Torstensson, B.A. 1977 "Geodrain and Other Vertical Drain Behavior", *Proc. 9th ICSMFE*, Vol. 1, Tokyo, 533-540.

Holtz, R.D., and Broms, B. 1972. "Long-Term Loading Tests at Ska-Edeby, Sweden". Proc. Special Conf. on Performance of Earth and Earth-Supported Structures". Vol. 1 pp 435-464. Purdue Univ. Lafayette, Indiana

Jamiolkowski, M., and Lancellotta, R., 1981. "Consolidation by Vertical Drains - Uncertainties Involved in the Prediction of Settlement Rates", Panel Discussion, Proc. X ICSMFE, Stockholm.

Jamiolkowski, M., R. Lancellotta, and W. Wolski. 1983. "Precompression and Speeding Up Consolidation." VIII ECSMFE 1983. Helsinki. pp. 311-314.

Jamiolkowski, M., Ladd, C.C., Germaine, J.T., and Lancellotta, R. 1985. "New developments in field and laboratory testing of soils: Theme Lecture 2." *Proc. 11th Int. Conference on Soil Mech. And Found. Engrg.*, San Francisco, 1, 57-153.

Koutsoftas, D.C. (2001). Personal Communication.

Ladd, C.C., J.J. Rixner, and D.C. Gifford. 1972. "Performance of Embankments with Sand Drains on Sensitive Clay." Performance of Earth and Earth-Supported Structures. ASCE, Purdue University, Lafayette, Indiana. Vol. 1, pp. 211-242.

Ladd, C.C. 1989. "Effects of Drain Spacing on Performance of Driven Sand Drains and Wick Drains". Internal MIT Memo. August

Lo, D.O.K. and Mesri, G. 1994. "Settlement of Test Fills for Chek Lap Kok Airport". Vertical and Horizontal Deformations of Foundations and Embankments. ASCE Geotechnical Special Publication 40. pp 1082 - 1099.

Long, R.P., Fontaine, L.F., and Olmstead, B. 1994. "Performance of Wick Drains Installed by Vibration". Vertical and Horizontal Deformations of Foundations and Embankments. ASCE Geotechnical Special Publication 40. pp 1193 - 1201.

Nicholson, D.P., and Jardine, R.J. (1982) " Performance of Vertical Drains at Queenborough Bypass." Vertical Drains Thomas Telford Ltd. London. pp 67-90

Onoue, A., Ting, N.H., Germaine, J.T., and Whitman, R.V. 1991. "Permeability of Disturbed Zone Around Vertical Drains". ASCE Geotechnical Engineering Congress, No. 27, Vol. 11, pp. 879 – 890.

Orleach, P. 1983. "Techniques to Evaluate the Field Performance of Vertical Drains." MS Thesis, Department of Engineering, MIT. 159 p.

Rixner, J.J., S.R. Kraemer, and A.D. Smith. 1986. Prefabricated Vertical Drains. Volume 1, Engineering Guidelines. FHWA/RD-86/168. August.

Sarkar, S.K. and R.J. Castelli. 1988. "Performance of Wick Drains in New Orleans Clays." Prefabricated Vertical Drains and Pavement Drainage Systems. Transportation Research Record No. 1159.

Saye, S.R., Easton, C.N., Smith, W.D., Nass, K.H., and Ladd, C.C. 1988. "Experience with Wick Drains in Highway Construction over Soft Clay - Storz Expressway, Omaha, Nebraska." Prefabricated Vertical Drains and Pavement Drainage Systems. Transportation Research Record No. 1159.

Saye, S.R., Ladd, C.C., Gerhart, P.C., Pilz, J. and Volk, J.C. 2001. "Embankment Construction in an Urban Environment: the Interstate 15 Experience". Proc. ASCE Geo-Institute Specialty Conference, 2001: A Geo-Odyssey – Foundations and Ground Improvement, Virginia Tech, June 9-12, 2001. pp 842-857.

Schmidt and Gould (1968) "Consolidation Properties of an Organic Clay Determined from Field Observations." Highway Research Board Record No. 243. pp 38-48.

Tanaka, H. 1994. "Consolidation of Ground Reclaimed by Inhomogeneous Clay". Vertical and Horizontal Deformations of Foundations and Embankments. ASCE

GSP No. 40, Vol. 2, pp 1262 – 1273.

URS Greiner Woodward Clyde. 1999. "Ground Improvement Feasibility Study Woodrow Wilson Bridge". Unpublished report. December.

Woodward-Clyde Consultants. 1989a. "Geotechnical Investigation Arthur C. Storz Expressway - Omaha, Nebraska." Unpublished Report, M81-1, Volume III. October.

Woodward-Clyde Consultants. 1989b. "Analysis of Instrumentation Data, Arthur C. Storz Expressway, Omaha, Nebraska." Unpublished Report, 86C8568. August.

Woodward-Clyde International-Americas. 1998. "PV Drain Installation Disturbance". Unpublished Project Design Memorandum PR-18. February 10, 1998.

TABLE 1:
DISTURBANCE EFFECTS OF DISPLACEMENT SAND DRAINS AND SAND WICKS

Project	Type of Drain	d_e (m)	Reported Mandrel/Anchor	d_m (m)	D_e/d_m	Description	Thick (m)	Index Properties	Lab c_v (m²/day)	$c_h(e)$ m²/day	$c_h(e)$/Lab c_v	Reference
										Field Observations		
Highway Embankment Wilmington, Delaware	Driven Sand Drain	2.24	Pipe	0.405	5.5	Organic Silt (OH)	7.5	WC = 84 LL = 84 Ip = 38	0.003	0.0017 ± 0.0004	0.63	Schmidt and Gould (1968) HRB REC 243
Fore River Test Section Portland, Maine	Driven Sand Drain	3.2	Pipe	0.46	7	Organic Clay (CH-OH)	7.5	WC = 65 LL = 65 Ip = 34	0.0056 to ±0.003	0.003 ± 0.0006	0.55	Haley and Aldrich (1970)
		4.48			9.8					0.0048 ± 0.0006	0.85	Report 2 to MSHC Aldrich and Johnson (1972)
Interstate 15 1960's Construction	Driven Sand Drain	5.6	Assumed 0.46m Pipe	0.46	12.1	Lake Bonnevile Clay (CH)	23	WC = 50-60 LL = 60-80 Ip = 30-60	0.015	0.125	8.2	Woodward – Clyde (1998)
		4.2			9.1					0.12	7.6	
		3.5			7.5					0.038	2.5	
		5.6			12.1	Lake Cutler Clay (CH)		WC = 25-60 LL = 45-60 Ip = 25-40	0.008	0.038	4.8	
		4.2			9.1					0.019	2.4	
		3.5			7.5					0.022	2.8	
Ska Edeby, Sweden	Driven Sand Drain dw = 0.18m	0.945	18cm Diameter Lid with 0.16m Pipe Mandrel	0.18	5.25	Glacial and Post-Glacial Soft Clay	10	WC > LL LL = 70-140 Ip = 50-110	0.0004	0.0004	1.0	Hansbo And Torstenssen (1977)
		1.58			8.8					0.0007	1.8	
		2.31			12.8					0.0008	2.1	
		2.31		0.18	12.8					0.0006	1.4	Holtz and Broms (1972)
		1.58			8.8					0.001	2.4	
		0.945			5.25					0.00035	0.86	

TABLE 1:
DISTURBANCE EFFECTS OF DISPLACEMENT SAND DRAINS AND SAND WICKS

Project	Type of Drain	d_e (m)	Reported Mandrel/Anchor	d_m (m)	d_e/d_m	Description	Thick (m)	Index Properties	Lab c_v (m²/day)	$c_h(e)$ m²/day	$\frac{c_h(e)}{\text{Lab } c_v}$	Reference
						Soil Characteristics				Field Observations		
Orebro Sweden	Driven Sand Drain dw = 0.165m	1.24 1.58 1.81	16.5cm Mandrel 18.5cm Bottom Plate	0.185	6.7 8.5 9.7	Soft Clay (CH)	8	WC =68-106 LL = 46-103 I_p = 27-56	0.0008	0.001 0.0011 0.0014	1.3 1.4 1.8	Eriksson And Ekstrom (1983)
Chek Lap Kok Test Section Hong Kong	Driven Sand Drain dw = 0.5m	3.15	Pipe Bottom Plate Not Reported	0.5	6.3	Marine Clay (CH)	6	LL = 85 I_p = 55 $I_L \geq 1$	0.0036 ± 0.0015	0.0024	0.69	Foott et al. (1987)
West Hiroshima	Driven Sand Drain dw = 0.5m		Pipe Bottom Plate Not Reported	0.5	4.4 5 5.5 6 6.5 11.5	Data Not Reported					0.45 0.61 0.7 0.8 1 4.8	Aboshi and Inoue (1984)
East Hiroshima	Driven Sand Drain				5 10						0.55 3	
Fukuyama	Driven Sand Drain				6.5						1.5	
Back Cove Test Section Portland, Maine	Driven Sand Drain	3.2 4.48	Pipe Bottom Plate Not Reported	0.46	7 9.7	Organic Clay (CH)	16	WC = 47 LL = 44 I_p = 20		0.0037 0.0037		Haley and Aldrich (1970) Report 2 to MSHC Aldrich and Johnson (1972)

TABLE 1:
DISTURBANCE EFFECTS OF DISPLACEMENT SAND DRAINS AND SAND WICKS

Project	Type of Drain	d_e (m)	Reported Mandrel/ Anchor	d_m (m)	d_e/d_m	Description	Thick (m)	Index Properties	Lab c_v (m^2/day)	Field Observations $c_h(e)$ m^2/day	Field Observations $\frac{c_h(e)}{Lab\ c_v}$	Reference
Queens-Borough Bypass, England	Sandwick dw = 0.065m	1.13	95mm Pipe Disposable Point	0.1	11.9	Silty Clay	10	WC =40-100 LL = 60-110 I_P = 20-40	0.0008	0.0014	1.67	Nicholson And Jardine (1982)
Sandwich Bypass England	Sandwick dw = 0.065m	0.9	95mm Pipe Disposable Point	0.1	9.5	Alluvial Clay	6	WC = 30-80 LL = 70-80 I_P = 40-60	0.0011	0.0009	0.78	Nicholson And Jardine (1982)
Porto Tolle	Sandwick dw = 0.07m	4.4	Reported as Equivalent Diameter	0.1	48	Alluvial Clay	40	LL = 50-60 I_P = 25-40	0.017	0.045	2.55	Jamiolkowski et al. (1985)

TABLE 2:
DISTURBANCE EFFECTS OF PREFABRICATED VERTICAL DRAINS

Project	Type of Drain	d_e (m)	Reported Mandrel/Anchor	Area Basis d_m (m)	Area Basis d_e/d_m	Perimeter Basis d_m (m)	Perimeter Basis d_e/d_m	Description	Thick (m)	Index Properties	Lab c_v (m^2/day)	Field Observations $c_h(e)$ m^2/day	Field Observations $\frac{c_h(e)}{Lab\ c_v}$	Reference
Chek Lap Kok Test Section, Hong Kong	Alidrain d_w=0.068m	1.58	Rhombic 75mmX166mm 100mmX100mm Plate Anchor	0.12	13.1	0.11	14.4	Marine Clay (CH)	6	LL = 85 I_P = 55 I_L > 1	0.0036	0.0049	1.4	Foott et al. (1987)
		3.15			26.3		28.6				± 0.0015	0.0085	2.4	Lo and Mesri (1994)
Storz Expressway Omaha, NE	Ameridrain 301 d_w=0.067m	1.04	5cm X 12.7cm Box 1.3cm X 2.5 cm Rib 0.95cm X 30cm Anchor	0.11	9.9	0.24	4.3	Oxbow Clay (CH)	12-15	WC = 55-85 I_P = 50 LL = 75-100 I_L = 0.65	0.002 ±0.0006	0.0011 ± 0.0003	0.55	Saye et al. (1988)
		1.68			16		7					0.0023 ± 0.0004	1.09	TRR 1159
Queens-Borough Bypass England	Colbond d_w = 0.152	1.52	165mm Flattened Tube 350mm Angle Plate Anchor	0.19	8	0.273	5.6	Alluvial Clay	10	WC= 40-100 I_P=20-40 LL-60-110 I_L = 0.85	AVG 0.0057 ±0.0008	0.0063 ±0.0009	0.91	Nicholson and Jardine (1982)
South Windsor, CT I-291	Wick Drain Type Not Reported	3.89	102mm X 150mm Anchor Not Reported	0.14	28	0.16	24.3	Conn. Valley Varved Clay	36	WC>LL LL=50-55 I_P=20-25	0.017	0.067 ±0.007	3.9	Long et at. (1994)
Ska – Edeby, Sweden	Geodrain	0.945	Not Reported	0.12	7.7	0.145	6.52	Post-Glacial Soft Clay	10	WC =60-100 I_P = 80-100 I_L > 1	0.0004	0.0033	2.4	Hansbo And Torstensson (1977)

TABLE 2:
DISTURBANCE EFFECTS OF PREFABRICATED VERTICAL DRAINS

Project	Type of Drain	d_e (m)	Reported Mandrel/Anchor	Area Basis d_m (m)	Area Basis d_e/d_m	Perimeter Basis d_m (m)	Perimeter Basis d_e/d_m	Description	Thick (m)	Index Properties	Lab c_v (m²/day)	Field Obs. $c_h(e)$ m²/day	Field Obs. $c_h(e)$/Lab c_v	Reference
Woodrow Wilson Bridge	Ameridrain 301 d_w=0.067m	1.26	Rhombic 89x152mm 190mmx9.5mm Bar Anchor	0.07	17.8	0.14	9.1	A 1 Clay	4	WC = 75 I_p = 50 LL=110 I_L = 0.65	0.0033	0.012	3.5	Unpublished
		1.58			22.3		11.5					0.020	6	
Interstate 15 Reconstruction Salt Lake City, UT	Ameridrain 301 d_w=0.067m	1.05	20cm 1.3cm Dia. Bar	0.105	10	0.178	6	Cutler Clay (CH)	23	WC = 50-60 LL= 60-80 I_p = 30-60	0.008	0.0086	1.1	Woodward-Clyde (1998) Saye et al. (2001)
		1.58	Plate Anchor	0.152	13	0.172	9.2					0.022	2.7	
		1.58	20cm 1.3cm Dia. Bar	0.105	15	0.178	9.2					0.026	3.2	
		2.1	20cm 1.3cm Bar (Note 1)	0.105	20	0.178	11.8					0.036	4.5	
Orebro, Sweden	Alidrain d_w = 0.071m	1.24	'Rhombic' A = 45cm² 150-200mm Section Drain For Anchor	0.076	16.3	0.069	17.97	Soft Clay (CH)	8	WC = 94 LL = 65 I_p = 38 I_L = 1.8	0.0008	0.001	1.3	Eriksson and Ekstrom (1983)
		1.58			20.8		22.9					0.0014	1.8	
		1.81			23.7		26.23					0.0017	2.1	
	Geodrain d_w = 0.071m	1.24	Rectangular Mandrel A = 120cm² 0.7mm thick 125mm X 150mm Plate	0.12	9.9	0.145	8.6					0.001	1.3	
		1.58			12.7		10.9					0.0016	1.9	
		1.81			14.5		12.5					0.002	2.4	
Porto Tolle	Geodrain	4	Reported as Equivalent Diameter	0.16	25			Silty Clay (CL-CH) Po River Delta	30	WC = 36 LL = 53 I_p = 30 I_L = 0.65	0.0017	0.035	2	Jamio-lkowski et al. (1985)
	Soil Drain d_w = 0.18	4		0.1	40							0.032	1.8	

1: Note: All pushed with a mandrel 12.7cm X 5.1cm Rib

TABLE 2:
DISTURBANCE EFFECTS OF PREFABRICATED VERTICAL DRAINS

Project	Type of Drain	d_e (m)	Reported Mandrel/Anchor	Area Basis		Perimeter Basis		Soil Characteristics			Lab c_v (m²/day)	Field Observations		Reference
				d_m (m)	d_e/d_m	d_m (m)	d_e/d_m	Description	Thick (m)	Index Properties		$c_h(e)$ m²/day	$\dfrac{c_h(e)}{Lab\ c_v}$	
Jourdan Road Terminal	Alidrain d_w = 0.067m	1.6	Rhombic 20.3 cm Bar Anchor	0.124	12.9	0.17	9.5	Louisiana Alluvial Clay (CH)	13	WC =80-140 LL=120-140 I_P = 80 I_L = 0.8	0.0028	0.0036	1.3	Sarkar and Castelli (1988)
Tokyo Airport	Data Not Reported										0.0065	0.004	0.62	Tanaka (1994)
												0.007	1.16	
												0.013	1.93	
												0.023	3.55	
												0.068	10.49	
												0.023	3.63	

TABLE 3
FIELD DATA FOR ASSESSMENT OF THE DISTURBANCE FACTOR, Fs

Test Location	Drain Type	d_e (m)	$c_h(e)$ m^2/day	Calculated Fs
Storz Expressway	PV	1.04	0.0011	
				(2.9)
		1.68	0.0023	
Hong Kong Airport	PV	1.58	0.0049	
				(5.2)
		3.15	0.0085	
Oredo, Sweden Alidrain	PV	1.24	0.0010	
				(3.3)
		1.58	0.0014	
				(3.8)
		1.81	0.0017	
Geodrain	PV	1.24	0.0010	
				(3.0)
		1.58	0.0016	
				(3.1)
		1.81	0.0020	
Interstate 15 Salt Lake City, UT	PV	1.05	0.0086	
				(2.3)
		1.58	0.026	
				(3.6)
		2.1	0.036	
Gould (1976) Trendline	Displacement Sand Drain			
		2.26	0.0025	
				(2.5)
		3.39	0.0049	
				(2.0)
		4.52	0.0094	
				(2.1)
		5.65	0.0017	
				(2.2)
		6.25	0.0025	
Back Cove Test Section	Displacement Sand Drain	3.2	0.0037	
				0.0
		4.48	0.0037	
Fore River Test Section	Displacement Sand Drain	3.2	0.0031	
				(2.9)
		4.48	0.0048	
Interstate 15 Salt Lake City, UT	Displacement Sand Drain	3.5	0.017	
				2.12
		4.2	0.019	
				(2.2)
		5.6	0.036	

( ) Denotes negative number

TABLE 3
FIELD DATA FOR ASSESSMENT OF THE DISTURBANCE FACTOR, Fs

Test Location	Drain Type	d_e (m)	$c_h(e)$ m^2/day	Calculated Fs
Storz Expressway	PV	1.04	0.0011	
				(2.9)
		1.68	0.0023	
Hong Kong Airport	PV	1.58	0.0049	
				(5.2)
		3.15	0.0085	
Oredo, Sweden Alidrain	PV	1.24	0.0010	
				(3.3)
		1.58	0.0014	
				(3.8)
		1.81	0.0017	
Geodrain	PV	1.24	0.0010	
				(3.0)
		1.58	0.0016	
				(3.1)
		1.81	0.0020	
Interstate 15 Salt Lake City, UT	PV	1.05	0.0086	
				(2.3)
		1.58	0.026	
				(3.6)
		2.1	0.036	
Gould (1976) Trendline	Displacement Sand Drain	1.7	0.0017	
				2.9
		2.26	0.0025	
				(2.5)
		3.39	0.0049	
				(2.0)
		4.52	0.0094	
				(2.1)
		5.65	0.0017	
				(2.2)
		6.25	0.0025	
Back Cove Test Section	Displacement Sand Drain	2.13	0.0037	
				0.0
		4.27	0.0037	
Fore River Test Section	Displacement Sand Drain	3.2	0.0035	
				(2.9)
		4.48	0.0048	
Interstate 15 Salt Lake City, UT	Displacement Sand Drain	3.4	0.017	
				0.16
		4.2	0.019	
				(2.2)
		5.5	0.036	

( ) Denotes negative number

TABLE 4:
DISTURBANCE EFFECTS OF NON-DISPLACEMENT SAND DRAINS

Project	Type of Drain	d_e (m)	Description	Thick (m)	Index Properties	Lab c_v (m²/day)	$c_h(e)$ m²/day	$\frac{c_h(e)}{\text{Lab } c_v}$	Reference
			Soil Characteristics				Field Observations		
Porto Tolle	Dutch Jetted Sand Drain	5.25	Po River Delta Silty Clay	30	WC = 35, I_P = 30, I_L = 0.65	0.017	0.042	2.4	Jamiolkowski et al. (1985)
Portsmouth, NH	Dutch Jetted Sand Drain		Sensitive Clay	10	WC = 50±5, LL = 35±5, I_P = 15±3, I_L = 1.8±0.3	0.014	0.02 to 0.03	1.5 to 2	Ladd et al. (1972)
Fore River Test Section Portland, Maine	Rotary Jetted Sand Drain	4.48	Organic Clay	8	WC = 65, LL = 65, I_P = 34, I_L = 1	0.0056 ± 0.0028	0.006	1	Aldrich and Johnson (1972)
	Augered Drain	4.48		8			0.006	1	
Battery Park City	Jetted Sand Drain	4.13	Organic Silt				0.02 to 0.03		Gould (1976)
Back Cove Test Section Portland, Maine	Rotary Jetted Sand Drain	3.2	Sensitive Clay	16	WC = 47, LL = 44, I_P = 20		0.006 to 0.008		Aldrich and Johnson (1972)
		4.48							
	Augered Drain	3.2					0.007 to 0.009		
		4.48							

EMBANKMENT DESIGN – THE EARLY DAYS

Joseph J. Rixner, P.E.[1]

Abstract

This paper describes in general terms the unique and innovative embankment design approaches developed and implemented by Dr. Charles C. Ladd in the late 1960s and early 1970s. Some of these approaches included 12.7 cm (5 in) diameter undisturbed tube sampling, in situ strength testing using the Geonor field vane shear device, innovative laboratory strength testing (Geonor Direct Simple Shear Tests), development and application of the SHANSEP approach to full scale test and production embankments, use of test embankments during design not only to verify soil properties but also to assess impacts of sand drain installation techniques on in situ soil properties, and first time use in the United States of the Dutch jet bailer (non-displacement) method of sand drain installation. Many of these approaches, particularly SHANSEP, have withstood the test of time and have been used on many other projects over the past 30 years.

Introduction

In the late 1960s and early 1970s, Dr. Charles C. Ladd consulted with Haley & Aldrich, Inc. on the design and construction of two major highway embankment projects in northern New England. These projects involved the construction of high embankment fills over deep deposits of soft clays along coastal regions of New Hampshire and Maine. He was responsible for initiating or developing numerous innovative design procedures and solutions to overcome extensive settlement and stability problems.

----

[1] Executive Vice President/COO, Haley & Aldrich, Inc., 200 Town Centre Drive, Suite 2, Rochester, NY, 14623; Phone 716-321-4201; jjr@haleyaldrich.com

This paper will describe in general terms the technical issues faced by the designers as a result of the adverse subsurface soil conditions and the solutions developed to allow the construction to proceed as planned. Many of the solutions and approaches developed for these embankments became a standard for future embankment design. A brief overview of how some of the techniques used in the 1970s have developed and evolved over the years will be discussed to show how these processes have withstood the test of time. Several of the techniques have naturally evolved along with advances in technology. However some of the underlying principles developed and used on these early embankments have also been applied to these new technologies, demonstrating the solid understanding and wisdom of the early concepts developed and used by Dr. Ladd.

This paper is not intended to be a detailed primer on technical procedures or a design manual. It will not present specific detailed design procedures, but will provide a general overview of design techniques and procedures developed over 30 years ago.

The intent is to show the innovations and foresight Dr. Ladd used and how that wisdom and knowledge has withstood the test of time, and evolved with changing technologies. To obtain details of the technologies and procedures discussed herein, the reader should refer to the list of references at the end of this paper, as well as to the references appended in those documents.

Background

Embankment Projects

The two projects that form the basis for the discussions in this paper are the I-95 Interchange project in Portsmouth, NH, and Highway I-295 through the Back Bay and Fore River areas of Portland, ME.

The I-95 Interchange with the Spaulding Turnpike is located in southern New Hampshire, near Portsmouth. The interchange included the construction of five bridges and approach embankments as high as 9.1 to 10.7 m (30 to 35 ft). A major portion of the interchange is underlain by 10.7 to 12.2 m (35 to 40 ft). thick deposits of highly compressible, soft marine clay with high sensitivity.

The I-295 project in Portland, ME consists of two approximately one mile long tidal flat areas, one crossing the Fore River Channel, and the other along the easterly edge of the Back Bay area. Embankments as high as 4.6 to 15.2 m (15 to 50 ft) were constructed over these tidal mud flats. Underlying each of the areas are extensive deposits of soft to medium consistency, sensitive, gray silty clays up to 30.5 m (100 ft) deep. Soft organic clays, 3 to 12.2 m (10 to 40 ft) thick, overlie the silty clays in the tidal marsh areas.

For both of these projects, settlements of up to 2.1 m (7 ft) were predicted, and extensive stability problems had to be overcome. Time constraints on construction schedules required accelerating consolidation of the compressible soils to limit post construction settlements.

Innovative Design Features/Approaches

In order to design stable embankments with acceptable post construction settlements within the geometric site constraints and construction schedules, current state of the practice approaches along with several innovative design and construction procedures were utilized. The innovative techniques included:

- Somewhat unique (12.7 cm (5-in) diameter undisturbed tube) field sampling methods
- In-situ strength testing with Geonor vane shear device
- Innovative (Direct Simple Shear) strength testing
- Analytical techniques, including development of SHANSEP approach
- Pre-design Test Embankments to not only assess in situ soil properties, but also to assess the impact of sand drain installation techniques on soil properties
- First time use in the United States of the Dutch jet-bailer (non-displacement) method of sand drain installation

Combining these innovative techniques with the state of the practice design procedures for preloading, surcharging, and staged construction pioneered an approach that was not only successful for these projects but also led to many successful future applications.

The details of some of these "early" applications are discussed in the following section.

Innovative Design Features

Field Test Sections

Field test sections were constructed to assess design parameters for each of the two projects. An Experimental Test Section (ETS) was constructed at the I-95 Portsmouth Interchange site, primarily to assess and verify subsurface strength parameters for use in embankment stability analyses. This test section was also instrumented to determine if any indications of impending failure were exhibited that could be used during actual embankment construction.

In Portland, test sections were constructed at both the Fore River (FRT) and Back Cove (BCT) areas to assess the impact of spacing and installation methods of vertical sand drains on the horizontal rate of consolation of the subsurface organic and silty clay deposits (Haley & Aldrich, 1969).

ETS at I-95 Portsmouth

After a comprehensive field and laboratory testing program was conducted on the soft marine clay deposits at the I-95 site, uncertainties still existed in the engineering properties, especially the undrained shear strength of the deposit.   A test section was purposely constructed to failure to define better the in situ strength parameters of the clay. Details of the ETS are shown on Fig. 1, including the location of field instrumentation. The slide area at the top of the embankment at failure was 67.1 m (220 ft) long by 9.1 m (30 ft) wide.   The ETS was originally predicted to fail at a height of 4.6 m (15 ft). Actual failure occurred at 6.6 m (21.5 ft).

Undrained shear strength of the clay was measured in the field using vane shear tests and in the laboratory using a variety of test methods, including:

- Unconfined compression tests (on both vertical and horizontal samples)
- Undrained triaxial compression tests (isotropically and anisotropically consolidated)
- K_o consolidated undrained Direct Simple Shear Tests (DSS) (Ladd and Edgers, 1971)
- Plane Strain Active and Plane Strain Passive tests

In addition, the undrained strength of the marine clay was estimated from the SHANSEP approach as discussed later in this paper.

The ETS failed at an elevation 2 m (6.5 ft) higher than predicted, as noted above. The failure occurred at night; it was massive and probably instantaneous. Measurements and photos after the failure demonstrated the failure surface was circular. It is depicted on Fig. 2. Measurements showed very little dissipation of pore pressures occurred before failure, indicating the failure occurred under undrained conditions.

Circular arc stability analyses simulating the failure surface were conducted using undrained shear strengths from the various field and laboratory test methods, as shown on Fig. 3. Strengths from both field vane tests and from the SHANSEP approach using DSS strengths and the average of the maximum past pressures from lab data gave the closest predictions of the failure (Ladd, 1971). Average of vertical and horizontal unconfined tests greatly underestimated the in-situ strength.

Vertical and horizontal deformations measured prior to failure were very small. Neither these deformations, nor pore pressure measurements made the day before the failure, provided any indications that a massive failure was imminent.

FRT and BCT at Portland

In the late 1960s, test sections were constructed at these two locations to evaluate the in situ performance of vertical sand drains. At that time, vertical sand drains constituted the method of choice to accelerate the rate of consolidation and gain in shear strength of soft cohesive clay soils. Such an approach was often required to reduce post construction settlements and to increase the rate of strength gain required for staged construction of embankments. Concern existed over the effect various installation methods had on soil disturbance, and hence the in situ coefficient of consolation. This was especially important at that time, since driven drains were the least costly and easiest to install.

The test section at Fore River (FRT) was built to evaluate the effectiveness of sand drains on a 7.6 m (25 ft) thick layer of soft organic clay. See Fig. 4 for details. At Back Cove (BCT), Fig. 5, sand drains were installed through 12.2 to 18.3 m (40 to 60 ft) of soft sensitive silty clay. Three methods of installation were used at each test section: a) Driven Closed-End Mandrel, b) Jetted, with open-end Mandrel, and c) Continuous Hollow Shaft Auger. All drains were 45.7 cm (18 in) in diameter and were installed in a triangular pattern. At BCT, drains were installed by each method at 3.0 m (10 ft) and 4.3 m (14 ft) centers to evaluate the effect of spacing. At FRT, the driven drains were installed at both these spacings, but, due to space limitations, the other two methods were installed only at the 4.3 m (14 ft) spacing.

Each test section was instrumented to monitor settlement and pore pressure dissipation. Coefficients of consolidation, c_h, were back calculated for each installation method and spacing to assess relative performance, as well as to compare with values from laboratory data. See Aldrich and Johnson (1972) for a detailed discussion of results.

In summary, for the sensitive silty clay at BCT results indicated that c_h for all three installation methods was significantly smaller than the one predicted from lab data. However, the jetted and augered methods were more efficient than the driven drains by a factor of two (Aldrich and Johnson 1972). Also, the closer spacing resulted in a 20 to 25% reduction in back-calculated c_h for the jetted and augered methods, while no difference was noted for the driven drains. Projected settlements from field measurements for all areas was larger than predicted, with settlements in areas with driven drains being almost double the predicted values, while the increase was only 15% in the area of augered drains.

At FRT, all three drain types produced an effective value of c_h for the organic clay less than that measured in the laboratory, but the reductions were not as large as at BCT where the clay was thicker and more sensitive. The driven drains were only about 80% as efficient as the other drains at the 4.3 m (14 ft) spacing. However, the driven drains at 3 m (10 ft) spacing produced a lower value of c_h, by approximately 45%. Projected consolidation settlements at the location of driven drains was 30% greater than projected, while only 15 to 20% for the other methods.

Results from these test sections confirmed the concern that displacement methods of drain installation can adversely impact the effectiveness of sand drains, especially in sensitive clays. The effect is most pronounced with driven drains, but can also be a factor in other installation methods. These tests indicated that further advances in sand drain installation techniques could lead to better performance. This fact was shown to be the case in the Portsmouth project where an improved method of installing jetted drains (Dutch Jet Bailer method, as will be described later in this paper) was developed that did not require the use of a continuous mandrel. Less disturbance and smear resulted and provided better performance effectiveness.

Over the years, advances in equipment and materials led to more cost effective and technologically superior types of drains. Currently prefabricated vertical (wick) drains consisting of combinations of plastic cores and surrounding porous membranes are used in lieu of sand drains. However, even with this drain type, the effect of disturbance and other design considerations associated with sand drains are important considerations in their performance. Once again, Dr. Ladd's perception of key factors influencing the soil properties and performance characteristics of soft sensitive clay has been shown to be relevant and applicable to newer technologies.

SHANSEP Approach

Ladd developed the Stress History and Normalized Soil Engineering Properties (SHANSEP) concept to estimate in situ undrained shear strengths of cohesive soils (Ladd and Foote, 1974). In this method, plots of $s_u/\bar{\sigma}_v$ vs log OCR ($\bar{\sigma}_{vm}/\bar{\sigma}_{vo}$), where $s_u/\bar{\sigma}_v$ are based on Direct Simple Shear Tests, are developed. As noted above, this method was applied to the ETS at Portsmouth and the SHANSEP strengths gave a very good estimate of the failure, using the average maximum past pressures from laboratory test data as shown on Fig. 6. However, field settlement data from actual embankment construction indicated higher maximum past pressures than predicted from laboratory data. Nevertheless, this approach gave as good an estimate of in situ strength as any other approach. Furthermore, measured strength increases with consolidation under the first stage construction fill, in Portsmouth, agreed quite well with those predicted from the SHANSEP method.

Based on data from the ETS and from numerous measurements made during construction, the SHANSEP method proved its worth for both predicting initial undrained shear strengths as well as strength increases during consolidation. This method became widely adopted and has been used on many subsequent projects.

Embankment Design, I-95 Portsmouth

Design of the embankments at the Portsmouth interchange was controlled by strength and consolidation characteristics of the soft sensitive clays. The results from the test sections discussed in the previous section were used to aid in selecting soil properties and sand drain installation methods.

Using the SHANSEP approach and the results from the ETS, appropriate undrained strengths were selected to analyze stability of embankments at design grades and conventional side slope geometries. It was not possible to construct the embankments to the design grades in one stage without causing failure of the underlying clay. Also, excessive post construction settlements would occur due to long term consolidation of the clay. The strength of the clay had to be increased to allow stable embankment construction, and the rate of consolidation of the clay accelerated to cause most of the compression to occur prior to final pavement. To accomplish these objectives, the design incorporated staged construction for the embankments combined with vertical sand drains to accelerate consolidation of the underlying clay.

The concept of staged construction was used to capitalize on the concept that the undrained strength of the clay is a function of the vertical effective stress and the Overconsolidation Ratio, OCR (i.e. the SHANSEP approach). In the staged construction approach, the first phase embankment construction fill (Stage I) was placed to heights that would produce stable embankments using existing undrained strengths of the foundation soils. As the clay consolidated under the weight of the Stage I fill, its strength would increase. The SHANSEP approach was used to predict these strength increases. Following these strength increases, additional fill could be placed. At Portsmouth, two stages of fill were placed-Stage I as described above, and Stage II to a height sufficient to surcharge the site as will be discussed below. Where permitted by site constraints, stabilizing berms were incorporated into the embankment geometry to provide more stable geometries at surcharge grades (see Fig. 7). These berms also allowed higher Stage I fill to be placed, which was beneficial in accelerating strength gains.

To reduce post-construction settlements to acceptable values, a surcharge scheme was adopted. Fill heights greater than final design grade were placed to precompress the foundation clay such that, when it was removed, settlements remaining under the final design grade would be acceptable. As noted above, staged construction was necessary to improve the strength of the foundation soils so they could support the surcharge fills.

In summary, consolidation of the foundation soils was essential for increasing their undrained shear strength to allow stable fills to be constructed to surcharge grades, and to preconsolidate the clay soil to minimize post construction settlements. If the clay consolidated under vertical drainage conditions only, it would have taken years more than was available to meet construction schedules. Vertical sand drains were selected to accelerate the consolation process by providing horizontal drainage paths in the clay.

Sand drains were selected since they were the most viable method in existence at that time to accelerate consolidation of subsurface soils. However, as was demonstrated by the test sections in Portland, the method of installation could have an adverse effect on sand drain performance. Driven methods could disturb the soil, thereby causing both the coefficient of consolidation and the undrained shear strength of the sensitive marine clays at Portsmouth to decrease. Performance specifications were therefore prepared that required the use of either jetted or augered methods of installation that would minimize disturbance and smear that could reduce their effectiveness. Ultimately a non-displacement jetted method using a Dutch jet-bailer was selected (Ladd et al, 1972)

Details of the design are shown in Fig. 7. Instrumentation was incorporated into the design to monitor the rate of settlement and pore pressure dissipation. These data were used to assess the degree of consolation that occurred during and following embankment construction. These data were used to assess when sufficient consolidation occurred under Stage I fills to allow Stage II construction to commence and when adequate consolidation occurred under surcharge grades to result in acceptable post construction settlement under final grades. Field vane shear tests were conducted after the Stage I fill achieved its required degree of consolidation to confirm that the predicted strength gains were occurring. See Ladd et al, (1972) for specific details.

It should be noted that a similar design approach was adopted for the I-295 embankments in Portland.

Embankment Performance, I-95 Portsmouth

In summary, the embankments were constructed safely in two stages, and the sand drains performed better than expected, allowing construction to proceed on schedule (Haley & Aldrich, 1972). Detailed evaluations of the field data and embankment performance are documented by Ladd, (1972) and Ladd et al, (1972). Some specific comments follow.

Geonor field vane tests were performed at numerous locations at the end of consolidation under Stage I fills. There was considerable scatter in the data, but average data agreed quite well with the $\Delta s_u / \Delta \bar{\sigma}_v$ of 0.20 as assumed in design, based on DSS test data. See Figure 8. Construction of Stage II fills proceeded using these data with no indications of embankment instability.

The observed settlements in the field exceeded those predicted from laboratory consolidation data by about $25\% \pm 15\%$. The primary reason for this discrepancy was the steep slope of the field compression curves just beyond the maximum past pressure. At higher stresses, the curves flattened out, approaching the laboratory values. The field curves did show a higher maximum past pressure than anticipated from lab data. However, the larger in-situ virgin compression curves just beyond the maximum past pressure outweighed the higher maximum past pressures, leading to the larger than anticipated settlements. This behavior demonstrated the sensitive nature of the clay. Lab data was likely impacted by sample disturbance which would have lowered the maximum past pressure and flattened the virgin compression curve near the maximum past pressure.

Measured values of the coefficient of consolidation, c_v, were determined at locations both with and without sand drains. The vertical c_v back-calculated from areas with no drains was about 60% higher than predicted from laboratory data. Coefficients of consolidation from sand drain areas verified that a c_h/c_v ratio of 2 was reasonable as assumed in design. However, the c_h values were higher than estimated by a factor of 1 to 2.

The fact that c_v and c_h values from field data exceeded those predicted from lab data was the opposite of the results obtained from the Portland sand drain test sections. This meant the drains at Portsmouth functioned more efficiently than anticipated. The installation method therefore must have had little or no effect on the performance of the sensitive clay (or at least less effect than sample disturbance had on the laboratory samples). This was remarkable considering the extreme sensitivity of the clay as demonstrated by the field compression curves. This performance clearly justified the care and concern for specifying and selecting non-displacement methods of drain installation.

Applications Following the Early Days

Subsequent Applications

Some or all of the innovative approaches developed and implemented on the Portsmouth and Portland embankments in the early 1970's, were used by Haley & Aldrich, as well as many other consultants, on later projects, and are still being applied today.

A few representative projects are presented and discussed below:

Charter Oak Bridge over Connecticut River

Subsurface conditions at this site consisted of loose alluvial silt and sand deposits overlying up to 42.7 m (140 ft) of soft varved clay. Dr. Ladd consulted on the design of approach embankments for this project. Extensive laboratory testing, including Direct Simple Shear tests, were performed to evaluate strength characteristics and to develop the $s_u/\bar{\sigma}_v$ ratio required for use in the SHANSEP approach. SHANSEP was used in conjunction with staged construction and surcharging. Wick drains were installed to accelerate consolidation of the varved clay, and lightweight fill was used to reduce settlements. Field instrumentation was installed to monitor the progress of consolidation of the varved clay, and to provide an early warning of impending failure.

Bangkok, Thailand Drainage Study

From 1968 to 1970, Dr. Ladd consulted on the evaluation of the stability of "klong" walls during proposed dredging. Klongs are open drainage ditches made by excavating below grade. The soils in the area of the study were very soft clays with shear strengths in the 9600 to 14,370 N/m^2 (200 to 300 psf) range. Extensive strength testing was undertaken, including Direct Simple Shear tests. The SHANSEP approach was applied (first time in Thailand) to assess side slope stability during dredging. A similar project in Bangkok was undertaken in 2000, and Dr. Ladd again consulted on the selection of soil parameters for evaluating stability of this soft clay deposit using the SHANSEP approach.

Maine Turnpike Interchanges (Jetport and Westport/Rand Road)

9.1 m (30 ft) high approach embankments were constructed over 12.2 m (40 ft) of soft marine clay. The SHANSEP approach was used to evaluate strength increases with consolidation. Strength values were determined from DSS tests from other nearby projects. Prefabricated vertical drains were installed to accelerate consolidation under surcharge loads to minimize post construction settlements. Field instruments monitored rate of consolidation during construction at the Jetport Interchange (1997 to 1999), and confirmed design parameters and embankment performance predictions. The other interchange is still under construction, and field observations indicate the soil has achieved a 60% degree of consolidation to date under surcharge loads.

Great Salt Lake Causeway

In the 1980's, the elevation of the railroad causeway over the Great Salt Lake had to be raised and erosion protection beefed up to accommodate rising lake levels and continuing embankment settlements. Extremely soft, plastic organic soils complicated the design. SHANSEP procedures were used to evaluate strength parameters. This required over 100 consolidation tests to assess maximum past pressures and numerous DSS tests to determine s_u/σ_v. This approach was successful and allowed design to proceed on this extremely difficult project.

Missisquoi Bay Bridge, Lake Champlain, VT.

As part of this 128 m (4200 ft) long lake crossing, 183 m (600 ft) of earth fill causeways were constructed over deep deposits of soft clay. Extensive strength and consolidation testing were performed to assess soil properties. Staged construction, strength increases with consolidation, wick drains and ground reinforcement were incorporated in the design to permit successful construction of these embankments.

Ashfill/Balefill Expansion, Scarborough, ME.

This landfill required construction of ashfill embankments as high as 9.8 m (32 ft) over 24.4 m (80 ft) of soft clay. Staged construction was required. 6.1 m (20 ft) high Stage I fills were placed, then raised to 9.7 m (32 ft) following strength increases with consolidation. The SHANSEP approach was used to evaluate these strength increases. DSS tests were used to assess strength conditions. DSS tests were also conducted under cyclic loading conditions to assess the potential for strength loss under dynamic loading conditions. Vertical drains would not be used to accelerate consolidation because the clay layer is relied on as an impermeable barrier for the landfill. Stabilizing berms of earth fill were required, which would later be replaced with ash fill as the strength of the clay increased with time. Piezometers, inclinometers, and settlement platforms were used to monitor subsurface and embankment performance.

I-95 Taylor River Crossing, Southern NH

Dr. Ladd assisted in selection of soil properties and design considerations for this project. 3 to 4.6 m (10 to 15 ft) high embankments were constructed over 4.6 to 7.6 m (15 to 25 ft) thick peat deposits in a tidal mudflat in southern NH. Design incorporated staged construction, and stabilizing berms. Successful construction of this embankment was controlled by monitoring lateral deformations of the peat, which ranged up to a meter or more in magnitude.

Other Advances to the State of the Practice

A significant amount of actual field data was obtained from the test sections and actual embankment construction on both the Portland and Portsmouth projects. These data not only provided a sound basis for design and construction of these projects, but also constituted a valuable data source for subsequent analyses by Dr. Ladd and many others. These analyses enhanced the profession's understanding of in-situ performance of soft sensitive clays with vertical drains and led to further enhancements in embankment designs.

For instance, Olson used field measured settlement and pore pressure data from the Fore River project to assess the validity of a finite difference method of analysis to predict the magnitude and rate of actual consolidation settlements in the field when sand drains were used to accelerate consolidation (Olson, et al, 1974). This finite difference method could specifically account for effective stress-dependent values of void ratio and coefficient of consolidation, stratified soils, time dependent loading, partially penetrating drains, and other variables. Other studies (Pelletier, et al 1979) used data from the FRT to evaluate the feasibility of less computer intensive closed form solutions involving extended equal strain theory to assess sand drain performance. This study also used the fill placement history at FRT to assess a theory on time dependent loading histories, which demonstrated the effectiveness of the proposed analytical method.

These are only some of many examples that could be cited of cases where data from the Portland and Portsmouth embankment projects provided a valuable source of documented field behavior which was used to assess the validity of analytical methods used to predict field performance.

Advances In Technology

Advances in new technologies and methods of analyses have occurred over the years. Many have affected the design and construction of embankments and methods of foundation soil stabilization. Two of these that are directly applicable to topics discussed in this paper are prefabricated vertical (wick) drains and deep soil mixing.

Prefabricated vertical drains are geocomposites consisting of a plastic drainage core surrounded by a permeable geosynthetic fabric. This drain type has replaced sand drains as a method of accelerating drainage of soft cohesive soils. Dr. Ladd was involved in developing design criteria and approaches for these drains. Even though the materials are different, many of the concerns, such as disturbance and smear effects identified by Dr. Ladd in the early days, apply to this drain type, and Dr. Ladd developed techniques and procedures for considering these efforts in their design.

Deep soil mixing, or the deep mixing method, is a relatively recent technique used to improve soil properties of soft cohesive soils. This concept is different from applying surcharges and accelerating consolidation to increase shear strength. Instead it involves a slow controlled mixing of cement grout into soft clayey soil deposits in the field using auger-like mixing tools. This process creates a soil-cement mixture that has much higher strength and lower compressibility than the original material. Again Dr. Ladd is involved with projects of this type and is playing a key role in assessing soil structure interaction issues associated with soil-cement mixtures. He also oversaw research on the mechanical properties of these mixtures in the mid-1990s.

As in the early days, Dr. Ladd remains in the forefront of technology as it evolves, especially as it relates to defining soil properties, and developing design solutions for embankment construction on soft cohesive soils.

Summary and Conclusions

- This paper summarizes design and construction procedures developed by Dr. Ladd in the late 1960s and early 1970s for highway embankments over deep deposits of soft clayey soils.

- Innovative procedures included 12.7 cm (5 in) diameter undisturbed sampling, in situ strength testing using the Geonor field vane shear device, innovative strength testing (Geonor Direct Simple Shear Tests), development and application of the SHANSEP approach to full scale test and production embankments, use of test embankments during design not only to verify soil properties but also to assess impacts of sand drain installation techniques on soil properties, and first time use in the United States of the Dutch jet bailer (non-displacement) method of sand drain installation.

- The Experimental Test Section (ETS) at Portsmouth was used to verify undrained shear strengths of the soft sensitive clay underlying the site. It helped establish the validity of results of the Direct Simple Shear (DSS) tests as a means of predicting strengths to be used for stability analyses.

- The SHANSEP Approach proved effective in estimating in-situ initial strengths as well as strength gains during consolidation. It is most effective when used in conjunction with the results from DSS tests

- Back Cove (BCT) and Fore River (FRT) Test Sections at Portland documented the negative effect driven and other displacement type sand drains had on the efficient performance of vertical sand drains, especially in sensitive marine clays. These results led to selection of alternate non-displacement drains for the Portsmouth (Dutch jet-bailer) and Portland production embankments that proved to be more efficient and cost effective.

- Design of production embankments incorporated staged construction and vertical drains to accelerate consolidation of the foundation soils. Consolidation of the soils was essential to precompress them to minimize post construction settlements. It was also required for strength gains due to consolidation. Without such strength gains, embankments could not have been constructed to either their final grade or to surcharge grades where required to reduce post construction settlements.

- Settlement and pore pressure measurements were monitored in the field during and following construction. These measurements were used to determine when sufficient consolidation occurred to proceed with the next stage of fill placement in staged construction areas, and when surcharge fills could be removed. These data also provided valuable field performance records that were used to verify soil properties assumed in design, and that could be used to assess the validity of analytical methods of predicting performance of compressible materials with vertical drains.

- The innovative approaches used on these early embankment projects have been used on numerous other embankments over the years, and are still being used today, demonstrating Dr. Ladd's ingenuity and foresight.

References:

Aldrich, H. P. Jr. and Johnson, E. G. (1972), "Embankment Test Sections To Evaluate Field Performance of Vertical Sand Drains at I-295, Portland, Maine", presented at the annual HRB meeting, Washington, D. C., January.

Haley & Aldrich, Inc. (1969) "Engineering Properties of Foundation Soils at Long Creek-Fore River Areas and Back Cove", Report No. 1 to the Maine State Highway Commission.

Haley & Aldrich, Inc. (1972), "Performance of the Jetted Sand Drains for The Approach Embankments, Interstate I-95, Portsmouth, N. H." Report to the New Hampshire Department of Public Works and Highways, March.

Ladd, C.C. and Foote, R. (1974), "New Design Procedure for Stability of Soft Clays", Journal of Geotechnical Engineering Division, ASCE, V. 100, GT7, pp. 763-786.

Ladd, C.C. and Edgers, L. (1971), "Consolidated-Undrained Direct-Simple Shear Tests on Saturated Clays", Research In Earth Physics Phase Report No. 16, Dept. of Civil Engrg. Research Report R71-30, M.I.T.

Ladd, C.C. (1972), "Test Embankment on Sensitive Clay", ASCE, SMFD Specialty Conference on Performance of Earth-Supported Structures, Purdue University, June.

Ladd, C.C., Rixner, J.J. and Gifford, D.G. (1972), "Performance of Embankments with Sand Drains on Sensitive Clay", ASCE SMFD Specialty Conference on Performance of Earth and Earth Supported Structures, Purdue University, June.

Olson, R. E., Daniel, D. E., and Liu, T. K. (1974), "Finite Difference Analysis for Sand Drain Problems," Analysis and Design in Geotechnical Engineering, Vol. 1, American Society of Civil Engineers, New York.

Pelletier, J.H., Olson, R.E. and Rixner, J.J. (1979), "Estimation of Consolidation Properties of Clay from Field Observations", ASTM Geotechnical Testing Journal, Vol. 2, No. 1, March.

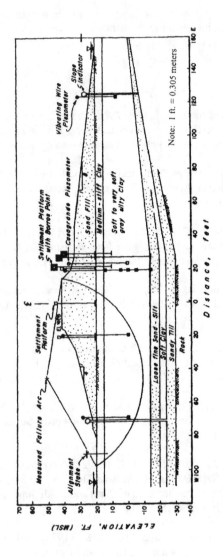

Figure 1 – Cross Section Of Experimental Test Section At I-95 Portsmouth. (Ladd, 1972)

TEST EMBANKMENT

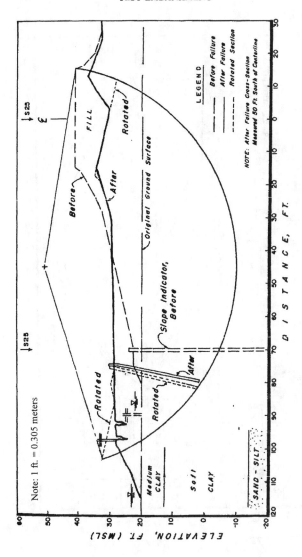

Figure 2 – Cross Section of ETS At I-95 Portsmouth Before and After Failure (Ladd, 1972)

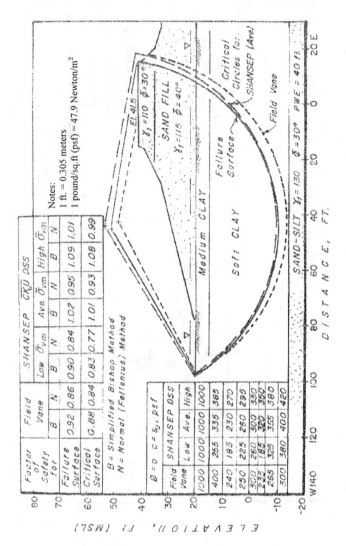

Figure 3 – Results of Stability Analysis For Experimental Test Section At I-95 Portsmouth (Ladd, 1972)

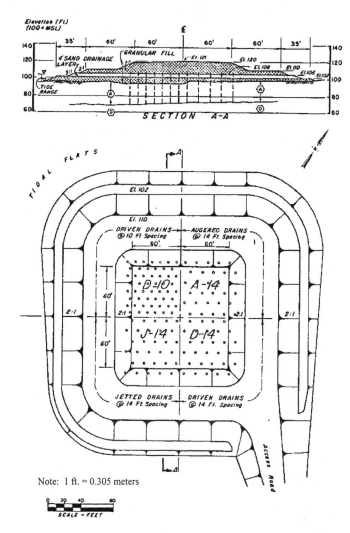

Note: 1 ft. = 0.305 meters

Figure 4 – Fore River Test Section At I-295 Portland, ME. (Aldrich and Johnson, 1972)

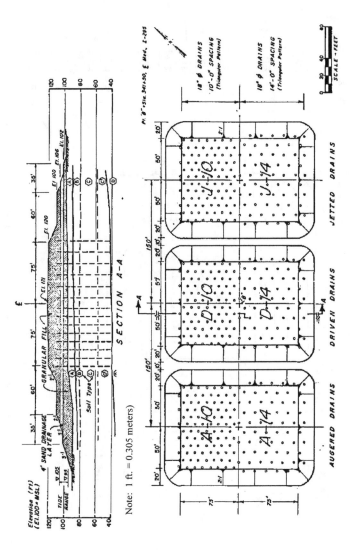

Note: 1 ft. = 0.305 meters)

Figure 5 – Back Cove Test Section At I-295, Portland, ME (Aldrich and Johnson, 1972)

Notes: 1ft. = 0.305 meters, 1 pound/sq.ft (psf) = 47.9 Newton/m²

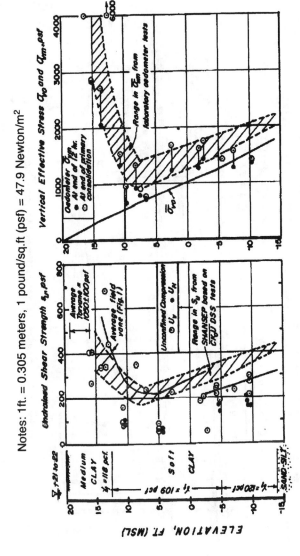

Figure 6 -Undrained Shear Strength and Stress History At Experimental Test Section At I-95 Portsmouth, NH

(Ladd, 1972)

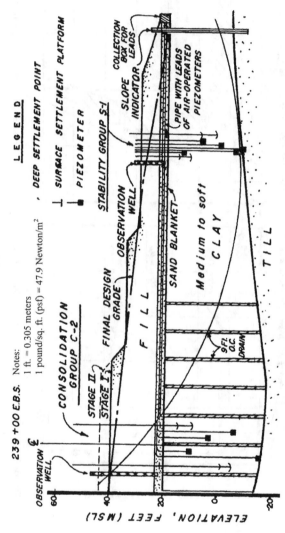

Figure 7 – Embankment Section At Typical Instrumentation Groups, I-95 Portsmouth, NH (Ladd et al, 1972)

SENSITIVE CLAY

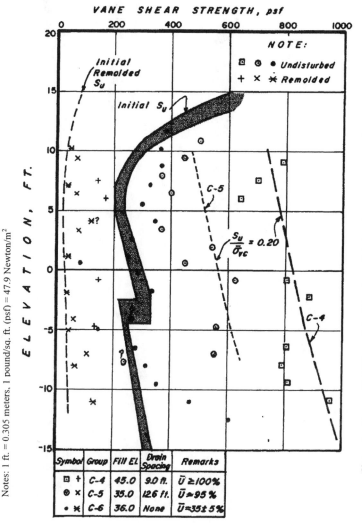

Notes: 1 ft. = 0.305 meters. 1 pound/sq. ft. (psf) = 47.9 Newton/m²

Figure 8 – Geonor Field Vane Data After Consolidation At I-95, Portsmouth, NH (Ladd, et al, 1972)

Subject Index

Page number refers to the first page of paper

Author Index

Page number refers to the first page of paper